INFORMATION TECHNOLOGY IN THE SERVICE ECONOMY

IFIP – The International Federation for Information Processing

IFIP was founded in 1960 under the auspices of UNESCO, following the First World Computer Congress held in Paris the previous year. An umbrella organization for societies working in information processing, IFIP's aim is two-fold: to support information processing within its member countries and to encourage technology transfer to developing nations. As its mission statement clearly states,

> *IFIP's mission is to be the leading, truly international, apolitical organization which encourages and assists in the development, exploitation and application of information technology for the benefit of all people.*

IFIP is a non-profitmaking organization, run almost solely by 2500 volunteers. It operates through a number of technical committees, which organize events and publications. IFIP's events range from an international congress to local seminars, but the most important are:

- The IFIP World Computer Congress, held every second year;
- Open conferences;
- Working conferences.

The flagship event is the IFIP World Computer Congress, at which both invited and contributed papers are presented. Contributed papers are rigorously refereed and the rejection rate is high.

As with the Congress, participation in the open conferences is open to all and papers may be invited or submitted. Again, submitted papers are stringently refereed.

The working conferences are structured differently. They are usually run by a working group and attendance is small and by invitation only. Their purpose is to create an atmosphere conducive to innovation and development. Refereeing is less rigorous and papers are subjected to extensive group discussion.

Publications arising from IFIP events vary. The papers presented at the IFIP World Computer Congress and at open conferences are published as conference proceedings, while the results of the working conferences are often published as collections of selected and edited papers.

Any national society whose primary activity is in information may apply to become a full member of IFIP, although full membership is restricted to one society per country. Full members are entitled to vote at the annual General Assembly, National societies preferring a less committed involvement may apply for associate or corresponding membership. Associate members enjoy the same benefits as full members, but without voting rights. Corresponding members are not represented in IFIP bodies. Affiliated membership is open to non-national societies, and individual and honorary membership schemes are also offered.

INFORMATION TECHNOLOGY IN THE SERVICE ECONOMY: Challenges and Possibilities for the 21st Century

IFIP TC8 WG8.2 International Working Conference
August 10-13, 2008, Toronto, Ontario, Canada

Edited by

Michael Barrett
University of Cambridge
United Kingdom

Elizabeth Davidson
University of Hawaii at Manoa
USA

Catherine Middleton
Ryerson University
Canada

Janice I. DeGross
University of Minnesota
USA

Information Technology in the Service Economy: Challenges and Possibilities for the 21st Century

Edited by Michael Barrett, Elizabeth Davidson, Catherine Middleton and Janice I. DeGross

p. cm. (IFIP International Federation for Information Processing, a Springer Series in Computer Science)

ISSN: 1571-5736 / 1861-2288 (Internet)

ISBN: 978-1-4419-3531-1 eISBN: 978-0-387-09768-8

Printed on acid-free paper

9 8 7 6 5 4 3 2 1

springer.com

Contents

Part 2: IT-Enabled Services in Industry Settings

Part 3: IT-Enabled Change in Public Services

Part 4: Outsourcing and Globalization of IT Services

Part 5: Panels

Part 6: Workshop Paper Contributions

PREFACE

This book represents the compilation of papers presented at the IFIP Working Group 8.2 conference entitled "Information Technology in the Service Economy: Challenges and Possibilities for the 21st Century." The conference took place at Ryerson University, Toronto, Canada, on August 10-13, 2008. Participation in the conference spanned the continents from Asia to Europe with paper submissions global in focus as well.

Conference submissions included completed research papers and research-in-progress reports. Papers submitted to the conference went through a double-blind review process in which the program co-chairs, an associate editor, and reviewers provided assessments and recommendations. The editorial efforts of the associate editors and reviewers in this process were outstanding. To foster high-quality research publications in this field of study, authors of accepted papers were then invited to revise and resubmit their work. Through this rigorous review and revision process, 12 completed research papers and 11 research-in-progress reports were accepted for presentation and publication. Paper workshop sessions were also established to provide authors of emergent work an opportunity to receive feedback from the IFIP 8.2 community. Abstracts of these new projects are included in this volume. Four panels were presented at the conference to provide discussion forums for the varied aspects of IT, service, and globalization. Panel abstracts are also included here.

We would like to thank the sponsoring and host organizations for their generous support. It was their involvement and endorsement that enabled this conference to take place. Ryerson University was instrumental in making this conference possible, through their generous financial support and the provision of the conference venue. We would like to extend our thanks to all of the following organizations:

- IFIP and in particular Working Group 8.2
- Ryerson University
- University of Hawaii at Manoa
- University of Cambridge

Finally, we would like to thank the graduate students at Ryerson for their assistance behind the scenes, developing the conference web site and maintaining the submission system.

Ojelanki Ngwenyama
Michael Barrett
Elizabeth Davidson
Catherine Middleton
Janice I. DeGross

CONFERENCE CHAIRS

General Chair
Ojelanki Ngwenyama
Ryerson University

Program Chairs
Michael Barrett
University of Cambridge

Elizabeth Davidson
University of Hawaii at Manoa

Organizing Chair
Catherine Middleton
Ryerson University

CONFERENCE SPONSORS

This conference was funded with the generous support of the Ted Rogers School of Management and the Office of the Vice President, Research and Innovation, at Ryerson University

In particular, we wish to acknowledge
Ken Jones, Dean,
Ted Rogers School of Management

Wendy Cukier, Associate Dean,
Ted Rogers School of Management

James Norrie, Director,
Ted Rogers School of Information Technology Management

ASSOCIATE EDITORS

PROGRAM COMMITTEE AND REVIEWERS

1 EXPLORING THE DIVERSITY OF SERVICE WORLDS IN THE SERVICE ECONOMY

Michael Barrett
Judge Business School
University of Cambridge
Cambridge, U.K.

Elizabeth Davidson
Shidler College of Business
University of Hawaii at Manoa
Honolulu, HI U.S.A.

1 INTRODUCTION

The transformation of modern economies into predominantly service-based economies is happening on a global scale. While services are not new, the scale and complexity of globally dispersed services are growing rapidly. These transformations are enabled by—and often dependent on—information technologies and are fueled by processes of globalization. Transformational change provides opportunities for innovation in business models, collaborations, and work practices but also presents challenges to established practices within industries and organizations. The aims of the IFIP Working Group 8.2 Conference on IT in the Service Economy were to better understand the possibilities and challenges of these transformations and to examine key implications for organizations, their employees, and IT professionals in the 21st century service economy.

2 WHAT IS DRIVING THE TRANSFORMATION TO THE SERVICE ECONOMY?

Growth in service sector employment and in the contribution from service sector industries to national economies has continued apace for some time. The shift from

Please use the following format when citing this chapter:

Barrett, M., and Davidson, E., 2008, in IFIP International Federation for Information Processing, Volume 267, Information Technology in the Service Economy: Challenges and Possibilities for the 21st Century, eds. Barrett, M., Davidson, E., Middleton, C., and DeGross, J. (Boston: Springer), pp. 1-10.

manufacturing-based to service-based economies began in the mid-20th century and is evident not only in highly developed economies but increasingly in developing nations. However, social, political, and economic trends are now interacting to amplify this shift (Bryson et al. 2004). For example, as per capita income rises and as individuals devote more time to work, demand for services, particularly personal and domestic services, increases. Demographic changes and increasing skills requirements for modern jobs have increased the demand for health and educational services. Most firms operate within extended production networks and employ sophisticated technologies in production processes; such networks require coordination services (such as accounting or legal services) and contribute to employment in professional service occupations. Requirements for interfirm coordination as well as outsourcing and externalization of service work (such as IT) contribute to the growth of specialized service providers (e.g., consulting firms). Governments influence service growth directly, as local, regional, national, and international governmental bodies extend their service offerings and payrolls, and indirectly through economic regulation, which stimulates growth in compliance-related services. International trade in services involves both provisioning of services remotely (e.g., with remote call centers) and relocation of clients to service areas (e.g., tourism).

When we consider these diverse factors, which contribute to growth in the level and variety of services in our modern economy, we begin to appreciate that the so-called service economy is much more than an economic shift; it is a transformation to a complex array of social, political, and economic *service worlds*. Bryson et al. (2004, p. 3) comment, “Service worlds are complex, evolving, heterogeneous, and long-established phenomena…a service world is also one in which there is a direct, even dialectical, relationship between service production and service consumption.”

3 INFORMATION TECHNOLOGIES AND SERVICE

Given this broad view of the service economy, we might question what we mean by “service” and whether services in today’s service worlds are essentially the same as services of the past. Of particular interest is how information technologies influence how we experience and how we innovate with services.

Services have been defined as intangible, heterogeneous, and perishable, and as requiring joint participation by the consumer and producer of the service using their skills and knowledge to produce (Gersung and Resengren 1973). Services have generally been viewed as more local and less subject to globalization than products. An example that illustrates this perspective is a consumer viewing a movie in a theater. The intangible service—entertainment—requires the consumer’s cooperation to sit through the movie and presumably to enjoy it. If a particular seat in a theater at a particular showing goes unfilled, that instance of the service perishes (unlike the movie popcorn, which can await future consumption). Each patron at the showing may experience the service differently. Similarly, an organizational example of the traditional view of services is a public accounting firm auditing a client firm, utilizing the accountants’ skills and knowledge and acting with the cooperation of the client firm, in order to satisfy regulatory requirements. The service is produced as the accountants’ time is expended and thus is not “storable,” although the artifacts of the service, such as the auditreport, are.

These traditional views of services are being challenged, particularly the differentiation between products and services. Gustafsson and Johnson (2003) suggest, for example, that "physical products are services waiting to happen." Vargo and Lusch (2004, 2008) similarly claim that all exchanges essentially are service exchanges, in which the customer and supplier co-create value with products or with services. Servitization, which generally involves adding value to products through services (often informational services) or offering products as services (e.g, software-as-a-service), further reduces the boundary between products and services.

This expanded understanding of the relationship between products and services can reveal opportunities for innovation and competitive advantage (Gustafsson and Johnson 2003). Moreover, embedding information and communication technologies in service offerings may alter other taken-for-granted characteristics of services. Using our movie example, "movies on demand" services available through home cable or the Internet enable the service provider to sell a viewing of the movie to consumers "24 by 7." Does this service actually perish? Is co-presence necessary for co-creation of value? Although the service experience (entertainment) is intangible, a complex array of technological artifacts is required to deliver the service, and the quality of the artifacts undoubtedly influences the quality of the entertainment experience.

Thus, not only do products take on service characteristics through servitization, but with extensive use of IT in services, services may take on some of the characteristics typically associated with products, that is, standardization, storability, and separation of production and consumption of the service. Expanding our understanding of these complex interrelationships can, we suggest, point to a variety of opportunities for service innovation across a wide variety of socio-economic activities.

4 GLOBALIZATION, IT, AND THE SERVICE ECONOMY

Contrary to the traditional expectation that service production is essentially local, in fact, globalization of services is a widespread socio-economic trend in modern economies. Nearly 20 percent of international trade is trade in services. The IT industry and its professionals have been at the leading edge of the trend to globalize services, a trend that now includes many other information services such as legal, accounting, educational, health care, and other professional services.

Globalization has been the subject of much controversy because of the disruptions to local economies that can result. By globalization, we mean the geographic separation (yet increasing interconnection) of service providers and service consumers, whether individuals or firms, the movement of service workers or service consumers around the globe to deliver services, and the development of capabilities around the globe to provide services. Globalization undoubtedly provides new opportunities for service innovation, particularly in the knowledge-based and informational services, enabled by information and communication technologies (ICTs). At the same time, there are significant barriers such as language, cultural, regulatory, environmental, and time zone differences, which pose challenges to the provision of global services. Thus, when we consider IT and the transformations in the service economy, globalization and its effects become a very important focus of the research of the IFIP Working Group 8.2 community.

5 OVERVIEW OF THE BOOK

The selection of completed research works, research-in-progress reports, panel discussions, and workshop papers presented at the 2008 IFIP Working Group 8.2 conference, "IT in the Service Economy," are documented in this volume. Collectively, these works address many of the themes outlined in our introduction: conceptualizing and theorizing about IT-enabled services, innovation in services through IT, growth of IT-enabled services in a broad array of service sectors, including government, healthcare, and other sectors, and outsourcing and globalization of IT services.

5.1 Conceptualizing and Theorizing about IT-Enabled Services

Recent developments in the services literature, particularly from the marketing and operations disciplines, portray an ongoing reflexiveness as to how services should best be conceptualized and theorized in today's service economy. However, there has been relatively little work in the IS discipline that has contributed to this growing debate. This is somewhat surprising since IT has not only played a significant role in aspects of service design and delivery, but calls into question previous conceptualizations of service, and suggests new conceptualizations (e.g., software as a service). It is therefore timely that the papers in this section take a step toward conceptualizing and theorizing IT-enabled services. Five papers challenged different conceptualizations of IT-enabled service and service work and provided novel ways of theorization and conceptualization. Three of these papers examine primarily global, macro-level issues while the other two are focused more squarely at the organizational and work system level.

In challenging populist conceptualizations, Walsham is critical of the "flat world" rhetoric that suggests there are essentially no barriers to globalization of service work. While the inappropriateness of this caricature may be well recognized by many in the 8.2 community, Walsham argues that we need to raise up a critical voice that constructively challenges the notion of a relationship-free world and highlights the continued importance of cultural diversity and local specificity amongst other unevenness. He outlines critical themes for further research, namely identity and cross-cultural working, globalization, localization, and standardization, and the explicit focus on power, knowledge, and control in understanding global working in a non-flat world.

Continuing the theme of cultural diversity, Vorakulpipat and Rezgui provide an intriguing illustration of the importance of local and cultural particularities through their study in a Thai IT research service organization. They explore the influence of shyness as a socio-emotional characteristic of the Thai culture on knowledge management practices in this organization and the need to consider shyness barriers for understanding team working, structure and culture, training, knowledge sharing, and motivation for knowledge management in this cultural setting.

Addressing the ways in which government regulations promote the need for and development of information services, Butler, Emerson, and McGovern offer a novel and timely conceptualization of compliance-as-a-service and as a viable value proposition. Drawing on institutional theory, they examine the different pressures faced by an IT manufacturing organization and show that regulatory influences have developed the deepest response in terms of innovative compliance-oriented procedures and protocols.

Alter challenges the service innovation literature for its predominant focus on large-scale topics of globalization and eco-systems of interacting suppliers and consumers, and argues the need to address more micro levels. Specifically, he suggests that less attention has been given to a systems perspective and the development of specific localized systems. To address these shortcomings, he critically evaluates the existing definitions of service and service systems and argues for simpler, broader-based definitions. He also draws on three interrelated frameworks, the work system framework, the service value chain framework, and the work system life cycle model to conceptualize service innovation across a wide range of service systems.

Drawing on Deleuze and Guattari's theory development of rhizomes as non-hierarchical networks, Atkinson and Brooks offer a novel conceptualization of rhizomatic informatics to explore IT-enabled service work to reinterpret a classic study of a problematic ERP implementation in a major university. In so doing, they not only illuminate the conceptual richness of such an approach for understanding the nature and complexities of IT-enabled service work and practice but they also question Deleuze and Guattari's overarching focus of information systems as instruments of corporate arboreal control.

Two research-in-progress papers contribute to our understanding of socio-cognitive services inherent in distributed software teams in global service work and in theorizing the service encounter. First, Shen and Gallivan's study examines how distributed software teams can develop capabilities to perform successfully across temporal, geographic, and cultural boundaries. Their novel theoretical perspective extends and applies transactive memory systems and faultline models to examine inter-subgroup dynamics of these teams on knowledge coordination. Second, Ramiller and Chiasson seek to conceptualize and theorize the role of sense giving and sense making in the creation of IT-enabled service encounters. They argue that in every service event one party's interpretations are being built upon the interpretations of others. They then develop a three-layer stratigraphy of sense-making and sense-giving in IT-enabled services to delineate the role of sense-making in the ongoing constitution of IT-enabled service events, the redesign of IT enabled service practices, and in the organizing vision discourse about the IT-enabled service.

5.2 IT-Enabled Services in Industry Settings

Traditional classifications of service industries have included retailing, wholesaling, transportation, financial services, healthcare, education, and so on. Classifying such diverse industries simply as service industries overlooks critical differences in the institutional environments that are likely to influence the ways in which information technologies are developed and applied, as well as the outcomes of IT use in these environments (Chiasson and Davidson 2005). The papers in this section address the specificity of industry context while drawing on and contributing to our knowledge of IT in service industries more generally.

Two papers examined the use of information technologies in healthcare settings, paying close attention to the array of institutional actors who participate in health IT systems. Andersen and Aanestad reported on the development and deployment of a community-wide support system for ambulatory psychiatric care for children in Norway.

Drawing on the concept of generativity and information infrastructure, they found that the existing installed base could not encompass user-driven innovations, but that users were able to overcome resistance from infrastructure supporters to promote development of a new and more effective design. The intent and design of this system was consistent with the movement to home-based, self-care practices and patient empowerment in the healthcare industry. Vuokko and Karsten examined the work practices of nurses in a pediatric unit of a hospital prior to the introduction of electronic medical records. They suggested that actor-network theory and complexity theory provide appropriate analytic lenses to investigate how the varied interests of actors in this work setting influence the reordering of time-place arrangements and enactment of changes in coordination, communication, and information sharing practices with the new technology.

Continuing the research focus on work groups but in a different industry setting, Van Daalen Fuente, Chiasson, and Devadoss investigated the ways in which a face-to-face community of practice of stock traders engaged in knowledge exchange and community learning in a colocation setting, despite utilizing technologies designed for remote working. Feller, Finnegan, Fitzgerald, and Hayes studied interfirm exchanges in an open source service network and examined the social mechanisms that facilitate coordination and safeguarding of exchanges of the Zea Partners network. They found that restricting firms' access to the network, establishing a macro-culture within the network, collectively sanctioning member firms' unacceptable actions, and building the network's reputation were social mechanisms that enabled the network of firms to deliver commercial products to exploit peer produced software.

Two research-in-progress papers report initial findings on how new information technologies may affect work practices and industry norms. Loebecke, Huyskens, and Gogan presented initial findings of a study of RFID (radio frequency identification) tags in retailing. This case study outlined how consumers' experience in a retail setting may be improved through use of the technology and how mangers may more effectively meet customers' needs with the information available through RFID systems. Sawyer and Yi utilized macro-economic data to explore how information technologies may have influenced employment in and productivity of the real estate industry. Their preliminary analysis suggests that IT investments have contributed to value produced in this industry but have not resulted in work force reductions, despite open access to real estate data by customers.

5.3 IT-Enabled Change in Public Services

Examining the intersection of IT, service innovation, and growth in the service economy is particularly interesting in the public service sector. We would normally expect innovation and change to be difficult here, due to bureaucratic institutions and lack of market incentives. However, the research studies reported here emphasize that change is underway, as governments expand their regulatory roles into the domains of IT services and as they incorporate IT-enabled innovations to be more responsive to constituents.

Petrakaki's study of e-government initiatives in Greece illustrated the interaction of these issues. The opportunities for service innovation such as "one-stop shopping" often conflict with bureaucratic practices. Implementing so-called best practices and technologies developed in commercial settings into the institutional environment of government may set in motion transformational changes in the nature of governmental service work; such changes are not uncontested by established interests and governmental actors and may lead to power shifts in the bureaucracy.

Tapia and Ortiz's study of municipal wireless network policy making processes explored how governmental services are expanding to encompass essential IT infrastructure services. They noted that unlike private enterprises, decision-making by governmental bodies necessarily involves public policy making, typically a top-down process. In the case of wireless broadband network services, a hotly contested area, local governments are jumping into the market, and the majority of policy making is now happening at the local level.

Three research-in-progress papers also considered themes of IT-enabled service innovation manifest in the actions and policies of governmental agencies. Maldonado and Tapia's assessment of Venezuela's policy for open source software adoption illustrates both the expansion of government regulation into the domain of IT and the growth in government services related to IT—in this case, promoting development of open source services and capabilities in the economy. Constantinides and Blackler's study of the National Program for IT in the United Kingdom similarly investigated the expansion of governmental services and oversight into IT activities. By examining the discursive activities of participants in a public review of this massive public healthcare IT project, the paper considered how attributions of success and failure are constructed and how co-orientation may be negotiated across diverse interest groups through discourse. Bernardi's study of the introduction of IT-enabled change in public health services in Kenya adopted a micro-level view of institutional change to examine how formal policy structures are interpreted and enacted in micro-level actions in this setting.

5.4 Outsourcing and Globalization of IT Services

Outsourcing and offshoring have been a key feature of the IT services landscape over the past decade. While globalization of services facilitated by ICTs continues apace, there is a need for service work to meet the many social, cultural, and political challenges that influence their successful growth and development. The papers in this section highlight key issues inherent in the unevenness of global service work, the importance of respecting cultural diversity, and the ongoing challenge of managing knowledge processes. These papers cover both IT and business process outsourcing, and highlight a number of conceptual themes including legitimacy and trust, power and self-knowledge, boundary objects, and knowledge gaps.

Barrett, Hinings, and Oborn explored the dynamics of the offshoring relationship between a multinational firm and its Indian vendors in the provision of global software development. They highlighted the value of legitimacy management in unearthing the strategies and activities in gaining, maintaining, and repairing legitimacy at different points in the evolution of the relationship, and discussed the role of trust in managing legitimacy.

Four research-in-progress reports explored a range of issues and themes on IT and business process outsourcing. Komporozos-Athanasiou proposes an alternative focus to relationship management in understanding the high failure rates in global outsourcing arrangements. He suggests a focus on self-knowledge in client-focused maturity maturity assessment to realize potential outsourcing benefits and draws on coevolutionary theory to better understand the asymmetry of knowledge and power dynamics. Gregory and Prifling focused on ways in which IS service providers can leverage their client relationships over multiple projects at multiple levels for expertise development and knowledge integration. At the more operational level, Aman and Nicholson developed a theoretical framework to examine the knowledge gaps in offshore software development work when teams are separated by time, distance, and culture. The framework emphasizes the different types of knowledge held by team members in analyzing the potential for knowledge gaps. Finally, in examining business process offshoring around call center service work, Devadoss and Chiasson examined and complicated the typical utopian and dystopian views of automation. The utopian view portrays workers as having pure and unrestricted agency to pursue a knowledge-based career, while the dystopian view is that of an "electronic panopticon," with restricted and imprisoned individuals having few options. In reality, the picture is much more complex with human and nonhuman participants being linked by an underlying complexity of boundary work and boundary objects in what they view as a stable industry.

5.5 Panels

Four panel presentations engage in the debates highlighted in paper presentations about the nature of service and the possibilities for IT-enabled service innovation. In our first panel, Ramiller, Davidson, Wagner, and Sawyer take issue with the predominant focus on servitization of products in the discourse on the service economy and suggest that services are also being transformed into something akin to products—structured, standardized, scalable artifacts that separate the service provider and service consumer in the co-creation process. Each panelist explores the contradictory implications of servicitization and productization in an industry setting and the tensions this engenders. In the second panel, Alter, Gal, Lipien, Lyytinen, and Russo examine the implications of servitization for the structures, processes, and competencies of service provider and service consumer organizations and the role of IT in such changes. The panelists focus on new types of collaborative relationships in service value chains and question how collaborative relationships affect organizational identity, goals, and missions. The panel considers practical approaches to sorting out the implications of servitization for system and organizational design.

Feller, Finnegan, Lundell, and Nilsson continue this discussion by addressing servitization of peer production networks, and using the example of open source software consider how profitable service offerings might develop around the process of open source software production. Each panelist brings their considerable experience in open source software development to bear as they consider how services are developing in this domain, and what we might learn more generally about servitization in peer production networks.

Our fourth panel delves into e-health initiatives to consider how IT-enabled innovation in health service delivery provides new opportunities for the inclusion of patients and the community into health care provisioning and reconsideration of institutionalized practices. Each of the panelists is actively engaged in an e-health program, and drawing on their experiences, they consider the practical and research challenges involved in developing systems for co-producing health care and health care delivery.

5.6 Workshop Paper Contributions

Turning finally to the workshop papers, we see that the research themes investigated in completed research and research-in-progress papers are evident in these emerging research projects, that is, conceptualizing and innovating with service, globalization of service work, and development of IT-enabled services in public service arenas. To conceptualize services, Germonprez and Hovorka posit the information services view as a shift from the defined and predetermined services to a user-enabled real-time production environment of *ad hoc* information systems. Carter, Takeda, and Truex suggest a novel theorization on services. They introduce a new process model as to how emergence arises and is manifest in organizational discourse, which they aim to illustrate by drawing on the discourse and narratives of IBM in transitioning to a service-oriented company. In the area of IT in public services, Miscione and Aanestad highlight the importance of free and open source software and organizational learning in public administrations of developing countries, using the case of a health information system being implemented in Kerala, India. Tan, Tan, and Teo develop a research model that identifies factors influencing a blogger's intention to participate in a commercial exchange, for example between buyers and sellers of a service. The last two workshop papers seek to contribute to the growing IT service management area. In their paper, Tong, Xu, and Pan aim to understand how dynamic capabilities are developed to facilitate software process improvement implementation in a small scale service-oriented company. Finally, Cuthbert, Pennesi, and McFarlane use a case study approach to develop a service information model whereby service information requirements are extracted to support the delivery of a service to a customer.

6 CONCLUDING REMARKS

The papers in this volume provide a rich milieu of approaches and draw on a wide range of theories to examine IT and transformations toward a service economy across four main themes. This diversity in "letting a thousand flowers bloom" is a strength of our 8.2 community, which has a lot to offer in understanding the social and political aspects of Bryson et.al.'s (2004) service worlds. It is timely that we as researchers at the intersection of information systems and organizations actively voice our contributions to this emerging research area. Not only are the trends of growth toward a service economy evident, but this growth in demand for service activity across industries has led to recent attempts to develop a new field of Service Science to integrate and bring coherence across silos of fragmented knowledge in a multidisciplinary approach (Chesbrough and Spohrer 2006).

Our conference has started some way down this track toward a multidisciplinary view on services. The diversity of scholars and the range of papers accepted for the conference reflects a breadth beyond what might traditionally be expected from an IS perspective on services, such as web services and service-oriented architectures (Spohrer and Riecken 2006), and indeed even the usual 8.2 community. For example, we welcome the perspectives on policy and government from colleagues in the information schools as well as those suggested by researchers coming from the manufacturing and management tradition.

Going forward, there are many opportunities for study in these service worlds. We are in favor of letting a thousand flowers bloom, but we also believe there is great value in cultivating research developments as they evolve and taking stock of and, where possible, categorizing the emerging concepts, theories, and methodologies for advancing our knowledge at both micro and macro-levels of research on IT and services. In doing so, we believe that future research on services should be embarked upon through innovative multidisciplinary research with scholars in a number of other fields, including marketing, strategy, operations, and organization theory. This research journey has only just begun, but it is one filled with a future of exciting challenges and opportunities.

References

Bryson, J., Daniels, P., and Warf, B. 2004. *Service Worlds: People, Organizations, Technologies*, London: Routledge.

Chesbrough, H., and Spohrer, J. 2006. "A Research Manifesto for Services Science," *Communications of the ACM* (47:7), pp. 35-49.

Chiasson, M., and Davidson, E. 2005. "Taking Industry Seriously in Information Systems Research," *MIS Quarterly* (29:4), pp. 599-606.

Gersung, C., and Resengren, W. 1973. *The Service Society*, Cambridge, MA: Schenkman Publishing Company.

Gustafsson, A., and Johnson, M. 2003. *Competing in a Service Economy: How to Create A Competitive Advantage through Service Development and Innovation*, San Francisco: John Wiley and Sons, Inc.

Spohrer, J., and Riecken, D. 2006. "Services Science," *Communications of the ACM* (49:7), pp. 31-32.

Vargo, S. L., and Lusch, R. E. 2004. "Evolving to a New Dominant Logic for Marketing," *Journal of Marketing* (68), pp. 1-17.

Vargo, S. L., and Lusch, R. E. 2008. "Service-Dominant Logic: Continuing the Evolution," *Journal of the Academy of Marketing Science* (36:1), pp. 1-10.

Part 1:

Conceptualizing and Theorizing about IT-Enabled Services

2 ICTs AND GLOBAL WORKING IN A NON-FLAT WORLD

Geoff Walsham
Judge Business School
University of Cambridge
Cambridge, U.K.

Abstract *This paper rejects the hypothesis of Thomas Friedman that ICT-enabled globalization is driving us toward a flat world. Instead, it is argued that the world remains uneven, full of seams, culturally heterogeneous, locally specific, inequitable, not well-integrated and constantly changing. This argument is supported by an analysis of three areas of ICT-enabled global working, namely global software outsourcing, global IS roll-out, and global virtual teams. The paper then builds on these analyses to put forward an agenda for future IS research on ICTs and global working based on three research themes: identity and cross-cultural working; globalization, localization and standardization; and power, knowledge, and control. The paper concludes that the area of ICTs and global working offers the IS field a major research opportunity to make a significant contribution to our understanding of a set of crucial issues in our more globalized world.*

Keywords Flat world, globalization, global software outsourcing, global IS roll-out, global virtual teams, IS research agenda, identity, cross-cultural working, standardization, power, knowledge, control

1 INTRODUCTION

The changes taking place in the global economy, including those in the burgeoning services component, are the subject of much debate by a wide range of commentators including journalists, practitioners, and academics. In the first of these categories, one book which has enjoyed remarkable success in terms of sales is *The World is Flat: A*

Please use the following format when citing this chapter:

Walsham, G., 2008, in IFIP International Federation for Information Processing, Volume 267, Information Technology in the Service Economy: Challenges and Possibilities for the 21st Century, eds. Barrett, M., Davidson, E., Middleton, C., and DeGross, J. (Boston: Springer), pp. 13-25.

Brief History of the Twenty-First Century by Thomas Friedman (2005). According to Friedman, information and communication technologies and globalization are driving us toward a flat world with a level playing-field for global competition, in which geography, distance, and even language become irrelevant. The thrust of this paper is to argue against Friedman that, although ICTs and globalization are indeed crucial to the global transformations that are taking place, the world is not flat.

Friedman's book has already produced much counter-argument although not, as far as I am aware, in the information systems field. For example, Leamer (2007) provides a detailed critique from the point of view of an economist. In addition to arguing that Friedman's definition of a flat world is vague, he takes issue with the notion of a relationship-free world in which every economic transaction is contested globally. Leamer argues that relationships matter and that the world is highly uneven in terms of advantages that individuals, groups, and societies enjoy. It is particularly ironic that Friedman uses India as a frequent example of the leveling of the playing field, a society of increasing wealth for the privileged, but where approximately one in two women are illiterate and one in two children under five are malnourished (UNICEF statistics, http://www.unicef.org/infobycountry/india_india_statistics.html).

Practitioners have also engaged with the nature of ICT-enabled global transformation and, unlike Friedman, the rhetorical hype tends to be less florid and the approach more down-to-earth. A good example is provided by Palmisano (2006), the chairman and chief executive of IBM. Palmisano argues that the term multinational has been superseded and should be replaced by the notion of a globally integrated enterprise. He says that such enterprises "have been made possible by shared technologies and shared business standards, built on top of a global information technology and communications infrastructure" (p. 19). Although, at first sight, this might sound unproblematic as a description of modern-day enterprises such as IBM, it is phrases like "globally integrated" and "shared business standards" that I wish to challenge in this paper.

In opposition to Friedman, and to some extent Palmisano, this paper argues that the world is uneven, full of seams, culturally heterogeneous, locally specific, inequitable, not well-integrated, and constantly changing. In the following sections, three IS research areas are used to illustrate this argument, namely global software outsourcing, the global roll-out of information systems, and global virtual communities. In each of these areas, the aim is not to produce a full literature review, but rather to use selected articles to illustrate and analyze the non-flat nature of ICT-enabled forms of global working. The paper then builds on these analyses to produce an agenda for future research on ICTs and global working based around three research themes. Finally, some conclusions are drawn on the high potential value of research in such areas for the IS field as a whole.

2 GLOBAL SOFTWARE OUTSOURCING

The outsourcing of software and other services on a global basis has become an emblem of globalization and, for Friedman and others, a prime example of the new flat world. However, any detailed analysis of particular global software outsourcing cases presents a much more uneven picture of complex and evolving relationships between the outsourcing partners, including issues of cross-cultural contradiction and sometimes conflict. For example, Walsham (2002) uses structuration theory as a basis for the analysis of two case studies of global software production and use. The first involved a team of Indian software developers working in a Jamaican insurance company, and the second the use in India of geographical information systems software originally developed in the United States. Walsham shows that cultural differences in areas such as detailed work patterns, and ways of conceptualizing space, created contradictions and tensions in such global relationships, resulting in the first case in outright conflict and, in the second case, in major problems of implementation and use. There is no feel of "flatness" in these cases, although the author discusses the concept of negotiated culture (Brannen and Salk 2000) as a possible way forward in trying to address cross-cultural issues and difficulties.

Krishna et al. (2004) based their analysis of ways of managing cross-cultural issues in global software outsourcing on a wide range of cases that the authors had carried out on outsourcing from North America, Western Europe, and Japan to software suppliers in developing countries, with a particular emphasis on India. The primary conclusion from their research is that working across culture when outsourcing software production is not a trouble-free process. They identify different work practices, different modes of communication and the cultural adaptation of bridgehead teams as examples of difficulties that arise. They suggest ways of approaching these difficulties through, for example, the strategic choice of which projects to outsource, the use of common systems and processes, attempts to move toward a negotiated culture of work practice, and staff education and training. Many of these approaches can be argued to have been reasonably successful since global outsourcing continues to grow, but the picture that is produced is certainly not that of the globally integrated organization based on shared business standards.

Nicholson and Sahay (2001) describe a longitudinal case study of outsourcing from a UK company to an Indian software supplier. In addition to discussing cross-cultural issues and the changing nature of the business relationship over time, as in the articles above, the authors also make an interesting set of comments on the nature of globalization itself. They argue that global software outsourcing can provide examples of how the local interacts with the global in a two-way manner. They illustrate this by an example of an embodied software methodology, developed in the West but taken up in India, which was then re-embedded into the British context by the Indian programmers. The point is that globalization should not be seen as some top-down behemoth which is flattening the world, but rather as a complex and evolving process in which global–local interaction is two-way with often unforeseen outcomes.

The book by Sahay et al. (2003) is perhaps the most substantial contribution to the literature to date on the nature of global IT outsourcing relationships. Much of the research underpinning the book looked at outsourcing to India, widely regarded as the leader in the field. However, it is clear from the book that India's relative success has not been achieved without lengthy and sometimes painful learning processes on the part of those involved on both sides of outsourcing alliances. The authors identify several micro-level themes, reflecting tensions in such alliances: shifting identity, the complexities of knowledge sharing, the limits and benefits of standardization, issues of power and control, and the challenges of cross-cultural communication. It is not possible to discuss all of these here, so I will focus on the single theme of shifting identity in the rest of this section.

One argument that is often made by those who see the world as flattening is that people across the globe are becoming more similar in their interests, attitudes, and perhaps, therefore, identity. If one believes this argument, then surely Indian software workers who travel and work across the globe would provide strong evidence of this. This makes the work of D'Mello (2005, 2006) particularly interesting, since her research involved an extensive and prolonged engagement with such "global workers" on the theme of "selves and identities of IT professionals in India." In D'Mello (2005), she argues that these workers are engaged in complex identity shifts, and tensions of identity. The latter are brought about by, for example, potential conflicts between Indian group and family orientation and the Western-oriented individualism encountered in their client organizations. The author nicely reverses the standard cliché by arguing that such workers often "think local but act global."

Now it could be argued that the Indian software workers are nevertheless nearer to their western colleagues in terms of identity than they were in previously less-globalized eras. However, processes of hybridization produce much diversity as history has surely shown us, and the new hybridized Indian software workers are one example of this. Putting it more strongly, and borrowing from Haraway (1991) on cyborgs, I would argue that we are all identity hybrids now. D'Mello (2006) expresses this as follows for the workers she studied:

> organizations, industries and marketplaces are milieus deeply imbued with personal, social and existential structures and processes which evoke a range of feelings and subjectivities, and influence workers' sense of self and identity. I argue that, in global work contexts, workers construct their selves and identities from both global and local elements' (p. ix).

At the level of the individual, therefore, the world again does not look flat.

3 GLOBAL INFORMATION SYSTEM ROLL-OUT

A second area concerned with ICTs and global working is that of the global roll-out of information systems in particular organizations. The term *roll-out* is worth noting. The image which is evoked is of a smooth process, like rolling out a carpet across a flat floor. As my subsequent examples will attempt to show, the global roll-out of IS is normally anything but smooth, and the floor is anything but flat. Perhaps the wide use of the term roll-out reflects wishful thinking on the part of management that the process will be smooth, reinforced by vendors of software or services who wish to present global IS development in this way?

Joshi et al. (2007) describe and analyze the global roll-out of an information system, called ISX, to a communications and media group across a global pharmaceuticals company. The concept was to provide group members, independent of where they were located or their specific role, with a standard means to support key aspects of their work such as handling media enquiries, maintaining contacts databases, and generating news and information about the company and its products. The system was not a failure in the sense that it was used by many communicators across the organization and provided some degree of global integration amongst the widely separated groups in different countries. However, the ISX system was less successful than its designers intended, since its use was less widespread than they had hoped, and its global reach was limited by factors such as local relevance and cultural fit. For example, the communicators group in Japan used a Japanese-language system to support their interactions with local media, and the global system written in English could not cope with this. Secondly, and more subtly, the new system was aimed to provide better lateral information flows across the global organization, but the structured and hierarchical nature of the organization inhibited this form of global knowledge sharing.

Rolland and Monteiro (2002) describe a case study of the global roll-out of an information system in a maritime classification company working in over a hundred countries. The IS was designed to support the surveying of ships in order to assess their condition, to be used, for example, for insurance purposes. Rolland and Monteiro discuss the need for a "pragmatic balance" between universal standards and local specificity. They describe how this was achieved in their case by largely invisible work on negotiations, work-arounds, tinkering, etc. The authors argue that these approaches are not compensations for poor design but are necessarily required in all global IS infrastructures. So, global systems, according to them, are not seamless, integrated systems that provide common standards and approaches in all locations. Rather, continuing local work needs to be done to mold the systems to local specificities and these, in turn, affect the evolving global approach. In the words of the authors, "The real issue, we argue, is to analyze how global, never-perfect solutions are molded, negotiated, and transformed over time into workable solutions" (p. 96).

One of the most ubiquitous of global systems is enterprise resource planning (ERP) systems. These have been rolled-out in many organizations over the last 15 years. The success of these endeavors has undoubtedly been mixed (see, for example, Davenport 1998). The literature on ERP systems typically looks at case studies of particular organizations and echoes many of the points made earlier in this section, for example, on the difficulties of local relevance and cultural fit, and on the need for local negotiations, work-arounds, and tinkering. Referring briefly to the theme of the current paper, this literature alone surely refutes the notion of a flat world served by seamless integrated systems.

Nevertheless, ERP systems are widely used around the world, so some success has been achieved in making global solutions fit local conditions. An interesting strand of research looks at this issue from the opposite end to that of the user organizations, namely that of the ERP supplier. Pollock et al. (2003) argue the need to understand the "biography" of software packages such as ERP and the way that suppliers try to provide a "generic" product, whereas users try to accommodate tensions between that and local needs. In a later paper, Pollock et al. (2007) describe this work of the suppliers as "generification" work. The authors provided a detailed description of this process in the case of an ERP system aimed at supporting the administrative functions of universities. They made the point that generic packages such as this can and do serve diverse contexts. This is undoubtedly true, but two qualifying comments can be raised. First, there is an interesting question as to which users get most say in the generification process. In other words, how do power relations influence the final design and make the system more suitable for some than others? Second, Pollock et al. (2007) argue that diverse organizations and standard technologies *can* be brought together, but the question remains as to what the consequences are for the organizations, positive and negative. The generification of systems does not have "flat" effects across different organizations.

4 GLOBAL VIRTUAL TEAMS

A third area of high relevance for contemporary forms of global working is that of global virtual teams. A precise definition of virtuality is hard to achieve (Crowston et al. 2007) but is normally taken to involve geographical dispersion, a dependence on electronic interaction, and sometimes a dynamism of structure (Gibson and Gibbs 2006). The addition of the qualifier global implies cultural diversity is also present. In terms of the non-flatness theme of this paper, the thesis in this section is not that global virtual teams cannot be made to work well in some cases, nor that global organizations should necessarily place less emphasis on them. Rather, the argument is that global virtual teams exhibit enormous variety and thus that any one-size-fits-all solution to their design, implementation, and management is unlikely to be successful.

An early article on global virtual teams by Dubé and Paré (2001), based on interviews with team leaders and members, pointed out difficulties brought about by the diversity of people in terms of culture, language, and IT proficiency. They also noted that accessibility to technology and its appropriateness in particular contexts were further problems that needed to be faced by global virtual teams. In another early article, Qureshi and Zigurs (2001) raise similar issues about the importance of culture and technology. They make the useful point that the nature of tasks faced by virtual teams differ greatly, and they say that their case study data supported the view that the teams that worked best involved tasks with a high level of structure and tasks requiring detailed teamwork.

Majchrzak et al. (2004) take a very positive view of virtual teams supported by technology based on interviews with 293 members of "successful" teams. It is worth noting that they did not obtain permission to talk to members of teams that were considered failures, so their sample is rather biased. Nevertheless, the article is a good counter to any technologically determinist view that virtual teams supported by ICTs are necessarily inferior to those supported by face-to-face encounters. Technologies found to be of high value included virtual work spaces, teleconferencing, and instant messaging. The authors emphasize that a lot of effort needed to be put in by the leaders to try to knit the teams together.

Kayworth and Leidner (2002) explore this theme of the importance of leadership in global virtual teams through a study of leadership effectiveness in multicultural student teams drawn from MBA students in Europe, the United States, and Mexico. The authors conclude that important attributes of effective leaders include being good mentors, having empathy with team members, prompt communication habits, and clear articulation of team roles and responsibilities. There is limited emphasis on technology in this study other than to note that web sites tailored to team members' needs seemed to work well.

The paper by Kankanhalli et al. (2007) is valuable in that it focuses on conflict episodes in global virtual teams rather than the somewhat rosier side of the picture that is painted by some of the authors above. Although based on work with student teams, which potentially limits the strength of their findings, it is clear that cultural differences between people from North America, Europe, and Asia were significant causes of team conflict. The article also discusses differences in task complexity affecting team outcomes but the results here are rather inconclusive. The discussion of technology is somewhat limited, other than noting, for example, that sending too many e-mails and slow responses can be problematic to teamwork.

Although a full literature review has not been provided in this section, the examples suggest that the sum total of research on global virtual teams is currently rather thin. This view is supported by a substantial survey of the literature on virtual teams by Martins et al. (2004). The authors conclude that research on virtual teams to date, while valuable, had been relatively limited in scope, and based mainly on student laboratory studies. They identify a wide range of areas for further research on inputs, processes, outcomes and moderators of virtual teams, mentioning global virtual teams as a subset. In a recent short article focusing specifically on global virtual communities, Tan (2007) echoes this need for further research: "Although some preliminary research findings on virtual communities have been reported over the past few years, there is still a clear lack of coherent theoretical development and systematic empirical investigation on this topic" (pp. i-ii).

I move now to consider the topic of future research, not just for global virtual teams, but also for ICTs and global working more generally.

5 FUTURE RESEARCH ON ICTs AND GLOBAL WORKING

A key aim of the previous three sections has been to use the results of published research to argue that the world is a complex, uneven place, where relationships matter whether at the individual or organizational level. Diversity is everywhere and global working processes must adapt to this reality. ICTs are deeply implicated in all such processes, and thus vital to the contemporary world. Others have made similar points at greater length (e.g., Avgerou2002; Walsham 2001), but it seems that that the theme of global diversity and unevenness needs to be reiterated in the face of the journalistic hyperbole of Friedman and others. A key implication for the IS research community is that ICTs and global working is an exciting research area with a relatively untapped potential for innovative and interesting research work.

Table 1 provides a summary of some common research themes across the three topic areas discussed above. These common themes are identity and cross-cultural working; globalization, localization and standardization; and power, knowledge, and control. Each of these themes is discussed below in terms of how future IS research could develop further understanding of the theme. In each case, some examples of specific research topics are identified together with a brief discussion of some relevant literature.

5.1 Identity and Cross-Cultural Working

Cross-cultural issues have been discussed in the IS literature to date as we have seen, but really the surface has been barely scratched with many countries not represented at all. What do we know in depth about outsourcing to China, IS roll-out in Nigeria, or global virtual team members in Brazil? In a related way, issues of identity in these contexts have received little attention in the IS literature to date. Now it could be argued that there are substantial non-IS literatures that deal with countries such as those above. These literatures can no doubt be used to help us understand issues of identity and cross-cultural working in such contexts, but future research is needed to show us how this can

Table 1. Common Research Themes Across the Topic Areas

Common Themes	*Identity and Cross-Cultural Working*	*Globalization, Localization and Standardization*	*Power, Knowledge, and Control*
Global software outsourcing	• Cross-cultural working not trouble free • Tensions of identity of software workers	• Two-way local/ global interaction	• Complexities of knowledge sharing • Power and control issues between outsourcers and vendors
Global IS roll-out	• Local relevance and cultural fit	• Pragmatic balance between universal standards and local specificity • Generification of software packages	• Hierarchies deterring global knowledge sharing • Power relations influencing who has most say in final design
Global virtual teams	• Fit between technology and culture • Cultural differences as source of conflict	• High variety of global virtual teams prohibits one-size-fits all solutions • Nature of task important in team success	• Role of leadership in making global virtual teams work

be done. As an example of such work, Ailon-Souday and Kunda (2003) discuss cross-cultural working in an Israeli high-tech corporation undergoing a merger with an American entity. Their paper demonstrates how concepts of Israeli national identity were drawn on as a resource in order to deal with some of the social struggles which arose in cross-cultural working.

Cross-cultural working normally involves more than one country but, even within countries, there are important differences in identity and culture, despite the dull homogeneity implied by Hofstede's (1991) simplistic indices of national culture. A good example of this in the recent IS literature is provided by Miscione (2007). The author describes telemedicine practices in the Upper Amazon in northeastern Peru and shows how these western-inspired practices were frequently mismatched with the local culture in these areas and, in particular, with the definitions of health and illness employed by local communities and healers. The increasing importance of cultural diasporas scattered around the world further emphasizes the need to study within-country identity and cultural difference. Appadurai (1996) discussed such differences using the concept of imagination. He argued that people draw on contemporary sources, such as the media, and their own cultural histories to "annex the global" into their own "practices of the modern" through their imagination.

Drawing on existing concepts, such as those of Appadurai, is desirable in trying not to reinvent the wheel of theories of identity and culture when carrying out IS studies. However, IS studies have the potential in turn to contribute to broader research areas than those of the IS field alone. For example, D'Mello and Sahay (2007) extend their work on identity of Indian global software workers, referred to earlier in the paper, to discuss how geographical, social, and existential mobilities of these workers shape their relationship to place and to their identity. D'Mello and Sahay argue that their theoretical concepts of "mobility–identity" contribute to the call of the sociologist Urry (2000) for the development of a "sociology of mobilities."

5.2 Globalization, Localization and Standardization

Turning now to the future research theme of globalization and localization, there is a large literature upon which IS researchers can draw (e.g., Beck 2000; Castells 1996, 1997, 1998) in order to inform our work. However, detailed IS studies can contribute a crucial dimension to our understanding of global/local processes in that technology is normally a key element of such processes, and detailed empirical studies are needed to extend our understanding of how this works. For example, there is much interest in the topic of knowledge sharing and knowledge management in global organizations. Although there is a sizeable literature on this topic area, most of it does not deal with the full diversity of global organizations with a few exceptions (e.g., Pan and Leidner 2003). This is perhaps understandable in that research access can be difficult when dealing with widespread global operations and research costs can be high. Nevertheless, we are getting a very partial picture of global knowledge sharing processes when we exclude large portions of the world from our studies.

A second example of a future IS research topic of high interest, with little published academic work to date, is the relationship between global working and the use of social networking technologies and other forms of user-driven web-based systems (McAfee 2006). One should not be carried away by the hype about Web 2.0 to believe, for example, that Facebook will solve the problem of cross-cultural collaboration, as one student suggested to this author recently. However, it would be equally foolish to neglect the potential of such technologies to support work in general, and global working in particular. Similarly, there is great interest (e.g., Jagun et al. 2007) in the potential for mobile technologies to provide relevant ICT-based services to relatively poor people in developing countries, to enable them to participate more fully in the globalizing world, where previous fixed technologies such as the landline phone and the fixed PC have not been widely available or used.

One important IS topic within the theme of globalization/localization is that of standardization and generification of software packages. Some work on this was discussed earlier in the paper, but again there is very little published work that extends beyond the context of a limited range of wealthy countries to the wider global arena, despite the fact that software such as ERP systems has spread across most of the world. An exception to this is the work known as the HISP program. Braa et al. (2007) describe some elements of this work on health information systems, which started in South Africa but has now spread to a wide range of developing countries. The authors describe the theoretical concept of flexible standards to support standardization processes in complex systems such as health care, and suggest an approach to implementing such standards in developing country settings that is sensitive to the local context, allowing changes to occur in small steps, but retaining a mechanism for scaling information systems across wider domains.

5.3 Power, Knowledge, and Control

The links between power and knowledge and their effects on control were present implicitly in much of the research literature discussed earlier. However, an explicit focus on such elements was often missing, but is surely crucial to an increased understanding of global working in a non-flat world. Future research is needed to address questions about the power relations between outsourcing suppliers and vendors. With respect to global IS roll-out and generification processes, who controls the processes and to what extent are local interests and local knowledge ignored or given low priority? How do power and politics play out in global virtual teams and who are the winners and losers?

A further broad area for future work on globalization and ICTs from the perspective of power and control would take a critical approach and ask research questions about silent voices, namely those often not considered at all in the networks of the powerful. For example, global software outsourcing may benefit the elites in India and elsewhere, but is there any evidence of spin-off effects in other areas such as the use of ICTs within India to benefit a wider range of stakeholders (Krishna and Walsham 2005)? Is the global roll-out of systems a form of neo-colonialism when imposed on particular developing countries (Adam and Myers 2003)? Who is marginalized by the use of ICTs in global working? For example, on the latter question, Thompson (2004) shows how the effects of the World Bank's global knowledge forum can be taken to exclude in-country inputs and grassroots participation in the development agenda.

IS researchers studying global issues involving power, politics, development, and marginalization have already drawn on a range of related literatures on these topics but this cross-discipline approach could be extended further. For example, institutional theory has been suggested as potentially valuable to the IS field (Orlikowski and Barley 2001) but its application has been limited to date.Although not an IS study as such, Khan et al. (2007) provide a good illustration of detailed work based on institutional theory and the "dark side" of even the well-intentioned exercise of power in the context of globally distributed working. The authors describe a case study of institutional entrepreneurship concerned with removing child labor from soccer ball production in Pakistan but with unintended negative consequences on women stitchers in that country. A second body of literature of relevance to issues of power and politics, which has received little attention in the IS field, is that of development studies. Thompson (2007) argues that work on ICTs and development should make more effort to connect to the development studies literature, and thus for IS researchers to become better engaged with issues of practice and policy in developing countries.

6 CONCLUSIONS

This paper has argued that the new ICT-enabled world is not flat. Analyses of three specific areas of ICTs and global working have shown the importance of unevenness in areas such as cultural difference, the heterogeneity of work processes, the significance of local specificity, and the asymmetry of power relationships. Organizations are not well-integrated as some senior practitioners claim, but instead struggle to make ICT-enabled global working effective. None of this argues against the importance of ICTs in the contemporary world, including the growing service-based component. Indeed, it could be argued that the increased emphasis on services makes the application of ICTs even more crucial since ICTs can be regarded as one of the primary transport systems for a globally-distributed service-based world.

There is an existing literature on ICTs and global working, but it remains relatively sparse to date. We can build on this research to further explore themes such as those identified in the previous section: shifting identity and cross-cultural working, globalization and standardization, and issues of power and control. We can try to connect better to other relevant literatures and disciplines such as those concerned with identity, culture, globalization, and development. Conducting in-depth research on ICTs and global working is not an easy task, due to issues such as time, cost, and complexity, but it offers the IS field a major opportunity to make a significant contribution to our understanding of crucial issues in our more globalized world.

References

Adam, M., and Myers, M. 2003. "Have You Got Anything to Declare? Neo-Colonialism, Information Systems, and the Imposition of Customs and Duties in a Third World Country," in *Organizational Information Systems in the Context of Globalization*, M. Korpela, R. Montealegre, and A. Poulymenakou (eds.), Boston: Kluwer Academic Publishers, pp. 101-116.

Ailon-Souday, G., and Kunda, G. 2003. "The Local Selves of Global Workers: The Social Construction of National Identity in the Face of Organizational Globalization," *Organization Studies* (24:7), pp. 1073-1096.

Appadurai, A. 1996. *Modernity at Large: Cultural Dimensions of Globalization,* New Delhi: Oxford University Press.

Avgerou, C. 2002. *Information Systems and Global Diversity*, Oxford, UK: Oxford University Press.

Beck, U. 2000. *What Is Globalization?*, Cambridge, UK: Polity Press.

Braa, J., Hanseth, O., Heywood, A., Mohammed, W., and Shaw, V. 2007. "Developing Health Information Systems in Developing Countries: The Flexible Standards Strategy," *MIS Quarterly* (31:2), pp. 381-402.

Brannen, M. Y., and Salk, J. E. 2000. "Partnering Across Borders: Negotiating Organizational Culture in a German–Japan Joint Venture," *Human Relations* (53:4), pp. 451-487.

Castells, M. 1996. *The Rise of the Network Society* (1st ed.), Cambridge, UK: Blackwell Publishers Ltd.

Castells, M. 1997. *The Power of Identity* (1st ed.), Cambridge, UK: Blackwell Publishers Ltd.

Castells, M. 1998. *End of Millennium* (1st ed.), Oxford, UK: Blackwell Publishers Ltd.

Crowston, K., Sieber, S., and Wynn, E. (eds.). 2007. *Virtuality and Virtualization*, Boston: Springer.

Davenport, T. H. 1998 "Putting the Enterprise into the Enterprise System," *Harvard Business Review*, July-August, pp. 121-131.

D'Mello, M. 2005. "'Thinking Local, Acting Global': Issues of Identity and Related Tensions in Global Software Organizations in India," *Electronic Journal of Information Systems in Developing Countries* (22:2), pp. 1-20.

D'Mello, M. 2006. *Understanding Selves and Identities of Information Technology Professionals: A Case Study from India,* unpublished Ph.D. Dissertation, Faculty of Social Sciences, University of Oslo.

D'Mello, M., and Sahay, S. 2007. "'I Am Kind of a Nomad Where I Have to Go Places and Places' …Understanding Mobility, Place and Identity in Global Software Work from India," *Information and Organization* (17:3), pp. 162-192.

Dubé, L., and Paré, G. 2001. "Global Virtual Teams," *Communications of the ACM* (44:12), pp. 71-73.

Friedman, T. 2005. *The World is Flat: A Brief History of the Twenty-First Century*, New York: Farrar, Straus and Giroux.

Gibson, C. B., and Gibbs, J. L. 2006. "Unpacking the Concept of Virtuality: The Effects of Geographic Dispersion, Electronic Dependence, Dynamic Structure, and National Diversity on Team Innovation," *Administrative Science Quarterly* (51:3), pp. 451-495.

Haraway, D. J. 1991. *Simians, Cyborgs, and Women: The Reinvention of Nature*, London: Free Association Books.

Hofstede, G. 1991. *Cultures and Organizations: Software of the Mind*, New York: McGraw-Hill.

Jagun, A., Heeks, R., and Whalley, J. 2007. "Mobile Telephony and Developing Country Micro-Enterprise: A Nigerian Case Study," Working Paper No 29, Institute for Development Policy and Management, University of Manchester.

Joshi, S., Barrett, M, Walsham, G., and Cappleman, S. 2007. "Balancing Local Knowledge Within Global Organizations Through Computer-Based Systems: An Activity Theory Approach," *Journal of Global Information Management* (15:3), pp. 1-19.

Kankanhalli, A., Tan, B. C. Y., and Wei, K-K. 2007. "Conflict and Performance in Global Virtual Teams," *Journal of Management Information Systems* (23:3), pp. 237-274.

Kayworth, T. R., and Leidner, D. E. 2002. "Leadership Effectiveness in Global Virtual Teams," *Journal of Management Information Systems* (18:3), pp. 7-40.

Khan, F. R., Munir, K. A., and Willmott, H. 2007. "A Dark Side of Institutional Entrepreneurship: Soccer Balls, Child Labour and Postcolonial Impoverishment," *Organization Studies* (28:7), pp. 1055-1077.

Krishna, S., Sahay, S., and Walsham, G. 2004. "Managing Cross-Cultural Issues in Global Software Outsourcing," *Communications of the ACM* (47:4), pp. 62-66.

Krishna, S., and Walsham, G. 2005. "Implementing Public Information Systems in Developing Countries: Learning from a Success Story," *Information Technology for Development* (11:2), pp. 123-140.

Leamer, E. E. 2007. "A Flat World, a Level Playing Field, a Small World after All, or None of the Above? A Review of Thomas L. Friedman's *The World is Flat*," *Journal of Economic Literature* (XLV), pp. 83-126.

Majchrzak, A., Malhotra, A., Stamps, J., and Lipnack, J. 2004. "Can Absence Make a Team Grow Stronger?," *Harvard Business Review*, May, pp. 131-137.

Martins, L. L., Gilson, L. L., and Maynard, M. T. 2004. "Virtual Teams: What Do We Know and Where Do We Go from Here?," *Journal of Management* (30:6), pp. 805-835.

McAfee, A. P. 2006. "Enterprise 2.0: The Dawn of Emergent Collaboration," *MIT Sloan Management Review* (47:3), pp. 21-28.

Miscione, G. 2007. "Telemedicine in the Upper Amazon: Interplay with Local Health Care Practices," *MIS Quarterly* (31:2), pp. 403-425.

Nicholson, B., and Sahay, S. 2001. "Some Political and Cultural Issues in the Globalization of Software Development: Case Experience from Britain and India," *Information and Organization* (11:1), pp. 25-43.

Orlikowski, W. J., and Barley, S. R. 2001. "Technology and Institutions: What Can Research on Information Technology and Research on Organizations Learn from Each Other?," *MIS Quarterly* (25:2), pp. 145-165.

Palmisano, S. 2006. "Multinationals Have Been Superseded," *Financial Times*, June 12, p. 19.

Pan, S. L., and Leidner, D. E. 2003. "Bridging Communities of Practice with Information Technology in Pursuit of Global Knowledge Sharing," *Journal of Strategic Information Systems* (12:1), pp. 71-88.
Pollock, N., Williams, R., and Procter, R. 2003. "Fitting Standard Software Packages to Non-Standard Organizations: The Biography of an Enterprise-Wide System," *Technology Analysis & Strategic Management* (15:3), pp. 317-332.
Pollock, N., Williams, R., and D'Adderio, L. 2007. "Global Software and its Provenance: Generification Work in the Production of Organizational Software Packages," *Social Studies of Science* (37:2), pp. 254-280.
Qureshi, S., and Zigurs, I. 2001. "Paradoxes and Prerogatives in Global Virtual Collaboration," *Communications of the ACM* (44:12), pp. 85-88.
Rolland, H. H., and Monteiro, E. 2002. "Balancing the Local and the Global in Infrastructural Information Systems," *The Information Society* (18:2), pp. 87-100.
Sahay, S., Nicholson, B., and Krishna, S. 2003. *Global IT Outsourcing: Software Development Across Borders,* Cambridge, UK: Cambridge University Press.
Tan, B. C. Y. 2007. "Leveraging Virtual Communities for Global Competitiveness," *Journal of Global Information Management* (15:3), pp. i-iii.
Thompson, M. 2004. "Discourse, 'Development' and the 'Digital Divide': ICT and the World Bank," *Review of African Political Economy* (31:9), pp. 103-123.
Thompson, M. 2007. "ICT and Development Studies: Towards Development 2.0," Working Paper 27/2007, Judge Business School, University of Cambridge (http://www.jbs.cam.ac.uk/research/working_papers/2007/wp0727.pdf).
Urry, J. 2000. "Mobile Sociology," *British Journal of Sociology* (51:1), pp. 185-203.
Walsham, G. 2001. *Making a World of Difference: IT in a Global Context,* Chichester, UK: Wiley.
Walsham, G. 2002. "Cross-Cultural Software Production and Use: A Structurational Analysis," *MIS Quarterly* (26:4), pp. 359-380.

About the Author

Geoff Walsham is a Professor of Management Studies (Information Systems) at Judge Business School, University of Cambridge. In addition to Cambridge, he has held academic posts at the University of Lancaster in the UK where he was Professor of Information Management, the University of Nairobi in Kenya, and Mindanao State University in the Philippines. His teaching and research is focused on the question: Are we making a better world with information and communication technologies? He was one of the early pioneers of interpretive approaches to research on information systems. Geoff is currently a senior editor of *MIS Quarterly*. He can be reached at g.walsham@jbs.cam.ac.uk.

3 EXPLORING THE INFLUENCE OF SOCIO-EMOTIONAL FACTORS ON KNOWLEDGE MANAGEMENT PRACTICES: A Case Study

Chalee Vorakulpipat
Yacine Rezgui
Research Institute for the Built and Human Environment
University of Salford
Salford, U.K,

Abstract *The objective of this empirical study is to explore the influence of socio-emotional factors, with a focus on shyness, on knowledge management practices in a Thai organization. The research adopts an interpretive stance and employs a case study approach involving multiple data collection methods. The paper is based on one author's personal expertise and close involvement, for over a decade, in the selected case study organization for over a decade. Using a grounded theory research approach, the study indicates that while shyness is overall perceived as a positive Thai cultural feature, it critically influences (1) the social network ties and relationship between employees within and across teams, (2) the resulting level of trust, including with management and senior staff, and (3) the ability to share and create knowledge effectively in the organizational socio-cultural environment. The study is limited to a Thai organization, but can be generalized to other organizations that exhibit similar characteristics. This empirical study provides a foundation to further the research and the validation of the summary of themes that emerged from this empirical study.*

Keywords Knowledge management, shyness, interpretive case study, grounded theory, Thailand

Please use the following format when citing this chapter:

Vorakulpipat, C., and Rezgui, Y., 2008, in IFIP International Federation for Information Processing, Volume 267, Information Technology in the Service Economy: Challenges and Possibilities for the 21st Century, eds. Barrett, M., Davidson, E., Middleton, C., and DeGross, J. (Boston: Springer), pp. 27-41.

1 INTRODUCTION

Knowledge management (KM) has become an important ingredient to sustain competitiveness in developing countries (Wagner et al. 2003). Very few articles, unfortunately, have reported KM implementations and strategies in developing countries. The ones that do include China (Burrows et al. 2005), Malaysia (Wei et al. 2006), India (Chatzkel 2004), and sub-Saharan Africa (Okunoye 2002). These studies have identified several distinctive features, including varying levels of expertise to adapt and adopt technologies, distinctive socio-cultural features, and lack of availability of human and financial resources to nurture KM practices (Okunoye 2002). A call has been made for further research to explore KM in different organizational and cultural (regional, national, and international) contexts in developing economies.

Although technology plays an important role in the successful implementation of KM initiatives (Koenig 2002), a number of distinctive socio-emotional features such as shyness have an equally important role and influence (Chaidaroon 2004), in particular in the cultural context of developing countries. Shyness in the study is defined as the presence of inhibition and discomfort (such as tension, worry, awkward behavior, and gaze aversion) in social situations (meeting strangers, for instance) (Cheek and Buss 1981).

Thailand is an example of a developing country where a number of distinctive cultural features have been identified, including shyness (Chaidaroon 2004). Therefore, it represents an interesting case to conduct a study on the influence of shyness on KM practices within an organizational context. While KM practices in Thailand have been reported on later than in other countries in the region, several private and public organizations have already initiated ambitious KM programs and initiatives (Vorakulpipat and Rezgui 2006). There is an interesting trend in the region to promote a competitive economy through technology and knowledge infused practices at a societal level. For example, the Ninth Malaysian Plan (2006-2010) has as one of its objectives to raise the capacity for knowledge and innovation, whereas the Ministry of Research and Technology (MRT) of Indonesia has identified information and communications technologies as a priority field to add value to its industries.

The aim of the paper is to explore the influence of shyness on KM practices in a selected Thai organization. As such, the core research question is, how does *shyness* (as a socio-emotional characteristic) influence, and how is it influenced by, knowledge management practices in an organizational context? The paper makes two main contributions. First, drawing on the rich data of a Thai IT research service organization, it generates a grounded understanding of the influence of shyness on KM. Second, the paper proposes a summary of themes developed from the grounded analysis of gathered primary data evidence from the case study, using social capital and related literature. This will allow researchers and practitioners to conceptualize and research further the influence of shyness on KM. The paper is organized into seven sections. Following this introduction, the paper presents related literature, and then the research methodology employed in this study, which involves a case study approach. The research results are then given, followed by the discussion. Finally, recommendations for further research are presented and conclusions are drawn.

2 RELATED RESEARCH

Previous research has highlighted the idea that technology adaptations in developed countries occur continuously in response to misalignments, gradually leading to a successful alignment (Leonard-Barton 1988). This is in contrast to developing countries, which tend to rapidly adopt technology created by developed countries, often in an *ad hoc* way (Archibugi and Pietrobelli 2003). In 2002, the National Science Foundation reported that more than 84 percent of the world's scientific and technological production is concentrated in developed countries. Developing countries have only marginally increased their participation, which emphasizes the scientific and technological gap that exists with the developed world. Also, in several of the information technology installations that were created and adapted for organizations in developing countries, local (regional and national) factors were not taken into account. This has resulted in outcomes that do not fit the needs of the direct beneficiaries in the developing nations (Cyamukungu 1996).

While the above is applicable to KM, the crucial issue might not relate only to technology but also include other factors, such as cultural-based resistance: "technology, designed and produced in developed countries, is likely culturally-biased in favor of industrialized socio-cultural systems, technology transferred to developing countries meets cultural resistance" (Straub et al. 2001). Moreover, it is reported that there is a significant gap in the understanding and maturity of KM between Asian developing companies and those in developed countries. This can be explained by the fact that American and European companies have had KM strategies and initiatives in place for over a decade, while Asian developing companies are still attempting to understand and apply the concept of KM (Yao et al. 2007).

Chaidaroon (2004) indicates that shyness is a national characteristic that distinguishes Thai culture and communication styles from Western (developed countries) counterparts. Thai people tend to place high value and responsibility in interactions through the process of receiving messages—in fact, Thai silence is a positive sign of respect (Knutson 2003). Chaidaroon (2003) argues that shyness or not speaking up can sometimes be strategically performed by Thai people to gain recognition from others. In addition, Thai culture is more hierarchical than in Western societies (McCampbell et al. 1999), which can engender a large gap between people due to personal barriers such as shyness. Based on this secondary evidence drawn from the literature, conducting an exploratory study in a national context (Thailand) is beneficial to further research and understanding of the role and impact of shyness on KM in a different socio-cultural context. This justifies the use of shyness as a national cultural variable as opposed to an individual characteristic in the context of this exploratory study, as argued by Choo (2003).

To be more general, research on human and organizational aspects related to KM has focused on understanding the socialization and organizational dimension of KM (Becerra-Fernandez and Sabherwal 2001). The concept of social capital has recently been adopted within the discipline (Huysman and Wulf 2006; Lesser 2000; Nahapiet and Ghoshal 1998), emphasizing the roles of trust, motivation, and social cohesion within the organization. Clearly, the higher the level of social capital, the more communities are stimulated to connect and share knowledge (Huysman and Wulf 2006).

3 RESEARCH APPROACH

The research aims at investigating the role of socio-emotional factors in the context of knowledge management in a Thai organization. The authors thus needed to gain an in-depth understanding of KM practices in the selected case study. Empirical studies that collect such data can be broadly classified as *interpretive case studies* (Walsham 1995). This type of approach has been selected since it aims to understand human thoughts and action in social and organizational contexts and to produce deep insights into information systems phenomena (Klein and Myers 1999). However, there are significant differences of methodology and theory under broad, interpretive case studies. The remainder of this section is devoted to describing the specific approach adopted in the research and the reasons for the choices.

The research methodology is based on grounded theory (Glaser and Strauss 1967). This is motivated by the facts that (1) grounded theory "is an inductive, theory discovery methodology that allows the researcher to develop a theoretical account of the general features of a topic while simultaneously grounding the account in empirical observations or data" (Martin and Turner 1986, p. 141), (2) grounded theory facilitates "the generation of theories of process, sequence and change pertaining to organizations, positions and social interaction" (Glaser and Strauss 1967, p. 114), and (3) there are few guidelines for analyzing qualitative data (Miles and Huberman 1994) and it has been argued that grounded theory approaches are particularly well-suited to dealing with the type of qualitative data gathered from interpretive field studies (Martin and Turner 1986; Oates 2005).

Site selection was guided by a technique of theoretical sampling (Strauss and Corbin 1998, p. 201): "data gathering driven by concepts derived from the evolving theory and based on the concept of making comparisons." BETA (the name of the organization has been disguised), a service organization in Thailand that conducts research in information technology was thus selected as a case study for the investigation as the authors believe that BETA exhibits a KM-rich environment to address the research question. Also, site selection depends on easy access for the case study. It is worth noting that one of the authors was not only employed by BETA, but had a close involvement and critical role, for over a decade, in BETA's IS and KM implementation initiatives. Indeed, the researcher has, over the years, acquired substantial personal knowledge of the organization's culture and work environment. Therefore, the organization welcomed the researcher to conduct this in-depth case study, and provided adequate support throughout the research. Also, because of personal involvement and role a in the organization, the researcher chose to analyze the collected data in an interpretive way based on his own experiences.

BETA was founded over 20 years ago. It employs more than 600 people, a majority of whom are highly educated and work in research and development production departments. BETA initially acted as a research supplier to Thai industries for over a decade. Following an increasing demand for R&D, BETA has transformed itself from a supply-driven to a demand-driven organization. This demand-focused strategy has helped BETA address and meet the needs of Thai organizations more effectively. In the late 1990s, management initiated a large KM program. In the first stage, a collaborative system was deployed and adopted to help staff collaborate more effectively while promoting knowledge-friendly practices. Also, physical and virtual social spaces have been provided for sharing knowledge. Later, management deployed a knowledge repository system to encourage staff to codify tacit knowledge and experience into a reusable form. A number of incentives have been introduced, including monetary rewards and recognition to motivate people to share and create knowledge. While KM initiatives have been underpinned by IT, it was found that the organization was not successful in achieving the objectives of those initiatives and the overall results were less than desired. It is believed that a socio-cultural feature, shyness, has critically influenced the achievement of BETA's KM initiatives. This justifies the need for research on the influence of shyness on KM in the context of BETA.

Data were collected through a variety of methods: semi-structured interviews, observation, and documentation. An interview guide was developed to collect critical qualitative data from all four "core" production (R&D) departments. The interview questions addressed a number of areas, including organization nature, teamwork and organization environment, information technology adoption, knowledge sharing and creation, and organizational change.

Before interviewing, the selected interviewees were encouraged to explain their knowledge background including their level of previous education, work experience, KM experience, etc. Twelve top management and key persons from the production departments were subjectively selected as interviewees as they were perceived to have permission to provide critical (or sensitive) data and constructive comments. Tape recording was used for nine interviewees, while the others requested not to be recorded during the interview.

Besides collecting sensitive data from management staff, additional data from employees were captured in the mode of direct observation (Yin 2003) throughout the entire study. The researcher was provided an opportunity to observe employees' working styles, environments, and reactions. Moreover, as the researcher had worked in BETA for over a decade, he had a chance to discuss informally some underlying issues among his colleagues and other employees. Also, documentation about the organization was examined in-depth.

The data triangulation technique was chosen to analyze data collected from multiple sources (Yin 2003), since "it is particularly beneficial in theory generation as it provides multiple perspectives on an issue, supplies more information on emerging concepts, allows for cross-checking, and yields stronger substantiation of constructs" (Orlikowski 1993, p. 312). The research involved multiple realities that were interpreted by different viewpoints, therefore, a broad variety of these have been particularly made in the field study. The BETA case was supposed to illustrate differing viewpoints among employees, managers, and the researcher. As an interpretive stance was adopted, the findings of this study would comprise the researcher's own interpretations and those of others (respondents: employees and managers) who were involved in the study. However, as the researcher controlled the study, the work is ultimately presented from the researcher 's perspective, a typical criticism of interpretive studies. Moreover, since the interpretive study may require sensitivity to possible biases and systematic distortions in the narratives collected from the participants (Klein and Myers 1999), the researcher might not take the informants' views at face value.

Finally, the process of data collection, coding, and analysis is iterative (Glaser and Strauss 1967). This iterative process only finishes when it becomes clear that further data no longer triggers new modifications to the data categories and emerging theory, that is, the research has reached *theoretical saturation* (Strauss and Corbin 1998). Eisenhardt (1989) notes that overlapping data analysis with data collection can allow researchers to take advantage of flexible data collection and make adjustments freely during the data collection process. In the research, this process of data collection, coding, and analysis was taken to validate that the researcher had captured the perspectives of the informants on the topic. Pattern coding techniques of qualitative analysis (Miles and Huberman 1994) were used to summarize segments of the data from interview transcripts and observation notes, and then to determine categories or pattern codes.

4 RESEARCH RESULTS

A number of categories emerged from the data analysis using pattern coding techniques of qualitative analysis (Miles and Huberman 1994). These are information and communications technology (ICT), team working, structure and culture, training, sharing knowledge, and motivation for knowledge management. The pattern coding here is processed iteratively. These categories are discussed below.

4.1 Information and Communications Technology (ICT)

Analysis of the interview transcripts suggest that ICTs help address shyness barriers as employees feel more at ease when communicating via electronic means as opposed to face-to-face in social or work contexts. Moreover, e-mail plays an important role and seems to be preferred to telephone and face-to-face interactions, in particular when these involve interactions between employees and their senior staff. Also, intranets and extranets are highly valued as they promote flexible work, including access to document repositories and knowledge, while minimizing physical social interactions between employees. However, it is observed that this can have the adverse effect of hindering social cohesion between employees, which is essential to develop trust and sustained relationships.

4.2 Team Working

Gathered data suggest that tasks and R&D in BETA are achieved through teamwork, and this emerges as the preferred mode of working across the organization. Hence, strong social relationships among employees are critical to promote effective working. It has been reported that a bureaucratic (hierarchical) organizational structure is perceived to inhibit positive social relationships among employees, in particular when teams involve members at various levels of the organizational structure. The interviewees have reported that a number of socio-emotional factors, including shyness (not speaking up) and seniority, inhibit teamwork effectiveness as employees usually believe that they should act in a receiver role in their team and should not elaborate and argue their own ideas against the ones of older or senior staff. Some respondents suggest that a more participative culture should underpin team working to gradually overcome the overall bureaucratic environment that characterizes work at BETA. Promoting the appropriate teamwork environment and atmosphere helps staff reduce their shyness and fears by encouraging them to contribute effectively through constructive comments to managers or team leaders. One interviewee confirms this and provides a suggestion to reduce shyness:

> *The problem is that my employees are very shy and thus often reluctant to present their ideas. This is a Thai behavior. But if I force them to do so, they can. In our meetings, I always allow my employees to ask and comment about everything. I hope they, especially young employees, will be at ease and brave enough to expose and present their innovative suggestions and ideas to me.*

Moreover, another interviewee states that shyness may make employees miss an opportunity to sustain ties with others:

> *When we've got some guests wanting to visit our offices, our colleagues tend to be shy and nervous to meet them. The problem is that they try to avoid discussing with them and present our work to them. I recognize that this is a Thai behavior, but it would be better if they could get to socialize and know others. Keeping up ties with significant people who visit us is always good.*

4.3 Structure and Culture

It is observed that the dominant bureaucratic (hierarchical) organizational structure within BETA may inhibit effective communication and employee participation and contribution, also leading to personal barriers. This is reflected in existing working procedures (e.g., reporting layers) and may generate conflicting corporate cultures. Although the hierarchical structure and culture in place is perceived to help increase trust and respect for senior staff, it inhibits self development as this amplifies employees shyness instead of helping it. One respondent confirmed this:

> *The ideas are always finalized by only senior members. Young members usually keep quiet, as they tend to be concerned about, and fear, the negative impact of criticizing ideas expressed by older members.*

Nevertheless, the respondents perceive that this bureaucratic structure is widely accepted in Thai organizations, and it is a non-changeable organizational feature. Despite this, they feel that promoting participation within a hierarchical organization can improve a sense of collectiveness and help remove shyness barriers.

4.4 Training

The interviewees perceive that fostering a learning organization culture promotes employee development. A variety of training methods such as formal training courses, on-the-job training, and learning from documents can encourage staff to develop their self-learning skills. Informal internal training (such as an informal forum during tea break) seems to be increasingly promoted as it allows people to exchange knowledge within a casual environment while addressing personal communication barriers due to shyness. While training is supposed to encourage lively interactions among trainers and trainees, and create strong relationships among employees during and after the training, in practice the impact tends to be limited, as trainees tend to be too shy, as reported by one interviewee:

> *Training would be ineffective if most trainees do not interact effectively with trainers such as not asking any questions and keeping quiet because of their reluctance and nervousness to speak in public.*

4.5 Sharing Knowledge

The respondents perceive that sharing knowledge by informal or traditional face-to-face interaction is preferred to virtual means (supported by technology) despite acknowledged shyness barriers. It has been reported that collective knowledge sharing in informal (face-to-face) contexts, such as discussion forums and coffee breaks, is highly valued as this method can break shyness barriers between employees and management by (1) establishing stronger relationships to develop trust among them, and (2) practicing and improving their presentation skills to gain confidence. One interviewee clearly explains that gaining confidence to remove shyness should be addressed before sharing knowledge in a team.

> *Firstly our objective is not knowledge sharing. We just want to practice their [employees'] presentation skills to reduce shyness and gain self-confidence. I'm quite sure they have good knowledge of their work, but this has always been difficult to share with others. They need to practice how to speak confidently in public and make the audiences understand them.*

The same interviewee adds a solution to improve presentation skills.

> *Another objective is to archive knowledge. I use a technique whereby I record their knowledge experience so that it becomes reusable. Then I invite my colleagues to watch the recording and be self-critical by identifying their weaknesses and thus improve further their presentation skills.*

On the other hand, virtual contexts also have been developed to break shyness barriers, as participants do not need to identify themselves and thus express their facial emotions. While this method seems promising, some interviewees believe that it is too informal and facial emotions cannot be expressed through this channel, as it is very important in formal interactions to avoid misunderstanding during communication.

4.6 Motivation for Knowledge Management

The respondents perceive that rewards are not always motivating them to share knowledge, as they do not feel that rewards impact their work performance. However, if reward systems are used in the organization, these should be reliable, reasonable, and fair enough to motivate people to share knowledge. It is observed that employees are still reluctant to share knowledge because they do not believe that the organization has sufficient financial resources to provide rewards. There is an overall feeling that the reward system/norm is unfair and does not factor in employees potential socio-emotional barriers that prevent them from adopting effective knowledge management practices. In addition, one interviewee has reported the following:

> *Perceived inappropriate criteria to get rewards lead to unfairness among staffs and cause them to feel embarrassed to not qualify for rewards. The criteria judge the employee performance based on the number of published research papers, and the same criteria are used for all employees. Therefore, these criteria encourage employees from R&D departments only, whereas it is very difficult for those from other non-research areas such as administrative departments to be rewarded. These are reported as unfair.*

Nevertheless, it is observed that although employees feel that this reward system is unfair, they are still shy enough to voice their concerns. They prefer to say nothing so as not to upset their hierarchy. This is because senior people are perceived to be the decision makers in Thai organizations, and it is very common that Thai junior employees are too shy to challenge their hierarchy. It is worth noting that management in BETA has initiated an anonymous informal online discussion to allow employees to comment or criticize anything they feel unsatisfied with in the organization. However, it has been reported that anonymity considerations are still an issue in this type of interaction. Senior staff should preserve the anonymity of participants in the discussion, and thus establish trust in relation to the use of such techniques.

5 DISCUSSION

As grounded theory facilitates "the generation of theories of process, sequence and change pertaining to organizations, positions and social interaction" (Glaser and Strauss 1967, p. 114), it is essential to base the discussion on existing relevant theories. This discussion will be based on social capital theory. Nahapiet and Ghoshal (1998) suggest that social capital should be considered in terms of three dimensions. First, the structural dimension refers to the opportunity to connect with each other. Second, the relational dimension refers to the character of the connection between individuals and motivation to share knowledge. This is best characterized through trust, norms, obligation, and respect. Third, the cognitive dimension refers to the ability to cognitively connect with each other in order to understand what the other is referring to when communicating and sharing knowledge. The discussion of shyness elaborates on how shyness is (1) affecting network ties and relationships, (2) affecting trust between people, and (3) affecting the ability to share and create knowledge. Thus, this section will involve the three dimensions of structural opportunity, relational motivation, and cognitive ability.

5.1 Structural Dimension

The concept of the structural dimension of social capital is used to refer to the overall pattern of connections between actors affected by shyness. The results confirm that shyness at work is perceived to inhibit the creation of strong social relationships among employees. It is observed that employees feel nervous when working with people they have not worked with previoiusly. They usually believe that they should act in a receiver role in their team and shyness can sometimes be strategically performed to gain recognition from others, as argued in related literature (Chaidaroon 2004).

To reduce shyness in a team, the results suggest that promoting adequate environments (informal, casual, and in smaller groups) and providing opportunities to staff to practice their presentation skills can yield promising results, as confirmed by findings related to the concept of open-mindedness (Al-Saggaf 2004). Employees are offered informal forums during tea break, leading to the opportunity of speaking with each other, especially visitors, without the potential of reducing shyness. It is suggested that a more participatory culture should be promoted in BETA.

Beyond pure communication, a virtual space that allows the creation and development of online collaboration may foster the structural dimension of social capital. The results suggest the adoption of virtual spaces (such as online discussion forums) to help boost confidence and reduce shyness barriers, leading to the initiation of effective online communities. Also, this virtual space can help employees direct themselves to strengthen existing social ties or build up new ones (Huysman and Wulf 2006). A substantial amount of research suggests that the use of the virtual space has the potential to break down some barriers to participation by removing certain psychological elements including shyness encountered by employees when expressing their views in public meetings (Al-Saggaf 2004). Additionally, Al-Saggaf (2004) notes that online communities help people become more open-minded in their thinking, and be more aware of the wider characteristics of individuals within their society.

5.2 Relational Dimension

The relational dimension here is based on a socially attributed characteristic of the relationship including trust and norm. This sub-section aims to discuss how shyness affects trust between people.

The results show that shyness may lead to a lack of motivation including trust, or simply a lack of awareness or lack of value being ascribed to sharing information with others as confirmed in Clayton and Fisher (2005). It is observed that employees are shy when they do not achieve the same work objectives as their colleagues. They sometimes believe that the organizational norm to evaluate them is unfair and this may have the result that they feel paranoid and do not trust others.

In terms of trust development among shy people, BETA has promoted an informal forum in physical social spaces to motivate employees to speak out publicly without shyness. This method leads them to establish good relationships, resulting in the development of trust. On the other hand, the relational dimension is also characterized through trust of others in virtual environments. Although the above discussion (structural dimension) emphasizes the advantages of a virtual space to reduce shyness, on the negative side, it is argued that participants in the virtual world may neglect trust within their family or social commitments and may become confused about some aspects of their culture and religion (Al-Saggaf 2004). This can be explained by the fact that human networks in Thai physical contexts (including face-to-face interactions) play the pivotal and strategic role of developing trust and relationships, as reported in related literature (Choo 2003; Thanasankit and Corbitt 2000).

5.3 Cognitive Dimension

The cognitive ability dimension of social capital here refers to the ability of the human actors to cognitively connect with each other with shyness. This subsection aims to discuss how shyness is affecting ability to share and create knowledge.

The results demonstrate that most employees in BETA are shy to participate in the discussion in a team because they are nervous when speaking in public. Also, shyness possibly is perceived to cause unwillingness to receive any assistance about information transfer and knowledge acquisition when it is offered even when needed (Nahl 2001). To foster the cognitive dimension of social capital, an appropriate representation of the history of knowledge-sharing activities may be useful since it allows the human actors to better understand and refer to past interactions (Huysman and Wulf 2006). As reported in the results, the idea of recording all presentations on digital archives for subsequent viewing by the presenters and other staff has been suggested. This idea is perceived to improve their shyness problems and thus their presentation performances, leading to improved knowledge sharing and creation.

5.4 Initial Theory Generation

The analysis of the case study for this research reveals how shyness influences KM in a Thai organization. While, the discussion portrays shyness as problematic, it has generated a number of interesting findings.

- Although shyness is a Thai positive cultural characteristic and communication competence notion, it seems to inhibit KM practices and creation of social relationships.
- ICT is perceived to help gain confidence and break shyness barriers, but it may not help develop trust among people, and may not be suitable in a bureaucratic culture.
- Trust and self-confidence are perceived to motivate shy people to share and create knowledge rather than any other rewards.
- Improving staff communication and presentation skills are a prerequisite before addressing effective knowledge sharing and creation.
- A bureaucratic culture as observed in BETA (or hierarchical structure), which causes seniority concerns, is likely to result in shyness barriers.

However, the discussion also opens up areas where shyness is very much subject to interpretation. The fact that it is perceived as difficult to change an organizational culture in Thailand does not necessarily equate with negativity. Some perceptions are accepted as a key and non-changeable feature of Thai culture. People in the organization prefer to preserve their culture (e.g., bureaucratic structure) regardless of the impact on KM.

Table 1. Summary of Themes Including Variables of the Influence of Shyness on KM Practices

Perspective	Variable
Technology	Impact of technology in breaking shyness barriers
Organization	Perceived influence of organizational structure and management in addressing shyness barriers
People	Perceived role of human networks and KM socialization activities in addressing shyness barriers

Using the concept of social capital, the study can characterize Thai people's experiences with KM. A summary of themes including variables of the influence of shyness on KM practices (Table 1) conceptualizes the thinking presented in this study in the general form, using three perspectives analyzed from the pattern codes and discussion: technology, organization, and people.

First, the variable "impact of technology in breaking shyness barriers" in the technology perspective refers to the level of impact of adopted technology in reducing shyness barriers. Technology, for example, includes virtual environment and knowledge management systems. Second, the variable "perceived influence of organizational structure and management in addressing shyness barriers" in the organization perspective refers to the level of influence of organizational issues including organizational structure, policy, and change in reducing shyness barriers. Last, the variable "perceived role of human networks and KM socialization activities in addressing shyness barriers" refers to the level of role of social networking, social capital, knowledge sharing and creation, and KM motivation in addressing and breaking shyness barriers. Each variable in the summary of themes helps researchers determine the level of KM influence in each perspective whereas practitioners may find it useful to take into account the levels measured by researchers to implement and adopt KM in their organization. The recommended future research is detailed in the next section.

6 RECOMMENDATIONS FOR FURTHER RESEARCH

Further investigation in other Thai organizations is highly recommended to validate and test the summary of themes, and then attempt to generalize it to Thailand. Further studies in Thailand may take into account the following issues:

- **Validation**: The need for the validation of the theory developed in this research is essential to determine the level of KM influence variables in each perspective. The levels can be measured by survey questionnaire using a scale. Each variable is assumed to develop a number of questions. For example, a variable "impact of technology in breaking shyness barriers" may involve many questions in relation to intention to use, perceived usefulness, voluntariness, client satisfaction, etc. Moreover, this further research aims to test whether the selected organization will represent the same culture as BETA, influencing or influenced by the KM practices, and whether the shyness feature is representative of the entire country.

- **Extension or simplification**: During an investigation, the variables in the theory could be extended or simplified if the researcher thinks them a subjective fit for the case, depending on many factors such as duration of field study, organizational culture, the researcher experience, appropriation, etc. For example, three perspectives (technology, organization, and people) could be subjectively extended to eight new attributes including technology, organizational structure and policies, change process, human network, social capital, knowledge sharing and creation ability, and KM motivation, where appropriate.

Besides shyness, further studies on different distinctive characteristics are important. Other distinctive cultural aspects such as collectivism, power distance, conscientiousness, and masculinity/femininity are reported to appear in Thailand and some developing countries. It would be interesting to further the research on the influence of these aspects on KM practices. In addition, further studies within the context of developing countries are highly recommended to manifest the status of KM practices in these countries, to test the extent to which there is a positive or negative trend toward KM awareness, and to investigate the need for the theory developed in this research. However, the extent to which richness of data can be captured about KM practices within an unfamiliar organization by case study method remains unclear. Further studies may be conducted by using alternative research methodologies such as action research and ethnography.

7 CONCLUSION

The research has investigated the influence of shyness on KM practices in BETA, a Thai organization. It demonstrates that an exploratory study on KM within a specific organization is far from being objective as the multiple realities associated with KM practices play out in various ways, resulting in the need for an interpretive case study to conduct the research. The use of the grounded theory approach has helped generate a set of insights, concepts, and interactions that address the critical organizational KM elements—elements from the cases in developing countries largely overlooked in the KM literature.

The summary of themes generated from the empirical findings suggests that shyness, a distinctive characteristic of Thai culture, critically influences, and is influenced by, a number of KM attributes in terms of technology, organization, and people perspectives. The research demonstrates how shyness influences, and is influenced by, KM practices which people in BETA have experienced. The study is limited to a Thai organization, leading to the recommendations for further research. Validation of the theory, and extension or simplification of the variables is suggested.

References

Al-Saggaf, Y. 2004. "The Effect of Online Community on Offline Community in Saudi Arabia," *The Electronic Journal on Information Systems in Developing Countries* (16:2), pp. 1-16.

Archibugi, D., and Pietrobelli, C. 2003. "The Globalisation of Technology and its Implications for Developing Countries: Windows of Opportunity or Further Burden?," *Technological Forecasting & Social Change* (70), pp. 861-883.

Becerra-Fernandez, I., and Sabherwal, R. 2001. "Organizational Knowledge Management: A Contingency Perspective," *Journal of Management Information Systems* (18:1), pp. 23-55.
Burrows, G. R., Drummond, D. L., and Martinsons, M. G. 2005. "Knowledge Management in China," *Communications of the ACM* (48:4), pp. 73-76.
Chaidaroon, S. S. 2003. "When Shyness Is Not Incompetence: A Case of Thai Communication Competence," *Journal of Intercultural Communication Studies* (12:4), pp. 195-208.
Chaidaroon, S. S. 2004. "Effective Communication Management for Thai People in the Global Era," paper presented at the International Conference on Revisiting Globalization and Communication in the 2000s, Bangkok, Thailand.
Chatzkel, J. 2004. "Establishing a Global KM Initiative: The Wipro Story," *Journal of Knowledge Management* (8:2), pp. 6-18.
Cheek, J. M., and Buss, A. H. 1981. "Shyness and Sociability," *Journal of Personality and Social Psychology* (41:2), pp. 330-339.
Choo, C. W. 2003. "Perspectives on Managing Knowledge in Organizations," *Cataloging and Classification Quarterly* (37:1/2), pp. 205-220.
Clayton, B., and Fisher, T. 2005. "Sharing critical 'Know-How' in TAFE Institutes: Benefits and Barriers," AVETRA, Brisbane, Australia (http://www.avetra.org.au/publications/documents/PA073ClaytonandFisher.pdf).
Cyamukungu, M. 1996. "Development Strategies for an African Computer Network," *Information Technology for Development* (7), pp. 91-94.
Eisenhardt, K. M. 1989. "Building Theories from Case Study Research," *Academy of Management Review* (14:4), pp. 532-550.
Glaser, B., and Strauss, A. 1967. *The Discovery of Grounded Theory*, Chicago: Aldine Publishing Company.
Huysman, M., and Wulf, V. 2006. "IT to Support Knowledge Sharing in Communities: Towards a Social Capital Analysis," *Journal of Information Technology* (21), pp. 40-51.
Klein, H., and Myers, M. 1999. "A Set of Principles for Conducting and Evaluating Interpretive Field Studies in Information Systems," *MIS Quarterly* (23:1), pp. 67-94.
Knutson, T. J. 2003. "Thailand as a Laboratory for Improved Intercultural Communication: Ethnographic, Metaphoric, and Social Scientific Implications," paper presented to the National Communication Association Convention, Miami, FL.
Koenig, M. E. D. 2002. "The Third Stage of KM Emerges," *KMWorld* (11:3), pp. 20-21.
Leonard-Barton, D. 1988. "Implementation as Mutual Adaptation of Technology and Organization," *Research Policy* (17), pp. 251-267.
Lesser, E. L. 2000. *Knowledge and Social Capital: Foundations and Applications*, Boston: Butterworth Heinemann.
Martin, P. Y., and Turner, B. A. 1986. "Grounded Theory and Organizational Research," *The Journal of Applied Behavioral Science* (22:2), pp. 141-157.
McCampbell, A. S., Jongpipitporn, C., Umar, I., and Ungaree, S. 1999. "Seniority-Based Promotion in Thailand: It's Ttime to Change," *Career Development International* (4:6), pp. 318-320.
Miles, M. B., and Huberman, A. M. 1994. *Qualitative Data Analysis*. Thousand Oaks, CA: Sage Publications.
Nahapiet, J., and Ghoshal, S. 1998. "Social Capital, Intellectual Capital, and the Organizational Advantage," *Academy of Management Review* (23:2), pp. 242-266.
Nahl, D. 2001. "A Conceptual Framework for Explaining Information Behavior," *Studies in Media & Information Literacy Education* (1:2), pp. 1-16.
Oates, B. J. 2005. *Researching Information Systems and Computing*, London: Sage Publications.
Okunoye, A. 2002. "Towards a Framework for Sustainable Knowledge Management in Organisations in Developing Countries," paper presented at the IFIP World Computer Congress Canada (WCC2002), Montreal, Canada.

Orlikowski, W. J. 1993. "CASE Tools as Organizational Change: Investigating Incremental and Radical Changes in Systems Development," *MIS Quarterly* (17:3), pp. 309-340.
Straub, D., Loch, K., and Hill, C. E. 2001. "Transfer of Information Technology to Developing Countries: A Test of Cultural Influence Modeling in the Arab World," *Journal of Global Information Management* (9:4), pp. 6-28.
Strauss, A., and Corbin, J. 1998. *Basics of Qualitative Research*, London: Sage Publications.
Thanasankit, T., and Corbitt, B. 2000. "Cultural Context and its Impact on Requirements Elicitation in Thailand," *The Electronic Journal on Information Systems in Developing Countries* (1:2), pp. 1-19.
Vorakulpipat, C., and Rezgui, Y. 2006. "A Review of Thai Knowledge Management Practices: An Empirical Study," paper presented at the IEEE International Engineering Management Conference, Salvador, Brazil.
Wagner, C., Cheung, K., Lee, F., and Ip, R. 2003. "Enhancing E-Government in Developing Countries: Managing Knowledge through Virtual Communities," *Electronic Journal on Information Systems in Developing Countries* (14:4), pp. 1-20.
Walsham, G. 1995. "Interpretive Case Studies in IS Research: Nature and Method," *European Journal of Information Systems* (4:2), pp. 74-81.
Wei, C. C., Choy, C. S., and Yeow, P. H. P. 2006. "KM Implementation in Malaysian Telecommunication Industry: An Empirical Analysis," *Industrial Management & Data Systems* (106:8), pp. 1112-1132.
Yao, L. J., Kam, T. H. Y., and Chan, S. H. 2007. "Knowledge Sharing in Asian Public Administration Sector: The Case of Hong Kong," *Journal of Enterprise Information Management* (20:1), pp. 51-69.
Yin, R. K. 2003. *Case Study Research Design and Methods*, Thousand Oaks, CA: Sage Publications.

About the Authors

Chalee Vorakulpipat is a Ph.D. student in the Research Institute for Built and Human Environment (BuHu) at the University of Salford. He has worked as a research assistant in the National Electronics and Computer Technology Center of Thailand. He has been involved in several projects in information systems development. His research interests include information systems, knowledge management, social and organizational studies, and software development. Chalee can be contacted at c.vorakulpipatt11@salford.ac.uk.

Yacine Rezgui is the (founding) Director of the Informatics Research Institute at the University of Salford, and a professor of Applied Informatics. For several years, he worked in industry as an architect and project manager on information technology research and development projects. In the last 15 years, he has led and been involved in a number of national (EPSRC) and European (FP4, FP5, FP6) multidisciplinary research projects. He conducts research in areas related to software engineering (including service-oriented architectures), information and knowledge management, and virtual enterprises. He has over 100 refereed publications in these areas, which appeared in international journals such as *Interacting with Computers, Information Sciences,* and *Knowledge Engineering Review*. He is a member of the British Computer Society. Yacine can be contacted at y.rezgui@salford.ac.uk.

4 COMPLIANCE-AS-A-SERVICE IN INFORMATION TECHNOLOGY MANUFACTURING ORGANIZATIONS: An Exploratory Case Study

Tom Butler
Bill Emerson
College of Business and Law
University College Cork
Cork, Republic of Ireland

Damien McGovern
Compliance & Risks
Cork, Republic of Ireland
Cork, Ireland

Abstract *In recent years, environmental concerns have led to a significant increase in the number and scope of compliance imperatives across all global regulatory environments. The complexity and geographical diversity of these environments has caused considerable problems for organizations, particularly those in high-technology industries. This paper first employs institutional theory to help understand the challenges for information technology manufacturing organizations that emanate from global institutional environments. While cultural–cognitive and normative influences from society-at-large and industry-based bodies have stimulated environment-oriented corporate social responsibility initiatives, it is undoubtedly regulatory influences that have generated the deepest responses in terms of the adoption of new compliance-oriented procedures and protocols. This paper first describes the general response from the organizational field in which high-technology firms operate and notes the extent of the response, with environmental compliance management systems being one of the institutional arrangements that organizations have adopted. The findings of empirical research based on*

Please use the following format when citing this chapter:

Butler, T., Emerson, B., and McGovern, D., 2008, in IFIP International Federation for Information Processing, Volume 267, Information Technology in the Service Economy: Challenges and Possibilities for the 21st Century, eds. Barrett, M., Davidson, E., Middleton, C., and DeGross, J. (Boston: Springer), pp. 43-59.

Compliance & Risks Ltd.'s compliance-to-product application and its deployment in Napa Inc., a Silicon Valley-based Fortune 500 company, are then offered and analyzed to illustrate the scale and scope of information systems support required to institute adequate compliance-oriented protocols and procedures in response to global regulatory influences, while also answering concerns raised by normative and cultural–cognitive sources.

Keywords Institutional theory, compliance, knowledge management, environmental compliance management systems

1 INTRODUCTION

Compliance with regulatory requirements of one sort of another has been with business enterprises for some time (Taylor 2005). In recent times, however, environmental concerns have led to a growing emphasis on compliance issues surrounding issues of energy consumption by, and the use of hazardous substances in, products across all industry sectors (Hristev 2006; Kellow 2002). The influence of environmentally oriented regulatory and social pressures are especially evident in the information technology sector (Murugesan 2007), particularly as shorter product life cycles, and longer product lines, have increased the use of materials that are deemed hazardous to the environment and, ultimately, to human health (Avila 2006; Brown 2006). Companies like Dell (2007) and Hewlett Packard (2006), for examples, increasingly advertise how they are exercising corporate social responsibility with respect to "green" pressures—both regulatory and social in origin.

Scott (2004) maintains that institutional theory has as its concern the forces that shape social structures, schemas, rules, norms, and routines and which, in turn, affect the behavior of social actors. It is clear from Scott (2001, 2004) that regulatory, normative and cultural–cognitive institutional influences shape organizational processes and structures, and help define what is effective performance or efficient operation in organizations (Powell 1991). Thus, institutional theory provides a suitable conceptual lens with which to examine the impact of institutional forces generated by environmental concerns in the organizational field, population, and environment in which high-technology manufacturing firms compete (Chiasson and Davidson 2005). This study, therefore, adopts institutional theory to first explore and understand the regulatory, normative, and cultural–cognitive forces shaping the environmental compliance imperatives confronting IT manufacturing organizations. It also illustrates that IT manufacturing organizations have responded to institutional pressures by deploying information systems to support internal compliance processes—with, it is argued, limited success (Avila 2006; Brown 2006). This paper focuses on how one software application and data services vendor, Compliance & Risks Ltd., had its compliance knowledge management system adopted by Napa Inc., a Fortune 500 high-tech manufacturer, in order to address the considerable challenge of global product compliance through IT-supported compliance-oriented protocols and procedures.

The research objective of this study is to explore how global environmental regulations have shaped the organization-level responses of firms in the IT manufacturing sector and to investigate how IT and data services are being employed to support new organizational protocols and procedures to address the challenges posed by green legislation.

2 APPLYING INSTITUTIONAL THEORY TO UNDERSTAND COMPLIANCE IMPERATIVES FACING ORGANIZATIONS

Institutions are "the humanly devised constraints that structure political, economic, and social interactions. They consist of both informal constraints (sanctions, taboos, customs, traditions, and codes of conduct), and formal rules (constitutions, laws, property rights)" (North 1991, p. 97). According to DiMaggio and Powell (1983, p. 143), an *organizational field* is comprised of "Those organizations that, in the aggregate, constitute a recognized area of institutional life: [it consists of] key suppliers, resource and product consumers, regulatory agencies, and other organizations that produce similar services or products." An organizational field is characterized by coercive (regulative and legislative) influences from government departments, state-sponsored agencies, the judiciary, and so on, in addition to normative and mimetic (cultural–cognitive; Scott 2001) influences from related organizations (suppliers, consulting organizations, distributors, professional bodies, etc.), competitors, stakeholders (nongovernment organizations, analysts, investment funds, for examples) and other related social entities. It is clear from DiMaggio and Powell, as with Scott (2001), that the external agencies which constitute an organizational field exert a significant influence over an organization's structures, policies, practices, and procedures. Following Chiasson and Davidson (2005), this study conceptualizes the organizational field as the electrical and electronics industry (although all organizations are now affected by environmental regulations of one kind or another) and the organizational population/environment as the high-tech IT manufacturing sector. However, we argue that, as recent regulatory developments in the European Union have illustrated, organizational fields exist within a global/societal environment rather than national environments.

2.1 Regulative or Coercive Influences on IT Manufacturing Organizations

Drawing on Scott (2001), the *emphasis* of regulatory institutional influences is on coercion, indicators of which are rules and laws, which agents such as governments and regulatory agencies legitimize using legal sanctions to ensure compliance. Institutional *carriers*, on the other hand, are social structures such as governance and power systems, which institute rules and laws, the organizational response to which is to institutionalize routines such as protocols and procedures. This section, therefore, focuses on governance power systems, rules, and laws that shape corresponding organizational compliance-oriented procedures and protocols.

High-tech IT manufacturing organizations operate in highly complex regulatory institutional environments internationally and, consequently, face a bewildering range of diverse regulations (Avila 2006; Hristev 2006). For example, recent European Union regulations such as the Restriction of Hazardous Substances Directive (RoHS), the Waste Electrical and Electronic Equipment Directive (WEEE), and the Registration, Evaluation and Authorisation of Chemicals (REACh) Regulation have enormous implications for diverse industry sectors operating globally (European Commission 2006; Hristev 2006). The implementation of the WEEE and RoHS directives resulted in highly complex legislation in EU member states, which does not lend itself to easy comprehension, application, and integration into an organization's research, development, manufacturing, and logistics processes (Pecht 2004). However, the task of maintaining compliance will become even more onerous for the IT manufacturing industry and related sectors, not only because of the recent moves to include 46 new hazardous substances on top of the original six under EU RoHS, but as the new REACh regulation came into force in June 2007. This new law requires organizations to specify the possible dangers of combinations of chemicals present in their products not only on disposal, but also while in use (Bush 2007). It will also place new disclosure requirements on companies under Article 33 by ensuring that customers, and also interested NGOs like Greenpeace, can insist on disclosure on a black list of substances. While the EU's environmental laws have received much attention, others are no less stringent. The Environmental Protection Agency (EPA) in the United States has issued a raft of legislation covering all hazardous substances across the whole range of manufacturing sectors, while Japan also has highly demanding laws (Hristev 2006). Over the last two years, Korea, Australia, Canada, and U.S. states such as California have introduced legislation similar to the RoHS and WEEE directives, while in China, a similar directive known as the China RoHS, or the Methods for the Control of Pollution by Electronic Information Products Directive, came into force in March 2007. The need to address compliance legislation in different geographical locations adds complexity for global IT manufacturing organizations; however, determining the applicable regulation for a given geographical area can be complicated by the problem of understanding which products are covered by, or are exempt from, sets of seemingly conflicting regulations (Kellow 2002).

The European Commission estimates that the cost of being in compliance with its new REACh legislation will be upwards of €5.2 billion ($7 billion) (European Commission 2006). Independent research also reports that the cost of compliance with RoHS and WEEE is approximately 2 to 3 percent of the cost of goods sold, a not insignificant amount given the size of the IT sector (Spiegel 2005). However, while the costs of ensuring compliance are considerable, the costs of not being in compliance are even more significant, with companies facing the risk of exclusion from key markets, stopped shipments, and product recalls, with a corresponding loss of revenue, and potentially disastrous consequences for brand image and/or corporate reputation (Avila 2006, Brown 2006; Goosey 2007). In cases of a serious breach of compliance, firms may also be faced with hefty fines and/or criminal prosecution (Brown 2006).

2.2 Normative and Cultural–Cognitive Influences on Environmental Compliance

While the forgoing illustrates the relatively strong incentives to comply with regulative imperatives, normative and cultural–cognitive influences increasingly come into play. Industry standards and professional bodies such as IEE and IEEE, customers, and suppliers are the agents bringing normative influences into play, while nongovernment organizations and other social stakeholders bring cultural–cognitive influences to bear (see Eisner 2004). In terms of the former type of influence, manufacturer associations, original equipment manufacturers, and firms that source components from suppliers in the industry mandate compliance to the regulations and standards in the institutional environments they face (Avila 2006; Pecht 2004). Take, for example, Eisner's (2004) report on the Ford Motor Company's decision to have all of its suppliers be ISO14001 compliant, thus forcing environmental social responsibility down the supply chain. In contrast, one recent example of cultural–cognitive influences comes from the nongovernment organization Greenpeace, which tests IT and electronic appliances for the presence of hazardous substances; Greenpeace publishes the results of its tests on the Internet in order to influence environmentally conscious customers and investors (Greenemeier 2007). It is clear, however, from recent practitioner analyses by the Aberdeen Group (Brown 2006) and McLean and Rasmussen (2007), that normative and cultural–cognitive influences aimed at generating corporate social responsibility are less influential in shaping organizational responses to environmental issues than regulatory influences, as it is hoped by organizations that addressing the latter will help address normative and cultural–cognitive influences as well.

2.3 Information Systems Support for Regulatory Compliance

Avila (2006) argues that information systems are required to help address the challenge facing the IT manufacturing industry in responding to the increasingly complex global regulatory environment. In terms of the organizational response of instituting suitable compliance-oriented protocols and standard procedures, Avila argues that, to be effective, information systems must, at base, possess material compliance analysis capabilities, the ability to reduce the total cost of compliance, and also to account for rapidly changing environmental regulations across multiple markets and geographies. However, in its recent study, The Aberdeen Group revealed that 80 percent of companies lacked a cohesive systems infrastructure to track, audit or manage product compliance. Most companies were relying on a variety of solutions that were not properly integrated, and which did not provide the necessary information needed to meet environmental regulations (Brown 2006).

Previous efforts to tackle corporate environmental auditing and reporting involved the implementation of environmental management systems (EMS)—which may, or may not, have been supported by IT (Eisner 2004). Eisner reports that EMS were an extension of total quality management, which helped companies define their corporate environmental policies and employed information on regulatory requirements and environmental impacts to determine quantifiable objectives and programs. It was only recently, however, that robust IT support for EMS emerged in the form of software-as-a-service tools that help companies manage environmental, health, and safety compliance; DuPont, Chevron, and Johnson & Johnson are three large multinational organizations using such systems (Brodkin 2007). While EMS grew out of the need to manage reasonably well-articulated regulatory requirements, and to help companies address corporate social responsibility, they did not address the information asymmetry problems that emerged with more complex, stringent, and highly differentiated global regulatory compliance imperatives—hence, the emergence of environmental compliance management systems (ECMS).

A brief industry analysis conducted by the authors of the various high-profile ECMS offerings currently available from vendors such as SAP AG, TechniData AG, E2open, and Synapsis Technology revealed that they can be deployed as

- stand-alone applications (off-the-shelf packages that are configured or customized)
- hosted solutions (e.g., compliance-as-a-service)
- either of the above, integrated with existing product lifecycle management (PLM) or enterprise resource planning (ERP) systems

It is significant that the features of these application only meet, in part, the criteria proposed by Avila (2006) and outlined above. Our analysis also revealed that such ECMSs emerged from extant vendors' PLM systems (e.g., EMARS from Synapsis Technology, Inc., and Product Governance and Compliance from Agile Inc.), enterprise resource planning (ERP) systems (e.g., CfP from SAP AG and TechniData AG), or supply chain management (SCM) systems (e.g., E2open Inc.'s Eco-Compliance solution). A recent report by the Forrester Group (McLean and Rasmussen 2007) on the adoption of governance risk and compliance (GRC) applications (of which ECMS would be a subset) by organizations across all sectors failed to register any of those mentioned above—indeed McLean and Rasmussen concluded that the absence of both SAP AG and Oracle Corp. from their analysis indicated that they were not servicing the compliance needs of organizations. Extensive (unpublished) market research of over 50 global organizations by one of the authors between 2006 and 2007 revealed that Fortune 500 organizations admitted to using a variety of internal solutions based on, for example, Excel spreadsheets, basic database systems, and point solutions from a range of vendors (see Brown 2006). The GRC executives interviewed indicated that none of these solutions in use supported the organizational protocols and routines needed to manage regulatory environmental compliance on a global basis. It was found that they were particularly poor in addressing information asymmetry and knowledge sharing problems between regulators and companies (and vice versa), between companies and their stakeholders, and between various organizational functions.

3 PARTICIPATORY RESEARCH ON COMPLIANCE-AS-A SERVICE

An exploratory, instrumental case study design was chosen for this two year longitudinal study (Stake 1995; Yin 2003), which is the result of an ongoing, participative study by the university researchers with Compliance & Risks Ltd. (C&R) on the compliance-to-product (C2P) application. It must be noted that the research study, which commenced in August 2005, did not meet the criteria demanded of action research (Baskerville 1999); however, in an effort to capitalize on the synergy between the university researchers knowledge of KM theory and IS design and C&R practitioners' "situated, practical theory" (Baskerville 1999, p. 17), a case-based participatory research strategy was chosen as the most appropriate approach.

Two university researchers participated in this study. Four practitioners from the company played an active role as coresearchers: the primary coresearcher was the founder of Compliance & Risks Ltd., while the secondary coresearchers included the California-based software team's project manager and the chief design architect of the C2P, a senior software engineer from C&R and the company's marketing and sales director. The majority (seven) of the development team were headquartered in northern California, with two being based in Europe. The company's legal data team and industry partners were primarily European based; however, it did have a number of lawyers working out of U.S. offices. The remaining participants included users of the pilot version of C2P at Napa Inc.

The data for the present study was gathered using semi-structured interviews during numerous meetings and on-site visits in Europe and the United States, spanning the period from August 2005 to April 2007; participant observation was also employed throughout (Yin 2003). It must be noted, however, that researchers had no access, at any time, to confidential client data, in accordance with C&R's nondisclosure and confidentiality obligations to its clients. Internet-based teleconferencing technologies were also employed to facilitate meetings, in addition to e-mails and instant messaging. The data was interpreted and analyzed on an ongoing basis and augmented by official company documentation, including C&R's business plan, training manuals, technology architecture documentation, and so on.

4 COMPLIANCE-AS-A-SERVICE: FROM CONCEPT TO SERVICE-ORIENTED ARCHITECTURE

The background to the emergence of the compliance-to-product application has its origins in C&R's CEO's experience in regulatory intelligence for Deloitte in Brussels. This provided him with a unique perspective on the need for, and potential of, ECMS, or compliance knowledge management systems (CKMS) as he conceptualized them, especially in industry sectors affected by environmental regulations being introduced by the European Union. He identified the electrical and electronics industries, and in particular the IT sector, as a potential niche market for his new business method and CKMS innovation. In order to help translate his innovative idea, he developed a conceptual model of the components and processes of an enterprise-wide CKMS. As a legal expert, he identified what he termed a *consumption problem*, in that an information asymmetry existed between regulatory institutions and agencies and the companies covered by the environmental legislation they produced. In terms of institutional theory, if companies did not know of, or understand, environmental regulatory compliance imperatives with respect to products/materials and substances of interest, how could they institute appropriate organizational responses in the forms of protocols and procedures to ensure compliance? Information-related problems were not, however, confined to the consumption of legislation, as high-tech manufacturers source an enormous range of sub-assemblies, components, etc. from diverse suppliers arranged in multitiered, geographically dispersed, supply chains. Knowing what is in a given product or model, and if it is (or will be) in compliance with global regulatory instruments, therefore, adds to the difficulty in arriving at suitably designed and effective protocols and procedures. The following section describes the conceptual model that emerged in C&R in response to regulatory institutional pressures identified by its founder and other legal experts.

4.1 The Compliance-to-Product Conceptual Model

Practitioners at C&R view CKMS for product compliance as being based on an integrative IT artefact that manages three high-level enterprise compliance processes viz. an external regulatory requirements gathering process, a compliance management process, and a knowledge management process (a fourth, the supply chain compliance process, was recently identified, but requires further elaboration). These processes concern themselves with assessing and managing the issues, risks, and tasks, communication, collaboration and knowledge sharing, document management, and disclosure activities, in addition to performing all of the product stewardship related to external requirements impacting an organization. The underlying activities or processes involve interpreting, tracking, and monitoring environmental regulations. All of this takes place in the context of evolving, strategic internal requirements. There are three primary categories of users—C&R's legal data team; the customer organization's compliance function; and managers and engineers in the product research and development, manufacturing, and logistics functions. The core of the system is the CKMS repository, which is underpinned by a highly sophisticated data model. Figure 1 presents a conceptual model of such a CKMS, called Compliance-to-Product.

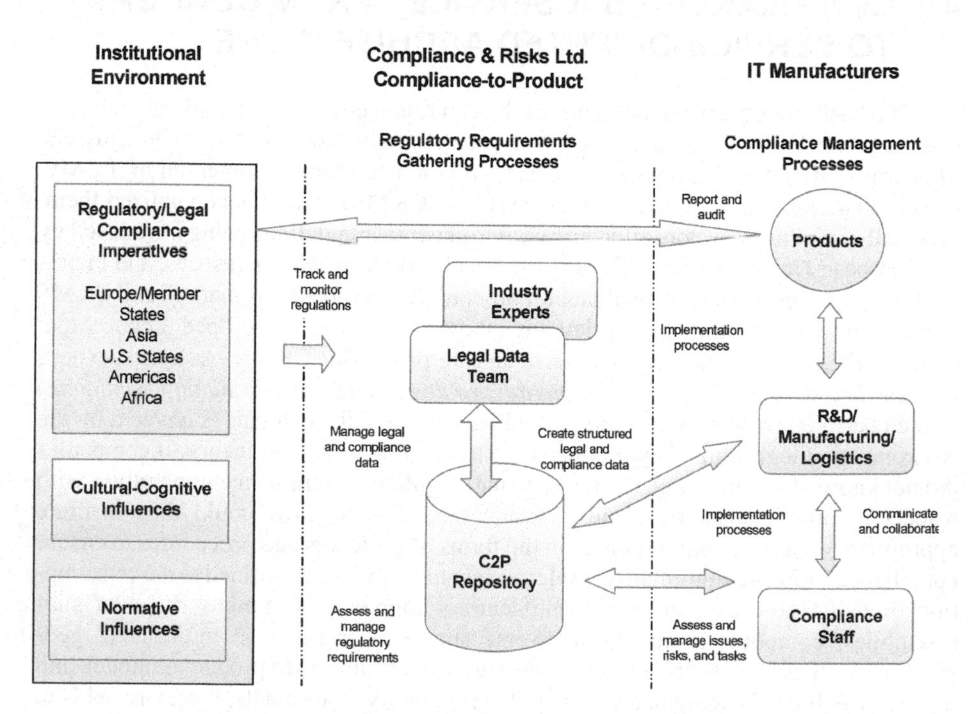

Figure 2. Compliance-to-Product Compliance-as-a-Service Conceptual Model

In this model, external regulatory requirements gathering processes are conducted by C&R, whose legal data team focuses on the jurisdiction, instrument type, and legal basis for compliance imperatives: as a result, their activities need to cover all of the business territories in which a company operates, in addition to juridico-political territories, future regional implementation areas, and so on. Identifying, managing, and tracking compliance imperatives is further complicated by the fact that they often fork into distinct regulatory requirements in "parent-child" configurations, in which the resulting child imperative may differ in terms of the criteria applied to business processes, behaviors, services, and products. Few, if any, organizations have the internal legal expertise to manage this complexity and to be able to track and monitor regulations, create and manage all legal and compliance data, while capturing the often subtle differences between, for example, an EU directive's local implementation in any of 27 member states, many of whom may "gold plate" legislation (i.e., make it more or less restrictive) to favor national industries or strengthen environmental protections.

Hence, in order to enable what are key processes in the model, C&R put together its legal data team, which is comprised of lawyers in Ireland, the United Kingdom, and the United States. The legal data team also includes paralegals, engineers and environmental specialists. Its primary function, therefore, is to convert legislation from global regulatory institutions to create C2P's knowledge repository. Table 1 provides an example of the policy areas covered by the LDT and corresponding compliance imperatives by region as of 2007; these will be increased significantly by 2008. Once the LDT has captured and properly formatted compliance imperatives, they are made available by C2P to all organizational users who manage the organization's enterprise compliance processes. The application's features therefore enable users to readily assess and manage regulatory requirements and to manage product-related risks and issues. These processes are enhanced by communication and collaboration features that enable users to automate information sharing (e.g., by triggering risk alerts when legislation is modified or compliance imperatives are modified or introduced) and to support the firm's compliance implementation process.

4.2 C&R's Business Model

The primary business model adopted by C&R saw the establishment of its application and data hub in northeastern California. Locating the application hub in California was a strategic move, as several of its customers are Fortune 500 companies are located in Silicon Valley. The company's U.S. and European offices are linked to the hub using encrypted virtual private network (VPN) technology. Thus, development and legal staff at the company's headquarters, in addition to legal experts located worldwide, access the application using standard, secure web technologies. C&R decided against hosting the client versions directly; instead, it outsourced the deployment of individually customized applications to a high profile application service provider in Sacramento, California, which implements a highly secure, industry-standard, service-oriented architecture. Figure 2 illustrates C&R's business model.

Thus as with initial customers such as Napa Inc., each client site will have a secure VPN link to its particular customized version of C2P. The only data common to all application instances will be database entries related to compliance imperatives and related

Table 1. Compliance-to-Product Knowledge Base of Global Product and Service Legislation

Regulatory Region Target	EU Legislation	Ireland	UK	EU AS	USA Federal	USA CA	USA NY	USA Other	Canada	South America	Asia, India,	International Standards
Batteries	10		3	2	5	2	1	15		5	8	1
Energy efficiency	22	1	4	6	48	6	1	12	2		13	4
Hazardous substances	69	3	19	66	41	30	4	64	5	9	38	23
Packaging	34	3	10	71	1	1	1	22	1		18	5
Disposal, take-back	8			14	12	3	11	75	6	9	9	6
Waste & waste mgmt..	23	5	7	81	2	6	1	7	6	1	9	8
Public & green proc.	10	5			5	1	2	1			2	
EMC low voltage	21					1						
ECO design	43		2		7						4	23
REACH	32		1									
Money laundering	10		12	5				1				
Anti-corruption	8	1	10	22	3							2
Health & safety	16	16		1	1	1					1	5
Dual use, export control	19	4	24	23	12						1	11
Data protection	20	17	2	1	1							1
Audit & governance	12	15	12	1	1	1						13
Product safety	12	1		4	2							9
Marketing, labeling	7			8	3	2	1	2	4		8	11
Medical regulation	8											
EU procedures	29											
Total sources	**413**	**71**	**106**	**307**	**144**	**54**	**22**	**199**	**24**	**24**	**111**	**122**

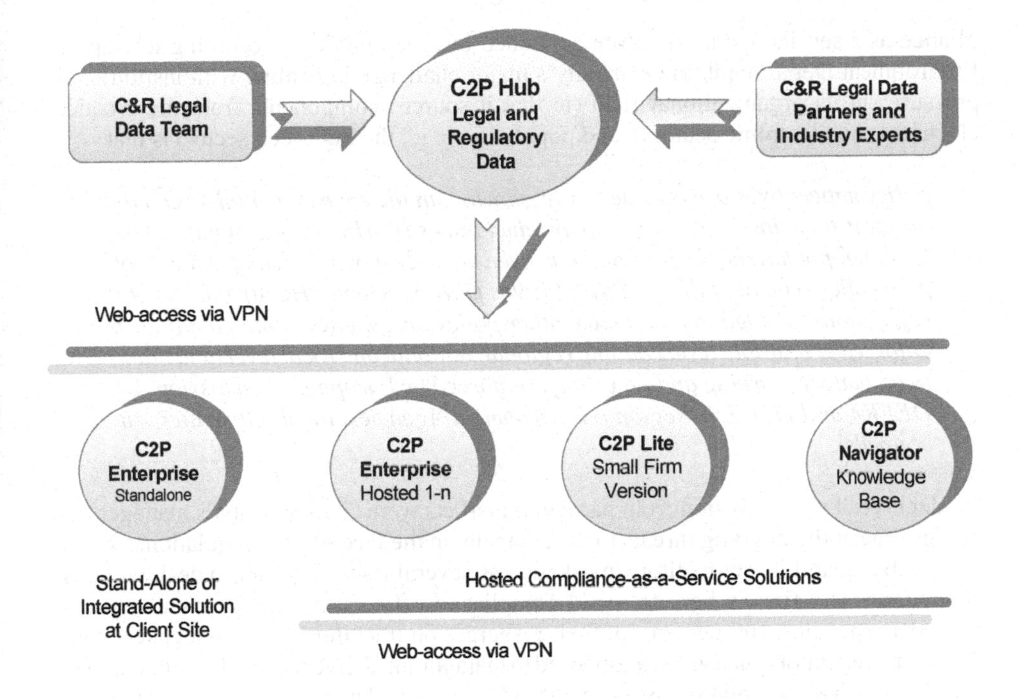

Figure 2. Compliance-as-a-Service Business Model

data, documents, and so on, which emanate from the C2P hub. This partitioning ensures that confidential customer data is secured from access by other customers, or indeed employees of C&R.

5 APPLYING COMPLIANCE-AS-A-SERVICE CONCEPTUAL AND BUSINESS MODELS AT NAPA INC.

The nub of the problem facing organizations in confronting environment-based institutional regulatory and corporate social responsibility pressures is that they generally do not have the necessary in-house legal capabilities to interpret, evaluate, capture, and store all relevant compliance-related information. These activities are usually out-tasked to legal experts who respond with voluminous reports that do little to lift the burden on compliance officers. Alternatively, in-house regulatory compliance requirements gathering demands strong legal competencies, as their focus must be on the geographic jurisdiction, instrument type, and legal basis for compliance imperatives. These were some of the challenges facing Napa Inc. prior to 2007.

Napa Inc. is a household brand name and a major player in the global IT industry worldwide. Due to the dearth of suitable environmental compliance management systems in the market in 2006, it piloted the use of C2P and subsequently adopted it as compliance-as-a-service using the aforementioned business model. According to Napa's Environment Department, the company's major challenge in dealing with institutional pressures in its organizational field (in that it sources components from the broader electrical and electronic sectors) and population (e.g., the high-tech sector) is that

> *policy imperatives are exponentially growing, in the environmental arena the policy is focusing increasingly on product issues (RoHS, Power management, labeling, packaging design) and has been steadily moving away from end-of-pipe policy typical of the 1980s and 1990s [Environmental Health and Safety] regulations. Added to this, unlike other policy areas, environmental policy is enforced at multiple levels adding regional, national and local level data points (e.g., battery marking and recycling is enforced by European Commission, UK DEFRA and DTI, UK Regional Environment Agencies, local authorities, city councils).*

Product recall and exclusion from particular markets were cited by Napa's managers as being some of the ongoing threats to the company in the face of such regulations. Napa argues that it, and its competitors, must address several issues in order to deal with this challenge: The first is how to avoid compliance officers and R&D manufacturing engineers spending 100 percent of their resources on tracking policy, associated standards, and regulatory measures, as opposed to maintaining delivery of higher value-added activities such as compliance assessments, addressing and managing issues, and implementing compliance with imperatives. The second issue concerns how to tie the actions and decisions taken at a product team level with the requirements, terms, definitions, and guidance provided by legislators. The third issue concerns the problem of how to reduce time and cost associated with getting independent guidance and assessment on regulatory definitions and requirements.

5.1 Napa's Experiences with Compliance-as-a-Service

Prior to its adoption of C2P, Napa adopted a five-stage process in order to help it manage compliance viz. (1) track and monitor all relevant regulations in the global marketplace; (2) assess all related risks in terms of process and product; (3) raise awareness across the organization, especially in engineering, design, R&D, and manufacturing, and enhance intra-organizational communications across all relevant functions; (4) implement compliance solutions in engineering and design functions at the earliest possible opportunity; (5) review the effectiveness of the steps taken and the level of compliance achieved. These steps were increasingly proving to be difficult to manage in terms of the scope and complexity of the global regulatory environments in which Napa operates. Like many other organizations, Napa was managing compliance using information gleaned from legal advisors and mapping this onto products, subcomponents, and so on, using Excel spreadsheets. Table 2 illustrates the support that C2P offers to Napa's compliance processes.

Regulatory Requirements Gathering Process: As indicated, C2P's regulatory requirements process is conducted by C&R's legal data team, with input from industry partners (such as law firms, consultants, policy area experts), and other sources. The

Table 2. Napa's Five-Stage C2P-Enabled Compliance Process

Napa Inc. Compliance Activities	Supporting C2P Features and Functions
1. Track and monitor	Structured, dynamic database of compliance imperatives
2. Assess risk	Detailed regulatory requirements, impact, issues, unambiguous terms and definitions, risk ratings
3. Raise awareness and communication	Alerts, comments, watches, and shared searches
4. Implement compliance solutions	Internal specifications, action plans, assignment and reports
5. Review 1 – 4	Historical reports, compliance indicators, and management summaries

scope of this process is that it encapsulates all of the business territories in which Napa operates, in addition to all juridico-political territories, future regional implementation areas, and so on. The complexity of capturing these types of data stands in stark contrast to the type of support provided by the majority of extant ECMS (e.g., SAP's compliance for product), where users create static "lists" of compliance requirements, rather than the multidimensional data structures needed to model compliance imperatives and requirements, and which also need to be updated in real-time. C2P, therefore, captures global regulatory imperatives, and represents them not as one-dimensional lists, but in complex data structures that Napa's compliance officers and other users can easily navigate and map onto internal specifications (in C2P format), products, subassemblies, parts, materials, and substances. Feedback from Napa's compliance and R&D teams indicated that C2P "adds value" to the company's compliance management processes, as the compliance-as-a-service application ensures that "source regulatory data is delivered pre-formatted, structured, and ready to use out of the box" (Compliance Officer, Napa Inc.). C2P was also argued to provide "a universal panoramic view of all relevant impacts to product or company operations" and it "gives instant, live snapshots on policy areas that are not adequately covered or understood—[C2P] helps guide management to allocate resources to appropriate risk areas."

Compliance Management Process: Napa's compliance officers and other users have full visibility into the status of its compliance activities using a simple dashboard and complex searching and reporting functionality. In most scenarios, manufacturing organizations will set down product specifications, including materials to be used in sub-components at the R&D or product design stage. An important feature of C2P is its ability to allow Napa to view the impact of regulations on, and issues around, products, subassemblies, parts, materials, and substances in real time. This means that once a regulatory requirement changes, or a new one emerges, or new data is fed into the system, the status of all related products can change, down to constituent materials, and on to parts provided by suppliers, communicated to "responsible" users using dashboard and email alerts. C2P's analysis features help users analyze compliance imperatives and requirements impacts on products, subassemblies, parts, materials, and substances. This feature encompasses facilities to map supplier data against compliance requirements so as to identify noncompliant parts, materials, or substances. As with other ECMSs, C2P features a personal dashboard, which supports access to the following features:

- An issue management feature helps Napa's compliance professionals, product design engineers, and others collaboratively evaluate, escalate, and address product and material compliance issues.
- A risk ratings feature displays a product's risk status for each compliance issue.
- The ability to allow Napa's compliance officer to delegate and monitor issues and responsibilities to relevant staff.
- An action plan feature, so that users can associate milestones with, and manage tasks for, each issue.
- An exception or alarm feature to remind users, and help them track the status, of assigned areas of responsibility.
- The ability to create custom reports according to their issues and products.
- The ability to have multiple views so that users can navigate between summary views and more detailed information.

Knowledge Management Process: Napa's user community felt that one of the strong features of the C2P was its knowledge-sharing capabilities and tools. The purpose of these features is to provide an additional dimension to enhance the understanding of users to enhance compliance-related decision making. Several of these features are accessible from C2P's dashboard, including

- Capturing discussion threads in instant messages and e-mail threads between users on any topic.
- Create contexts (i.e., background information) for classifying and reporting the evolving impact of compliance issues on products.
- Capturing a history of all changes to regulatory imperatives and requirements and associated changes to products, subassemblies, parts, materials, and substances.
- Automatic e-mail and/or instant messaging notification of any changes in the regulatory status of products, subassemblies, parts, materials, and substances.
- An attachment feature to provide links to, or to attach directly, related documents (e.g., legal interpretations or advice, industry journal articles, reports, etc.) that describe in-depth compliance imperatives, requirements, issues, impacts, or data on products, subassemblies, parts, materials, and substances.
- Search features to allow users to run queries and produce reports based on specific parameters.

Napa's compliance executive summed up comments from colleagues by stating that "C2P has shown us that the era of paying external organizations to dump information on our doorstep has come to an end." He maintained that the application "enables our compliance team to move away from the inordinate amount of time spent on tracking and monitoring activities and to focus on activities 2, 3, and 4 [see Table 2] which are the bits that really add value to the company." Another member of the Napa's compliance team supported this and stated that "C2P helps us to a specific risk assessment for all affected products, getting quickly to impacts and risks." Thus, the C2P application offers faster time to knowledge to Napa in managing product compliance and minimizing any risk to the company by being out of compliance.

6 SUMMARY AND CONCLUSIONS

This paper describes the changing institutional environment confronting high-tech manufacturing organizations in the IT sector. Growing societal concerns with the planet's environment has led governments and their agencies across the globe to institute highly complex and far-reaching legislation that has significant implications for all IT products, their production, use, and disposal. Thus, there is no place to hide for IT manufacturers, wherever they are based. This development has brought an important and new challenge to organizations to respond to such institutional pressures by instituting new compliance-oriented organizational protocols and procedures. As the problems facing organizations in achieving compliance with environment-oriented regulatory requirements are essentially information-based, then it is no surprise to find that a new breed of IS has emerged—environmental compliance management systems. While several such solutions are currently available in the marketplace, they have not been the focus of information systems research or industry analysis. This study's major contribution is that it identified a ground-breaking approach to solving the problem of organizational responses to increasing institutional and stakeholder pressures (often climate change related) through the provision of compliance-as-a-service by a small, innovative software company—Compliance and Risks Ltd.

The paper revealed that compliance-as-a-service emerged as a viable value proposition to organizations as it addresses: (1) the complexity of the global institutional regulatory environments in which they operate, and (2) the inability of such companies to form adequate responses to those challenges due to the dearth of in-house legal capabilities, and the resources required to provide adequate IT support for the application of new protocols and procedures. It is clear from marketplace assessments made by the authors of this study, and confirmed by practitioners in Napa Inc., and other Fortune 500 and 100 organizations, that, with few exceptions, both the organizations themselves and extant software vendors are not meeting the compliance and risks needs of the IT manufacturing and related sectors. This study illustrated how one innovative software company identified these requirements and responded by building an IT artefact that was tailored to help organizations institute appropriate protocols and procedures to meet the regulatory compliance needs of these sectors. The key differentiator of C&R's C2P application, and one that conveys a comparative advantage on it over other vendor offerings, is that it approaches the problem from a legal/compliance perspective and offers a combination of software, regulatory data, and data services. Alternative approaches from competing vendors are based on document management, supply chain management, or enterprise resource planning perspectives, none of which captures the multidimensional nature of regulatory compliance imperatives and their impact down the product hierarchy to component materials. This lacuna prevents adopting organizations using such ECMS for developing effective protocols and applying related procedures to address institutional challenges. It must also be noted that while the knowledge and skills of C&R's legal data team are important, so too is the conceptual model underpinning the C2P application in that that captures the legal essence of the institutional influences, and thereby enables the team to translate the complex regulations into basic compliance imperatives such that users can understand them. In conclusion, the practical success of the C2P application with a growing number of Fortune 100 and 500 companies provides support for its underlying concept and architecture in delivering compliance-as-a-service and underpins its position as an innovative solution in this emergent area of the IS field.

References

Avila, G. 2006. "Product Development for RoHS and WEEE Compliance," *Printed Circuit Design and Manufacture* (23:5), pp. 28-31.

Baskerville, R. L. 1999. "Investigating Information Systems with Action Research," *Communications of the Association for Information Systems* (2:19), October.

Brodkin, J. 2007. "Hosted Software Manages Environmental Compliance," *Network World*, August 1 (http://www.networkworld.com/news/2007/080107-hosted- software.html).

Brown, J. 2006. "The Product Compliance Benchmark Report: Protecting the Environment, Protecting Profits," Aberdeen Group, Boston, September.

Bush, S. 2007. "EU's REACH Directive Will Hit Electronics Firms," *Electronics Weely.com* February 28 (http://www.electronicsweekly.com/articles/2007/02/28/ 40856).

Chiasson, M., and Davidson, L. 2005. "Taking Industry Seriously in Information Systems Research," *MIS Quarterly* (29:4), pp. 591-605.

Dell. 2007. "Values in Action: Dell Sustainability Report, Fiscal Year 2007 in Review," Dell Inc. (http://www.dell.com/downloads/global/corporate/environ/report07.pdf).

DiMaggio, P. J., and Powell, W. W. 1983. "The Iron Cage Revisited: Institutional Isomorphism and Collective Rationality in Organizational Fields," *American Sociological Review* (48:2), pp. 147- 160.

Eisner, M. A. 2004. "Corporate Environmentalism, Regulatory Reform, and Industry Self-Regulation: Toward Genuine Regulatory Reninvention in the United States," *Governance: An International Journal of Policy, Administrations and Institutions* (17:2), pp. 145-167.

European Commission. 2006. "REACH in Brief," Enterprise & Industry, December 12 (http://ec.europa.eu/enterprise/reach/overview_en.htm).

Goosey, M. 2007. "Implementation of the RoHS Directive and Compliance Implications for the PCB Sector," *Circuit Design* (33:1), pp. 47-50.

Greenemeier, L. 2007. "Greenpeace: Apple iPhone More Brown than Green," *Scientific American.com News*, October 18 (http://www.sciam.com/article.cfm?id=iphone-more-brown-than-green).

Hewlett Packard. 2006. "HP FY06 Global Citizenship Report." March 17 (www.hp.com/hpinfo/globalcitizenship/gcreport/index.html).

Hristev, I. 2006. "Rohs and WEEE in the EU and US," *European Environmental Law Review*, March, oo. 62-74.

Hutter, B. 1997. *Compliance, Regulation and Environment*, Oxford, UK: Clarendon Press.

Kellow, A. 2002. "Steering through Complexity: EU Environmental Regulation in the International Context," *Political Studies* (50:1), pp. 43-60.

McLean, C., and Rasmussen, M. 2007. "The Forrester Wave™: Enterprise Governance, Risk, and Compliance Platforms, Q4 2007," *Forrester*, December 21 (http://www.forrester.com/Research/Document/Excerpt/0,7211,41751,00.html).

Murugesan, S. 2007. "Going Green with IT: Your Responsibility Toward Environmental Sustainability," *Cutter Consortium Business-IT Strategies Executive Report* (10:8), August (http://www.cutter.com/alignment/fulltext/reports/2007/08/index.html)

North, D. C. 1991. "Institutions," *Journal of Economic Perspective* (5:1), Winter, pp. 97-112.

Pecht, P. 2004. "The Impact of Lead-Free Legislation Exemptions on the Electronics Industry," *IEEE Transactions on Electronics Packaging Manufacturing* (27:4), pp. 221-232.

Powell, W. W. 1991. "Expanding the Scope of Institutional Analysis," in *The New Institutionalism in Organizational Analysis*, W. W. Powell and P. J. DiMaggio (eds.), Chicago: University of Chicago Press, pp. 183-203.

Scott, W. R. 2001. *Institutions and Organizations*, Thousand Oaks, CA: Sage Publications,.

Scott, W. R. 2004. "Institutional Theory," in *Encyclopedia of Social Theory*, G. Ritzer (ed.), Thousand Oaks, CA: Sage Publications, pp. 408-414.

Spiegel, R. 2005. "Cost of Compliance – 2 to 3 Percent of Cost of Goods," *Design News*, September 6 (http://www.designnews.com).

Stake, R. E. 1995. *The Art of Case Study Research*, Thousand Oaks, CA: Sage Publications.

Taylor, C. 2005. "The Evolution of Compliance," *Journal of Investment Compliance* (64), pp. 54-58.

Yin, R. K. 2003. *Case Study Research: Design and Method*, Thousand Oaks, CA: Sage Publications Inc.

About the Authors

Tom Butler is a Senior Lecturer in Business Information Systems, University College Cork, Ireland. A former IT professional, he worked for 27 years in the telecommunications sector. Tom's research focuses on investigating the origins of firm-level IT capabilities and the design, development, and implementation of information systems, particularly knowledge management systems (KMS). From 2003-2006, he was lead researcher and project manager on two major action research-based initiatives on the design, development, and deployment of IT-enabled KMS from the United Nations Population Fund Agency and the Irish government. Since 2005, Tom has been conducting research into the design of compliance knowledge management systems (CKMS). His work has been published outlets including *Information Systems Journal, Journal of Strategic Information Systems, Journal of Information Technology*, and the proceedings of major international conferences such as the International Conference on Information Systems, the European Conference on Information Systems, and IFIP Working Groups 8.2 and 8.6. Tom can be reached at TButler@afis.ucc.ie.

William Emerson is a College Lecturer in Business Information Systems at University College Cork. A former IT professional, he worked for 15 years in financial services organizations as an analyst programmer and system architect. His current research interests focus on the design, implementation, and use of compliance knowledge management systems, particularly in the financial services sector. Bill's research has been published in journals, including *Journal of Strategic Information Systems*, and in the proceedings of conferences such as the European Conference on Information Systems. Bill can be reached at b.emerson@ucc.ie.

Damien McGovern is founder and CEO of Compliance & Risks Ltd. Damien is a lawyer and spent a considerable portion of his career working for Deloitte & Touche in Europe. It was in this capacity that he identified the need for a dedicated compliance knowledge management system (CKMS). Consequently, he began to draft a blueprint for his Compliance-to-Product application in 2001. His CKMS concept has been tested and validated by Compliance & Risks' clients and is now fully operational in several sites. C&R has its headquarters in Cork, Ireland, while its software development team is located in northern California and in Ireland. Damien's team of governance, risks, and compliance domain experts include lawyers in Ireland, the United Kingdom, Europe, and the United States. Damien can be reached at d.mcgovern@complianceandrisks.com.

5 SERVICE SYSTEM INNOVATION

Steven Alter
University of San Francisco
San Francisco, CA U.S.A.

Abstract *Service innovation has been discussed by many authors, but usually not from a system perspective. Recent literature about service systems and service innovation stresses multi-partner commercial offerings, ecosystems of interacting suppliers and consumers, globalization, and the changing nature of advanced economies. Interesting as these large-scale topics are, discussions at that level tend to overlook operational and organizational issues that service innovators must address in order to create or improve specific, localized systems that deliver services to internal and/or external customers.*

This paper builds on three interrelated frameworks that describe fundamental aspects of service systems at the level at which they are designed, operated, and improved:

- *The work system framework identifies nine elements that should be included in even a rudimentary understanding of any work system. (Service systems are work systems.)*
- *The service value chain framework incorporates characteristics often associated with services, such as coproduction by providers and customers.*
- *The work system life cycle model treats the system's life cycle as a set of iterations involving planned and unplanned change.*

This paper shows how each of those three frameworks provides insights that apply to service innovation across a wide range of service systems.

Keywords Service innovation, service system, definition of service, work system framework, service value chain framework, work system life cycle model

Please use the following format when citing this chapter:

Alter, S., 2008, in IFIP International Federation for Information Processing, Volume 267, Information Technology in the Service Economy: Challenges and Possibilities for the 21st Century, eds. Barrett, M., Davidson, E., Middleton, C., and DeGross, J. (Boston: Springer), pp. 61-80.

1 INTRODUCTION

Over half of the revenues of technology companies come from services (Wood 2007). Noting the increasing importance of services in the economy in general and for technology companies in particular, IBM joined with other technology companies in a major initiative to develop university courses of study related to service and to develop "service science" (Chesbrough and Spohrer 2006; Spohrer et al. 2007). Motivations for this initiative included the long term need to hire employees who can succeed in a service economy and the belief that many practical and theoretical ideas related to service are not yet well developed.

This paper presents a broadly useful view of service system innovation (SSI). Service innovation is mentioned often in discussions of service, and many examples of service innovation have been described. However, service innovation is rarely presented as a type of business initiative that can be analyzed and implemented based on an organized set of ideas that apply to almost all services, not just hospitality or retail or e-commerce or e-government. Similarly, the term *service system* appears frequently but is rarely defined clearly or used as an analytical concept.

The term *service system innovation* is especially rare in the service innovation literature, which contains many discussions of service innovation, service systems, and service system engineering. For example, a Google search on "service system innovation" on March 9, 2008, returned only 66 hits, many of which were tangentially relevant or irrelevant. Many the relevant hits involved *Methodology for Product Service System Innovation: How to Implement Clean, Clever, and Competitive Strategies in European Industries* (van Halen et al. 2005). While that book certainly deals with system-related issues, it focuses on strategic issues and partnerships between firms. For example, it says "product service systems theory brings companies to a new strategic level and provides new industrial perspectives. It opens new approaches on value creation, social context and industrialization visions" (p. 18). In contrast, our discussion of service system innovation applies to both internally and externally directed services.

The goal in this paper is to provide a system-oriented view of service innovation that is based on service system concepts and is both useful and broadly applicable. Going beyond defining service innovation or providing accounts of service innovation in particular settings, this paper couches service innovation in terms of an organized set of ideas about service system innovation that should help in describing, analyzing, and researching how service systems are created, how they operate, and how they evolve over time through a combination of planned and unplanned change.

This paper proceeds as follows. After reviewing existing definitions of service, it defines service, service system, and service innovation. It explains that the proposed definitions differ from corresponding definitions reflecting very recent thought in the service science community. It presents three frameworks related to systems and services and shows how each framework provides an analytical lens and a set of insights that managers and service innovators can use for describing, analyzing, and implementing specific service system innovations. Each framework has been discussed in considerably more depth, and would yield many additional implications in a longer paper or book. The main contribution of this paper is its demonstration of basic implications of viewing service innovation from a service system innovation (SSI) perspective organized around the three frameworks.

2 Basic Definitions

There is surprisingly little agreement about definitions of service and service innovation. Debates about the definition of service stem from difficulties distinguishing between goods (often called products) and services. Most marketing books bypass this issue by saying that offerings to customers often combine product and service features.

Existing definitions of service. Researchers in marketing, operations, and computer science discuss and analyze services from vastly different viewpoints. Table 1 lists a number of definitions, four of which were cited by Rai and Sambamurthy (2006), who said,

> there is reasonable triangulation on what services are…in general, the definitions emphasize a simultaneous or near-simultaneous exchange of production and consumption, transformation in the experience and value that customers receive from engagement with providers, and intangibility in that goods are not exchanged (p. 328).

Rai and Sambamurthy's "reasonable triangulation" is open to question because many service situations fit neither their triangulation nor the definitions in Table 1. Many views of service emphasize the customer's subjective experience of person-to-person service encounters, for example, Carlzon's (1989) term "moments of truth" and Teboul's (2006) book, *Service Is Front Stage.* Subjective experience of service encounters is less important (although still present) for highly automated services (e.g., web-based mortgage brokers or semi-automated translation services), back office services (producing tax returns or generating market research reports), computer-to-computer services (e.g., remote monitoring of equipment or automatic backups), and self-service situations (e.g.,

Table 1. Typical Definitions of Service

"A service is a change in the condition of a person, or a good belonging to some economic entity, brought about as a result of some other economic entity, with the approval of the first person or economic entity" (Hill 1977, p. 318).
"A service is any act or performance that one party can offer to another that is essentially intangible and does not result in the ownership of anything" (Kotler and Keller 2006, p. 402).
"A service is a time-perishable, intangible experience performed for a customer acting in the role of a coproducer" (Fitzsimmons and Fitzsimmons 2006, p. 4).
Services are "the application of specialized competencies (knowledge and skills) through deeds, processes, and performances for the benefit of another entity of the entity itself" (Vargo and Lusch 2004, p. 2).
Service is "the application of competences (knowledge skills, and resources) for the benefit of another entity in a mutually agreed and mutually beneficial manner" (IfM and IBM 2007, p. 16).
"Service [is] the application of resources for the benefit of another" (Vargo and Lusch cited in Spohrer et al. 2008, p. 1).
An article called "Foundations and Implications of a Proposed Unified Services Theory" found one unique characteristic that differentiates services from products: "with service processes, the customer provides significant inputs into the production process" (Sampson and Froehle 2006, p. 331).

using an ATM). Subjective experiences are of no importance in computer scientists' views of services, such as "a service is generally implemented as a course-grained, discoverable software entity that exists as a single instance and interacts with applications and other services through a loosely coupled (often asynchronous), message-based communication model" (Brown et al. 2005, p. 728).

Definition of service. We adopt a simple, dictionary-like definition by which the following are all services: performing surgery, installing networks, producing customized software, providing Internet-based search capabilities, accepting orders through an e-commerce web site, building houses, producing televisions, providing leisure opportunities on golf courses, performing legal work, and selling groceries.

Services are acts performed for others, including the provision of resources that others will use.

This definition applies to a wide range of services:

- services for external customers and for internal customers
- automated, IT-reliant, and nonautomated services
- customized, semi-customized, and non-customized services
- personal and impersonal services
- repetitive and non-repetitive services
- long-term and short-term services
- services with varying degrees of self-service responsibilities

It applies to the three types of value configurations discussed by Stabell and Fjeldstad (1998): value chains, value networks, and value shops. It covers special cases such as self-service and automated services for people. In self-service, service providers provide resources that are used by customers performing self-service activities. In automated services for people, machines perform the service activities.

This definition is similar to the Vargo and Lusch definitions in Table 1, but differs from other definitions in a number of ways. Contrary some of the definitions in Table 1 and to Rai and Sambamurthy's triangulated definition (mentioned above), it says nothing about service characteristics such as intangibility, customization, simultaneity of production and consumption, time-perishability, or involvement of customer interactions or experiences. Such characteristics are treated as continuous design variables that apply to different services in differing degrees and can be set to different levels depending on the goals of the service situation. Similarly, while coproduction of value is often associated with services, coproduction can also be viewed as a continuous design variable whose different levels include

- The customer provides a request for service (minimal level of coproduction).
- Customers participate in some aspects of service fulfillment processes (beyond specifying requirements).
- The service occurs largely through service interactions including direct participation by customers.
- A self-service approach is used, whereby the service provider creates and provides the means by which the customer performs self-service processes and activities.

Two other aspects of the definition are worth mentioning. The definition says nothing about provider–customer interactions because many services involve minimal customer interaction. In self-service, for example, provider–customer interaction devolves to the customer's use of the provider's technical artifacts and procedures. The fact that such usage might be viewed as an interaction, especially from the viewpoint of actor–network theory, is not sufficient to include interaction as part of the definition. In addition, the definition does not assume services are provided "in a mutually agreed and mutually beneficial manner," a phrase included in a definition in Table 1 that appeared in a recent discussion paper produced by an international symposium of service research leaders and others committed to developing service science (IfM and IBM 2007). That phrase is normatively attractive but inconsistent with practice, as noted in a recent *Harvard Business Review* article (McGovern and Moon 2007) that discusses how companies in cell phone service, airlines, banking, health clubs, car rental, and credit cards take advantage of customers through increasingly opaque, provider-centric strategies that are not designed for customer benefit.

Probably the most controversial aspect of our dictionary-like definition is that it ignores the traditional distinction between goods and services because a precise boundary is of little practical importance in marketing, operations, or information systems. The "service-dominant logic" (contrasted with "goods-dominant logic") proposed by Vargo and Lusch (2004) addresses the goods versus service issue by proposing that "goods are distribution mechanisms for service provision" (p. 8). This idea mirrors Levitt's observation that people who buy quarter-inch drills actually want quarter-inch holes (Christensen et al. 2005, p. 76), whereby the extent to which customer value is delivered is more important than whether it is delivered through goods or services. Instead of pondering the precise dividing line between product and service (e.g., the point where a restaurant meal or a new home flips from product to service), we assume that purposeful activity performed for someone else is a service.

Finally, unlike most other definitions, the definition of service needs only a slight revision to cover service computing (the computer science view). In that realm, which this paper otherwise ignores, *another entity* replaces *others* and the definition becomes

> *Services are acts performed by one entity for another, including the provision of resources that another entity will use.*

Definition of service system. A service system is a work system that produces services. "A work system is a system in which human participants and/or machines perform work using information, technology, and other resources to produce products and services for internal or external customers" (Alter 2003, p. 368). Based on the discussion of the definition of service, the distinction between products and services is of little importance when evaluating, analyzing, designing, or implementing systems in real-world situations. If, as mentioned earlier, one views the characteristics commonly ascribed to services as continuous design variables, the important practical issues involve identifying the level of each design variable that is the best compromise between provider and customer needs and interests.

As with the definition of service, the definition of *service system* is not obvious. For example, a recent paper by service research leaders defined a service system as

> a dynamic value co-creation configuration of resources, including people, organizations, shared information (language, laws, measures, methods), and technology, all connected internally and externally to other service systems by value propositions....Every service system has a unique identity, and is an instance of a type or class of service systems (e.g., people, businesses, government agencies, etc.). The history of a service system is a sequence of interaction episodes with other service systems, including interaction episodes with itself....Service systems have a beginning, a history, and an end (Spohrer et al. 2008, p. 5).

Our definition of service system does not assume that service systems are dynamic because many typical service systems (such as payroll systems and mass education systems) exist in essentially the same form over extended periods. Also, it says nothing about interacting through value propositions because many services are not voluntary for customers (e.g. elementary education and driver registration). Most important, it views service systems as work systems. Typically one would not view a person as work system; from the other side, most businesses and government agencies of more than minimal size would not be viewed as service systems. Rather they would be viewed as entities consisting of multiple service systems.

Definition of service innovation. Many, and perhaps most, published discussions of service innovation address broad concerns that are somewhat distant from the realities of local decision making and action related to systems that actually produce services. For example, the call for papers for this IFIP 8.2 Working Conference, "IT and Change in the Service Economy," reflects broad, high-level concerns and does not even contain the word *system*. It sets the context by mentioning transformations within the service economy, globalization, new business models, and new regulatory, normative, and socio-cultural institutions at many levels, including societal, industry, profession, organizational, group, and individual.

Consistent with Schumpeter's (1934) distinction between invention and innovation, we define *service innovation* as the implementation of new services and/or better ways to produce existing services. Consistent with the discussion of innovation in Wikipedia (2008), service innovation may involve any combination of service system changes:

- changes generating incremental value for customers (small innovations that tend to be less interesting as innovations)
- changes generating substantially greater value for customers (large innovations)
- changes that improve the provider's efficiency, cost, quality, or reliability, whether or not customers observe the improvements directly.

Within that view, service innovation involves creating or improving specific service systems such as a firm's systems of hiring employees, finding sales prospects, delivering products and services, performing corporate planning, or providing customer services. It also applies to systems that cross organizations, such as supply chains.

Our definition of service innovation is related to a particular situation and says nothing about whether others may have created similar innovations elsewhere and whether innovation is linked to competition, competitiveness, or globalization. With this situated view, service innovation is fundamentally about creating beneficial improvements and is not about demonstrating that the resulting service system differs from or is better than other service systems elsewhere. (Nonetheless, competitively significant innovations are often among the most interesting.)

In contrast, the glossary of the service symposium discussion paper mentioned above (IfM and IBM, 2007) defines service innovation as "the combination of technology innovation, business model innovation, socio-organizational innovation, and demand innovation to improve existing or create new service value propositions (offerings or experiences) and service systems." Our definition does not assume that technology innovation, business model innovation, socio-organizational innovation, and demand innovation are all necessarily involved in service innovation. To the contrary, a service innovation may involve little more than installing newer technology or terminating service for customers who absorb excessive resources. Since service innovations might involve internal processes that are invisible to a service system's customers, our definition does not assume that value propositions will necessarily change.

3 APPLYING THREE FRAMEWORKS

Our definitions of service, service system, and service innovation form the basis of a system-oriented view of service innovation. Our view of service system innovation (SSI) stems from these definitions, which conform to three interrelated frameworks that describe service systems from a business viewpoint (Alter 2008).

Each framework serves as an analytical lens and provides a set of insights that are useful for describing, analyzing, and implementing specific service system innovations. In contrast to typical analysis and design concepts related to data, workflows, and technology, these frameworks provide concepts that reflect the semantics and business context of services, rather than just computing or engineering concepts. These frameworks can be used to organize many additional concepts related to each element of the frameworks.

Taken together, the three frameworks provide a rich and broadly applicable model of how services operate and evolve. They create a platform for comparing service situations, identifying important special cases of services, and describing service design strategies. In turn, these ideas can contribute to research about relative advantages and disadvantages of different types of innovations in the presence of specific situational characteristics.

4 THE WORK SYSTEM FRAMEWORK

The work system framework (Figure 1) uses nine basic elements to provide a system-oriented view of any system that performs work within or across organizations. This framework is the basis of the work system method, a systems analysis method that incorporates many additional concepts related to the individual elements and work systems as a whole. Basic ideas for describing and analyzing work systems are applicable to service systems because service systems are work systems that produce services.

As an illustrative example of a service system, Table 2 shows a work system snapshot related to a hypothetical loan application and underwriting system that combines

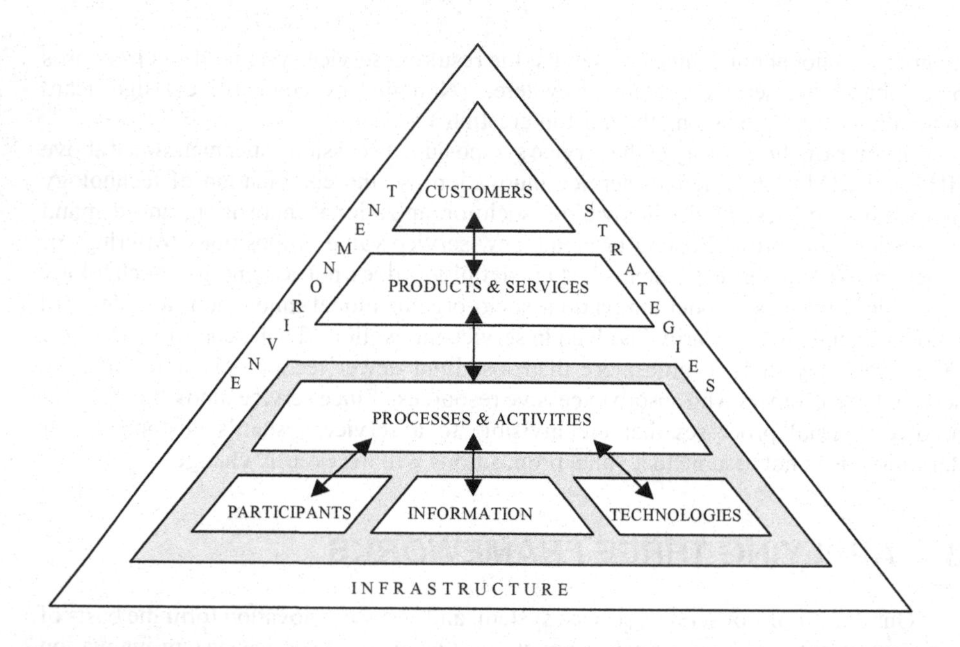

Figure 1. The Work System Framework (slightly updated; Alter 2006, 2008)

functional characteristics from a number of different real-world systems. A work system snapshot is a tabular, one-page summary of a work system based on the six central elements in the work system framework. A service innovator wishing to improve this work system might consider changing any or all of the six elements (and possibly the other three that are not included in the one page summary).

The work system framework illuminates service innovation at two levels, by providing a holistic view of SSI and by providing SSI-related insights about specific work system elements.

At the holistic level, the framework implies that service innovation necessarily involves SSI and is not just about changing whatever customers see. The framework's linkage between specific elements implies that SSI typically involves mutually aligned changes in multiple elements of a service system.

The many facets of SSI (as shown through the work system framework) imply that attempts to design and implement SSI around a single goal for a single performance indicator (e.g., amount of service produced per person-hour of work) are usually short-sighted. On the one hand, potential SSIs should be evaluated based on improvements in efficiency, consistency, employee satisfaction, and other internally directed metrics. On the other hand, for both internal and external customers, performance from the customer perspective is related to total cost borne by the customer (including time and energy expended) and to customer perceptions of quality, responsiveness, reliability, the overall customer experience, and conformance with standards and expectations.

Attributes of the various work system elements form the basis of SSI design decisions at a big-picture level. For example, big-picture SSI decisions related to processes and activities involve issues such as the degree of structure, complexity, and rhythm of

Table 2. Work System Snapshot for a Loan Application and Underwriting System for Loans to New Clients (Alter 2006)

Customers	Products and Services
• Loan applicant • Loan officer • Bank's risk management department and top management • Federal Deposit Insurance Corporation (FDIC) (a secondary customer)	• Loan application • Loan write-up • Approval or denial of the loan applicaiton • Explanation of the decision • Loan documents

Work Practices (Major Activites or Processes)
• Loan officer identifies businesses that might need a commercial loan. • Loan officer and client discuss the client's financing needs and discuss the possible terms of the proposed loan. • Loan officer helps client complete a loan application including financial history and projections. • Loan officer and senior credit officer meet to verify that the loan application has no glaring flaws. • Credit analyst prepares a "loan write-up" summarizing the applicant's financial history, providing projections explaining sources of funds for loan payments, and discussing market conditions and applicant's reputation. Each loan is ranked for riskiness based on history and projections. Real estate loans all require an appraisal by a licensed appraiser. (This task is outsourced to an appraisal company.) • Loan officer presents the loan write-up to a senior credit officer or loan committee. • Senior credit officers approve or deny loans of less than $400,000; a loan committee or executive loan committee approves larger loans. • Loan officers may appeal a loan denial or an approval with extremely stringent loan covenants. Depending on the size of the loan, the appeal may go to a committee of senior credit officers, or to a loan committee other than the one that made the original decision. • Loan officer informs loan applicant of the decision. • Loan administration clerk produces loan documents for an approved loan that the client accepts.

Participants	Information	Technologies
• Loan officer • Loan applicant • Credit analyst • Senior credit officer • Loan committee and executive loan committee • Loan administration clerk • Real estate appraiser	• Applicant's financial statements for last three years • Applicant's financial and market projections • Loan application • Loan write-up • Explanation of decision • Loan documents	• Spreadsheet for consolidating information • Loan evaluation model • MS Word template • Internet • Telephones

the processes, and the incentives and expected skill level for the participants. In combination, design decisions related to specific elements result in holistic system characteristics such as centralization/decentralization, capacity, leanness, scalability, resilience, agility, and transparency.

Customers. SSI is equally applicable to externally facing processes and inwardly facing processes and activities that are services, such as processes for hiring and managing employees, planning, organizing, maintaining facilities and equipment, and qualifying sales prospects. In contrast, discussions of services often assume that services necessarily face outward, as in "our sales revenue increased after we introduced new services for our customers."

Speaking of innovations related to "the customer" of a service system is often misleading. A work system approach quickly reveals that many service systems have multiple customers with different needs and concerns. Those customers often include the direct beneficiaries of whatever a service system produces, plus other customers whose interest and involvement is less direct. For example, direct beneficiaries of a service may not be paying customers (e.g., medical service paid by insurance or employee counseling paid by the employer). SSI directed toward one group of customers may have no effect on other customers or may result in better or worse service for other customers.

Products and Services. SSI often involves combinations of products and services because the actions performed for customers might include the creation and transfer of physical things. The innovator's goal is to design or improve a service system to attain the right combination of internal efficiency and customer satisfaction. Thus, classifications related to products versus services are unimportant for SSI even though these classifications are important for analyzing industry structure and macroeconomic trends.

Processes and Activities. SSI-related changes in processes and activities can support a variety of goals simultaneously. For example, innovations may improve the products and services experienced by customers and/or may improve the provider's efficiency, cost, quality, or reliability whether or not customers observe the improvements directly.

SSI design is about more than changing process details that might be captured in flow charts or business rules. The vocabulary for process-related aspects of SSI includes process design characteristics such as degree of structure, degree of integration, complexity, variety of work, degree of automation, rhythm, time pressure, amount of interruption, error-proneness, and formality of exception handling. Each of these big-picture decisions can be discussed at a management level to identify situation-specific guidelines that will help in making detailed choices that will appear in flow charts or business rules.

SSI often involves automation of information handling and physical processes. For example, IndyMac Bank's SSI in the mortgage industry converted largely manual processes to highly automated processes, partly by permitting web-based self-service (Krogh et al. 2005). In such situations, the arrows in the work system framework point to a series of corresponding changes that must occur in other work system elements.

Also, SSI sometimes needs to deal with the treatment of workarounds and exceptions, another common work system topic. The SSI question is the extent to which processes and activities should be structured, at least in theory, and the extent to which workarounds should be allowed or even encouraged.

Participants. The success of SSI depends on many factors related to participants, including skills, knowledge, incentives, and presence of a service mindset.

SSI often involves changing roles of participants. Increasingly, SSI incorporates self-service, whereby customers perform functions that were previously performed by the producer's employees. SSI through automation (as in the IndyMac case mentioned above) often eliminates roles of some system participants (such as certain mortgage brokers). In such cases, SSI usually includes replacing former participant roles by technician roles focused on maintaining the service system's technology and external infrastructure.

Information. SSI often involves using different information or achieving improvements related to information accuracy, accessibility, timeliness, and so on. As indicated by the arrows in the work system framework, however, better information contributes to SSI only if that information is incorporated into the system's processes and activities.

Technologies. Almost all major service systems rely on IT. However, the headline for understanding and analyzing SSI in specific situations involves the entire system of providing service, rather than the parts that happen to use IT in interesting ways.

Environment. A service system's environment includes organizational culture and relevant regulations, policies and procedures, competitive issues, organizational history, and technical developments. Even within the same business function, company, and industry, a successful SSI in one situation may be unsuccessful in another due to differences in the environment surrounding the service system. Consequently, the common claim that specific practices built into commercial software packages are "best practices" is often misleading for SSI.

Infrastructure. SSI initiatives sometimes succeed or fail based on the presence and operation of human, information, and technical infrastructures that the service system shares with other work systems and does not manage or control. Consequently, SSI design and evaluation should not stop at the boundaries of the service system, and should identify expectations related to external infrastructure.

Strategies. SSI sometimes involves changes in a service system's strategy, which includes its value proposition to its customers and its production strategy. Its value proposition summarizes how customer needs will be met, and at what implicit or explicit cost to customers. Its production strategy concerns how the provider organization will execute the steps it is responsible for performing. SSI projects sometimes encounter problems when the strategies designed into the innovation conflict with strategies of the firm or organization.

5 THE SERVICE VALUE CHAIN FRAMEWORK

All of the foregoing observations are equally applicable to all work systems, regardless of whether they are considered service systems. The service value chain framework (Figure 2) augments the work system framework by introducing ideas that are associated specifically with services regardless of whether the customers are internal or external. The entire service value chain for a particular service might be viewed and analyzed as a single work system. Alternatively, different subsystems in Figure 2 (such as provider or customer preparation) might be analyzed as separate work systems.

The two-sided structure of the service value chain framework provides the conceptual motivation for a flexible systems analysis tool called a *service responsibility table* (SRT) that can be used in various forms throughout the analysis and design of a service system. Table 3 presents an SRT for the example in the work system snapshot in Table 2. The first two columns identify activities and responsibilities of service providers and service consumers. The third column identifies problems and opportunities at each

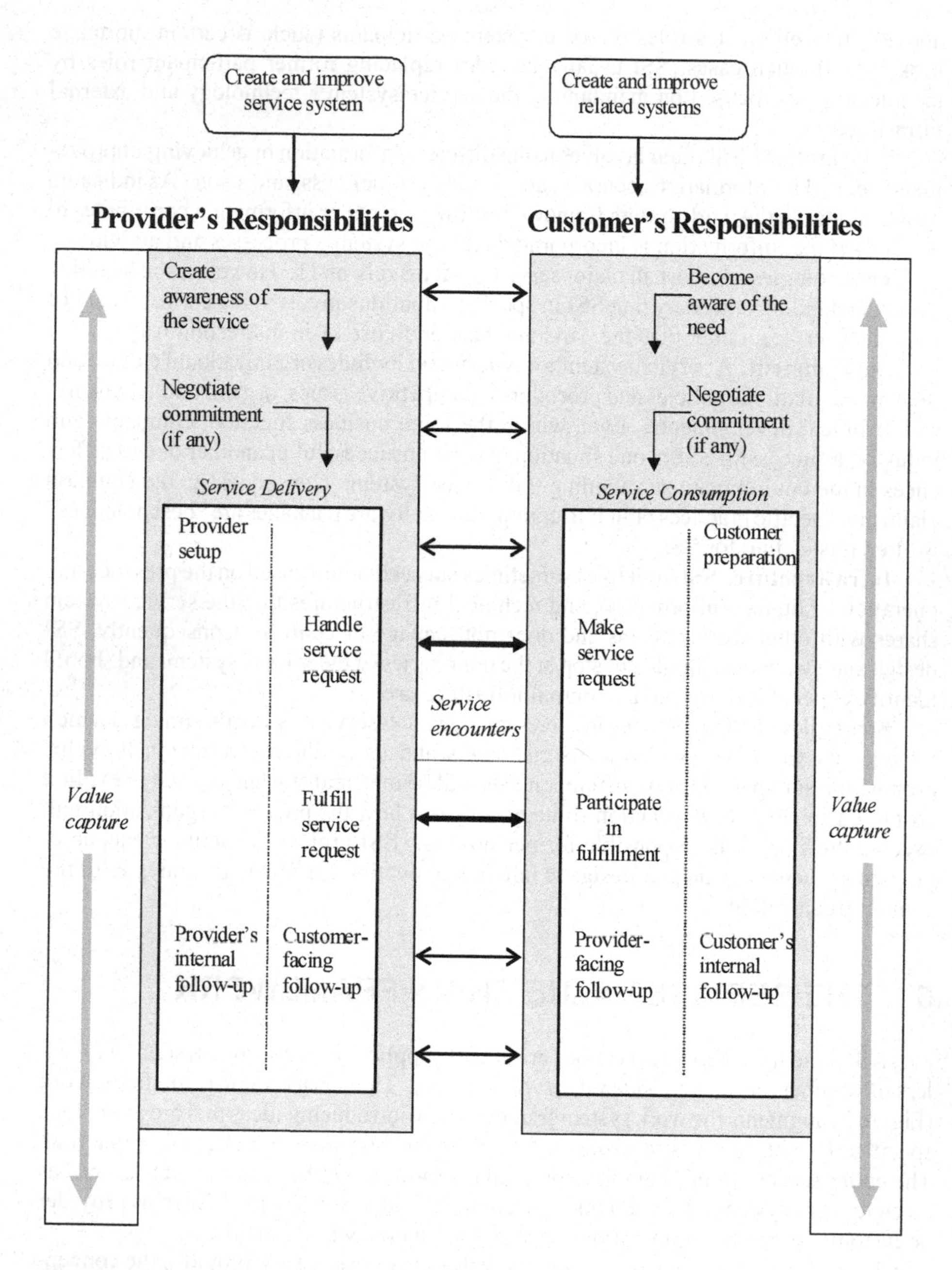

Figure 2. Service Value Chain Framework (updated slightly from Alter 2007, 2008)

Table 3. Three-Column Service Responsibility Table Including a Column for Problems and Issues (Alter 2007, 2008)

Provider Activity or Responsibility	Customer Activity or Responsibility	Problems or Issues
Loan officer identifies businesses that might need a commercial loan.		• Loan officers are not finding enough leads.
Loan officer contacts potential loan applicant.	**Potential loan applicant** agrees to discuss the possibility of receiving a loan	
Loan officer discusses loan applicant's financing needs and possible terms of the proposed loan.	**Potential loan applicant** discusses financing needs.	• Loan officer is not able to be specific about loan terms, which are determined during the approval step, which occurs later.
Loan officer helps loan applicant compile a loan application	**Loan applicant** compiles loan application.	• Loan applicant and loan officer sometimes exaggerate the applicant's financial strength and prospects.
Loan officer and **senior credit officer** meet to verify that the loan application has no glaring flaws.		• 20% of loans applications have glaring flaws.
Credit analyst prepares a "loan write-up" summarizing the client's financial history, providing projections of sources of funds for loan payments, etc.		• 10% rate of significant errors, partly because credit analysts use an error prone combination of several spreadsheets and a word processing program. • Much rework due to inexperience of credit analysts.
Loan officer presents the loan write-up to a senior credit officer or loan committee.		• Meetings not scheduled in a timely manner. • Questions about exaggerated statements by some loan officers.
Senior credit officer or **loan committee** makes approval decision.		• Excessive level of nonperforming loans. • Rationale for approval or refusal not recorded for future analysis.
Loan officer informs loan applicant of the decision	**Loan applicant** accepts or declines an approved loan.	• 25% of refused applicants complain reason is unclear. • 30% of applicants complain the process takes too long.
Loan administration clerk produces loan documents for an approved loan that the client accepts		

step. Alter (2007, 2008) identifies many other topics that could be included in SRTs, such as information generated or used, preconditions, important exceptions, typical errors, and even recommendations for improvements related to specific steps.

The service value chain framework is useful in SSI because it identifies both a number of steps and a number of general service topics that could be the focus of service innovations in specific situations. Topics related to the general form of the service value chain framework include coproduction of value, value capture, service interactions, front stage and back stage, and the customer experience. We summarize these topics first, and then look at SSI opportunities in the various steps.

Coproduction of value. The two-sided structure of the service value chain framework is based on the common observation that services are typically coproduced by service providers and customers (Fitzsimmons and Fitzsimmons 2006; Sampson and Froehle 2006; Vargo and Lusch 2004). Accordingly, service design requires attention to activities and responsibilities of both service providers and service customers. (Beyond this paper's scope, Vargo and Lusch suggest that the term coproduction should be replaced by cocreation of value because production brings connotations of manufacturing.)

Value capture. Customers may experience benefits as the service is produced and/or may experience benefits later. Value capture, represented by the left-most and right-most portions of the service value chain framework, includes the customer's experience of attaining value from the service and the provider's experience of attaining value in exchange for the customer's value. Since service systems exist to provide value for customers, SSI efforts should focus substantial attention on augmenting value capture, rather than just improving internal processes or providing a better form of service interactions.

Form and substance of service interactions. The service value chain framework shows that service interactions occur throughout the service value chain, not just when the service is being provided directly. Opportunities for SSI include improving provider–customer interactions during individual steps and creating new ways to integrate activity and information across the various steps.

Assignment of front stage and back stage. Services often involve front stage and back stage activities by both service providers and customers. The front stage versus back stage distinction raises a design challenge about the proper balance between front stage and back stage activities: Which service interactions are necessary, which are desirable, and which processes and activities related to the service are best done back stage? For example, the trend toward self-service is basically about transforming service interactions with the supplier's employees into what the providers might see as back stage activities performed by customers in self-service mode. Providing tools such as web-based self-service affords the provider more visibility of self-service activities.

Customer experience. Although the service fulfillment step in the service value chain model is typically viewed as the core of the service, SSI opportunities may transform or facilitate activities related to any part of the entire experience that typical customers associate with acquiring, receiving, and benefiting from a particular service.

Each of the steps in the service value chain framework affords SSI opportunities.

Prerequisite systems. Systems in organizations are always related to other systems, both for service providers and for their internal and external customers. SSI initiatives should recognize that most service innovations are based on assumptions about complementary systems elsewhere. For example, a hospitality system that provides enjoyable resort experiences for vacationers is related to the travel systems that deliver the vacationers to the resort. Travel to the resort may provide SSI opportunities that improve the overall customer experience, such as better ways to deliver vacationers to the resort or providing tourism opportunities on the way.

Awareness. Value creation does not occur if customers are unaware of available services. Consequently, SSI opportunities for a specific service start with better ways to create awareness of the service.

Negotiated commitments. Many service situations involve delivery of services based on negotiated commitments under which the service may be requested and delivered repeatedly. Complex, detailed negotiations related to IT outsourcing represent one extreme of negotiated commitments; simple negotiations leading to verbal agreements about personal services represent the other extreme. Across the entire spectrum of negotiation complexity, SSI can sometimes reduce costs and increase quality of the results for both customers and providers.

Customer and provider preparation. Preparation by providers and/or customers prior to each instance of service delivery is often essential for service efficiency and effectiveness. SSI may improve aspects of the preparation that occur before the service is produced for a specific customer, such as better scheduling and clean-up of examination rooms in a clinic so that patients feel comfortable while waiting for doctors and don't experience excessive waits.

Service request. For many services, each instance of service delivery includes an explicit or implicit service request. This is another area for SSI because handling of the service request is an important part of service delivery and often affects customer satisfaction.

Service fulfillment. Often viewed as the core of services, service fulfillment processes can be improved or facilitated through SSI related to automation of the services, automation of guidelines and control, self-service, enhanced communication and collaboration, and many other approaches.

Follow-up. Some services require follow-up by providers and/or customers. Follow-up may be related to a single service instance (Was the installation OK?) or to multiple service instances (How responsive is your account manager?). In these cases and others, SSI can lead to greater efficiency and effectiveness for providers and customers.

6 THE WORK SYSTEM LIFE CYCLE MODEL

The *work system life cycle* (WSLC) model looks at how work systems (including service systems) change and evolve over time (see Figure 3). It treats a system's life cycle as a set of iterations involving planned and unplanned change. Unlike system development life cycle (SDLC) models that describe programming projects, this framework makes no assumption about whether information technology is involved or whether IT-related changes occur. After creation of the first version of the service system, each iteration of planned change goes through the stages of initiation, development, and implementation, thereby creating the next version of the system. Unplanned change occurs through experimentation and adaptation during any phase of the life cycle.

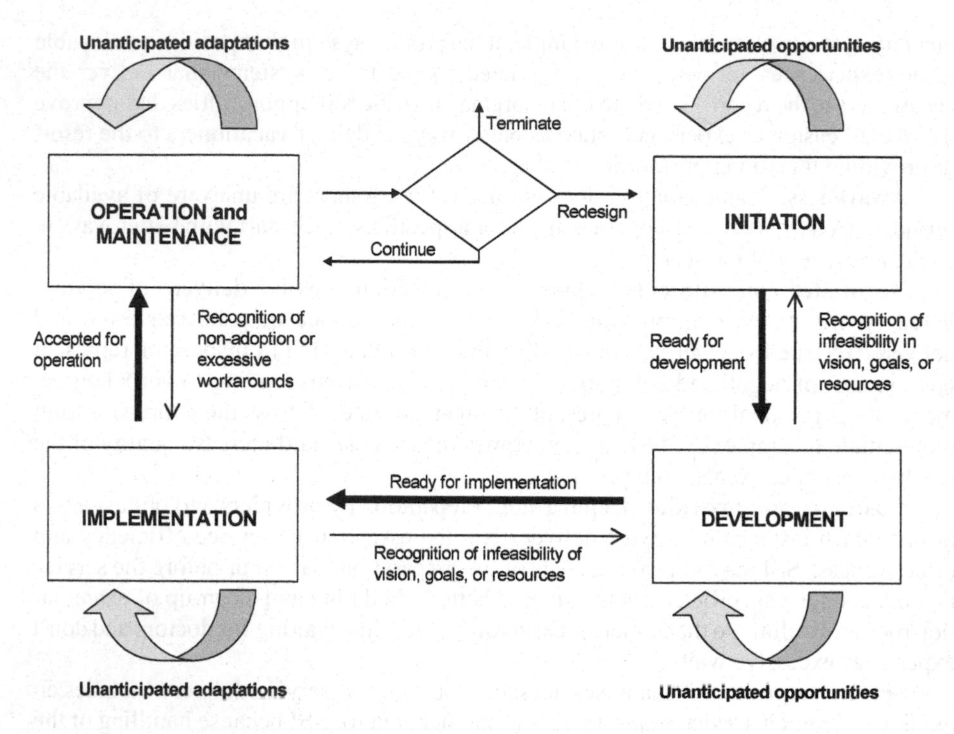

Figure 3. The Work System Life Cycle Model (Alter 2003, 2006, 2008)

The form of the WSLC indicates that SSI is often iterative. Consequently, the innovation strategy for a particular service system might involve frequent incremental changes interspersed with occasional leaps forward. In contrast, note how the SDLC, even in its iterative forms that deliver software in phases, is basically a project model that goes from project initiation to project completion. In other words, the WSLC provides a better long-term picture of how service innovation unfolds as service systems evolve.

SSI opportunities occur during each phase of the work system life cycle model.

SSI during the operation and maintenance phase. SSI during the operation and maintenance of existing systems typically involves unanticipated changes that can be viewed as small innovations. Such changes often involve using software differently or changing procedures. Sometimes they also involve small modifications of software that do not require significant projects.

SSI during the initiation phase. This phase is important for SSI because this is where the direction for planned innovations is negotiated. Discussions during a well-executed initiation phase may find that the changes that launched the initiation phase are only part of a larger set of changes that would be required to obtain better results.

SSI during the development phase. The detailed analysis during the development phase often finds additional issues and opportunities that must be dealt with. In some instances, the SSI becomes broader. In others, realization about complexity and capacity limitations during the development phase may result in curtailment of the extent and impact of the innovation.

SSI during the implementation phase. The implementation phase is another opportunity for SSI that goes beyond the initially intended changes that launched the project. In many cases, participant training and initial usage of software demonstrates that procedures must be changed in unanticipated ways and/or that the software itself must be changed.

A more detailed view of the WSLC (beyond this paper's length restrictions) would reveal many other SSI opportunities. It would also discuss process guidelines, issues related to division of responsibility between business and IT professionals, issues related to change management, and common caveats and pitfalls related to each stage. Suffice it to say that a discussion of SSI tends to be incomplete if it focuses on the service process or outcome and overlooks the service system life cycle.

7 CONCLUSION: VALUE OF A SERVICE SYSTEM VIEW OF SERVICE INNOVATION

Business enterprises, government organizations, nonprofits, and departments within larger organizations often need to create service innovations in order to meet their goals, succeed, and even survive. The ideas about SSI presented in this paper could be of substantial use to most organizations because service innovation almost always involves innovation related to the service systems that produce services. Given the importance of service system innovation, rather than service innovation in general, it is perhaps surprising that so little of the marketing, operations, and IS literatures focus on that topic.

We have seen how three interrelated frameworks illuminate important aspects of service system innovation. Individually or in combination, these frameworks can help in describing, analyzing, and researching the nature of service innovations and the processes through which service innovation occurs. A longer paper or a book could have explained those ideas in much greater depth and could have mentioned many other implications for describing and understanding service system innovation.

Moving beyond definitions of service and innovation. This discussion of SSI tried to take a step beyond the ongoing debates about the definitions of service and innovation. It used a simple, dictionary-like definition of service in order to avoid focusing on product versus service distinctions that are of little practical value to managers or IS practitioners, all of whom need to focus on whatever combination of products and services their customers want and need. It also defined innovation in a simple, situated way that reflects management concerns rather than the concerns of economic researchers. These simple, broad definitions facilitated using three frameworks to visualize aspects of service system innovation.

Identifying opportunities for service system innovation. This paper summarized how each of the three frameworks can be used for identifying opportunities for innovation. It identified separate opportunities related to nine elements of the work system framework, five concepts (coproduction of value, front stage versus back stage, and so on) and seven generic processes expressed in the service value chain framework, and four phases of the work system life cycle model. In other words, it identified at least 25 areas of potential service innovation that apply across a vast range of service situations.

Challenge for other service innovation models. The three frameworks presented in this paper are applicable to all service innovations in organizations. The system-oriented approach to service innovation poses several implicit challenges for other frameworks or models related to service innovation. The first challenge involves breadth: Do the alternative frameworks or models have the same breadth, or are they relevant to specific types of service situations, such as hotels, hospitals, law firms, software as a service, or IT consulting? The second challenge involves utility: This paper's three frameworks lead directly to 25 directions for possible service innovation. When comparing the view of service innovation presented here with any other view, one would immediately think about which types of innovation opportunities or possibilities are implied by alternative models, how broad are the relevant service areas, and which areas of service simply aren't covered.

What is next? The ultimate test of the ideas presented here is whether they help SSI practitioners and researchers understand and analyze real-world situations, and whether they help instructors teach about service and service innovation. Steps in that direction include

- Developing detailed comparisons between the three frameworks and other frameworks that might be used to understand and guide service innovation efforts
- Using the three frameworks to summarize existing accounts of service innovation, identify omissions, and draw lessons for the future
- Developing guidelines for using the three frameworks during service innovation projects
- Testing those guidelines through action research and other research methods
- Developing educational and training material that help people use the three frameworks in delivering and managing services and in creating service innovations

Each of these steps is a substantial project that could generate valuable results.

References

Alter, S. 2003. "18 Reasons Why IT-Reliant Work Systems Should Replace the IT Artifact as the Core Subject Matter of the IS Field," *Communications of the Association for Information Systems* (12:23), pp. 365-394.

Alter, S. 2006. *The Work System Method: Connecting People, Processes, and IT for Business Results*, Larkspur, CA: Work System Press.

Alter, S. 2007. "Service Responsibility Tables: A New Tool for Analyzing and Designing Systems," in *Proceedings of the 15th Americas Conference on Information Systems*, Keystone, CO, August 9-12.

Alter, S. 2008. "Service System Fundamentals: Work System, Value Chain, and Life Cycle," *IBM Systems Journal* (47:1), pp. 71-85 (http://www.research.ibm.com/journal/sj/471/alter.html).

Brown, A. W., Delbaere, M., Eeles, P., Johnston, S., and Weaver, R. 2005. "Realizing Service-Oriented Solutions with the IBM Rational Software Development Platform," *IBM Systems Journal* (44:4), pp. 727-752 (http://www.research.ibm.com/journal/sj/444/brown.pdf).

Carlzon, J. 1989. *Moments of Truth*, New York: Harper Collins.

Chesbrough, H., and Spohrer, J. 2006. "A Research Manifesto for Services Science," *Communications of the ACM* (49:7), pp. 35-40.

Christensen, C. M., Cook, S., and Hall, 5. 2005. "Marketing Malpractice: The Cause of the Cure," *Harvard Business Review* (83:12), pp. 74-83.

Fitzsimmons, J. A., and Fitzsimmons, M. J. 2006. *Service Management: Operations, Strategy, and Information Technology* (5th ed.), Boston: Irwin/McGraw-Hill.

IfM and IBM. 2007. *Succeeding through Service Innovation: A Discussion Paper*, University of Cambridge Institute for Manufacturing (IfM) and International Business Machines Corporation (IBM), October (http://www.ifm.eng.cam.ac.uk/ssme/documents/ssme_discussion_final.pdf).

Hill, T. P. 1977. "On Goods and Services," *The Review of Income and Wealth* (23:4), pp. 315-338.

Kotler, P., and Keller, K. 2006. *Marketing Management* (12th ed.), Upper Saddle River, NJ: Prentice Hall.

Krogh, E., El Sawy, O. A., and Gray , P. 2005. "Managing Online in Perpetual Perfect Storms: Insights from IndyMac Bank," *MIS Quarterly Executive* (4:4), pp. 425-442.

McGovern, G., and Moon, Y. 2007. "Companies and the Customers Who Hate Them," *Harvard Business Review* (85:6), June, pp. 78-84.

Rai, A., and Sambamurthy, V. 2006. "Editorial Notes: The Growth of Interest in Services Management: Opportunities for Information System Scholars," *Information Systems Research* (17:4), December, pp. 327-331.

Sampson, S. E., and Froehle, C. M. 2006. "Foundations and Implications of a Proposed Unified Services Theory," *Production and Operations Management* (15:2), pp. 329-343.

Schumpeter, J. A. 1934. *The Theory of Economic Development*, Boston: Harvard University Press.

Spohrer, J., Maglio, P. P., Bailey, J., and Gruhl, D. 2007. "Steps Toward a Science of Service Systems," *IEEE Computer* (40:1), pp. 71-77.

Spohrer, J., Vargo, S. L., Caswell, N., and Maglio, P. P. 2008. "The Service System Is the Basic Abstraction of Service Science," in *Proceedings of the 41st Hawaii International Conference on Systems Sciences*, Los Alamitos, CA: IEEE Computer Society Press.

Stabell, C. B., and Fjeldstad, O D. 1998. "Configuring Value for Competitive Advantage: On Chains, Shops, and Networks," *Strategic Management Journal* (19), pp. 413-437.

Teboul, J. 2006. *Service Is Front Stage: Positioning Services for Value Advantage*, New York: Palgrave Macmillan.

Van Halen, C., Vezzoli, C., and Wimmer, R. 2005. *Methodology for Product Service System Innovation: How to Implement Clean, Clever, and Competitive Strategies in European Industries*, Assen, Netherlands: Royal Van Gorcum.

Vargo, S. L., and Lusch, R. F. 2004. "Evolving to a New Dominant Logic for Marketing," *Journal of Marketing* (68), pp. 1-17.

Vargo, S. L., and Lusch, R. F. 2006. "Service-Dominant Logic: What It Is, What It Is Not, What It Might Be," in *The Service-Dominant Logic of Marketing: Dialog, Debate, and Directions*, R.F. Lusch and S. L. Vargo (eds.), Armonk, NY: M. E. Sharpe, pp. 43-56.

Wikipedia. 2008. "Innovation" (http://en.wikipedia.org/wiki/Innovation, viewed on March 17, 2008).

Wood, J. B. 2007. "Major Service Challenges Facing Technology Companies in 2007," presentation at the 16th Annual Frontiers of Service Conference, San Francisco, CA, October 4-7.

About the Author

Steven Alter is a professor of Information Systems at the University of San Francisco. He received a Ph.D. from MIT, taught at the University of Southern California, and was vice president of the manufacturing software firm Consilium before joining USF. His research for over a decade has concerned developing systems analysis concepts and methods that can be used by typical business professionals and can support communication with IT professionals. His latest book, *The Work System Method: Connecting People, Processes, and IT for Business Results*, is a distillation and significant extension of ideas in four editions (1992, 1996, 1999, 2002) of his information system textbook. His articles have been published in *Harvard Business Review, Sloan Management Review, MIS Quarterly, IBM Systems Journal, Communications of the Association for Information Systems*, and other journals and conference proceedings. Steve can be reached at stevenalter@comcast.net.

6 RHIZOMATIC INFORMATICS: The Case of Ivy University

Chris Atkinson
Manchester Business School
University of Manchester
Manchester, England

Laurence Brooks
Department of Information Systems and Computing
Brunel University
Uxbridge, England

Abstract *While the debate over information systems and their role in society persists, the discipline continues to seek approaches to better understand IS, and its complex interactions with organizations and people. This paper draws ideas from Deleuze and Guattari's work on rhizomes (a nonhierarchical network) and its opposing tree (arboreal structure), to develop better insights (Deleuze and Guattari 2004). The paper specifically revisits the case of Ivy University, in which a major introduction of an ERP application was attempted by the powerful central administration. When this centralized and arboreal structure collapses, it is eventually replaced by a much more organic system, which emerged from the localized and ad hoc software developments that had already taken place. The rhizomic interpretation of this case enables both a conceptualization of information systems within social and organizational settings and offers a set of principles on which the concept of a "rhizomatic informatics" may be based and IS case studies interpreted.*

Keywords Rhizomatics, arboreal, territorialization, nomad, ERP, case study

Please use the following format when citing this chapter:

Atkinson, C., and Brooks, L., 2008, in IFIP International Federation for Information Processing, Volume 267, Information Technology in the Service Economy: Challenges and Possibilities for the 21st Century, eds. Barrett, M., Davidson, E., Middleton, C., and DeGross, J. (Boston: Springer), pp. 81-101.

1 INTRODUCTION

Over the past 20 years, information systems research has drawn on a number of conceptual frameworks, from a variety academic traditions (Klein and Myers 1999; Walsham 1993). One of the earliest examples of this was the sociotechnical frameworks and methodology of Land (1979) and Mumford (1995). More recently, IS has drawn on Giddens' structuration theory (Giddens 1979, 1984; Jones 1999; Jones and Karsten 2008) from sociology and the traditional actor network theory (ANT) of Latour (2005), Callon (1986), and Law (1987, 1992) from science studies. Continuing this tradition of drawing on external theory and deploying it within IS, this paper outlines and explores the potential of Deleuze and Guattari's *rhizomatics* (1983, 1996, 2004) as an overtly post-modern ontology and analytical framework for IS interpretive studies, which also has the capacity to inform practice.

Information systems as a discipline has many development methods, with a heterogeneity of research schools, subjects, approaches and epistemologies; despite attempts to technologically ground it (Lee et al. 2001). It also shows eclecticism, practice focus, development processes, and tools, all with divergent schools of thought as a critical component. When contrasted to the "noble" sciences, such as Physics and Chemistry, IS is (to use a Deleuze and Guattari term) an "ignoble science," a post-modern pragmatics, complete with a number of divergent and competing perspectives and practices—from the managerial to the critical. This ignobility for the authors is a term of esteem rather than disapproval; even more so for IS, it is argued and explored here, when it enables a rhizomatic interpretation of an IS case study. In doing so, this paper seeks to challenge Deleuze and Guattari's own view of information or computer systems as being essentially and necessarily arboreal. Rather, to argue, that IS can also be rhizomatically critical.

These explorations start with the identification of core concepts and questions as to the relevance of rhizomatics to IS research and practice. It then proceeds to offer an interpretation, or rather a reinterpretation, of a case study surrounding the problematic introduction of an enterprise resource planning (ERP) system into a North American Ivy League University. These outcomes lead to the proposition that rhizomatics can be used to better understand information systems.

2 RHIZOMATICS

A rhizome

> connects any point to any other point, and its traits are not necessarily linked to traits of the same nature; it brings into play very different regimes of signs, and even nonsign states... .It is composed not of units but of dimensions, or rather directions in motion. It has neither beginning nor end, but always a milieu from which it grows and which it overspills... .The Rhizome proceeds by variation, expansion, conquest, capture, offshoots....the Rhizome pertains to a map that...is always detachable, connectable, reversible, modifiable, and has multiple entryways and exits (Deleuze and Guattari 2004, p. 23).

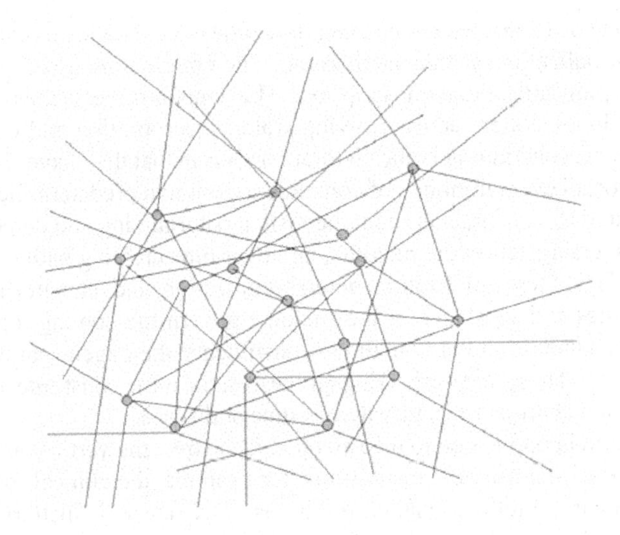

Figure 1. A Representation of the Rhizome. There is no center, but each node can link to an infinite number of other nodes (Seto 2006).

The rhizome, for Deleuze and Guattari, is a nonhierarchical network; it is root-like, spreading over space and time, growing incessantly as long as there are nutrients and energy available (see Figure 1). The rhizome has the potential to be universally ubiquitous, neither beginning nor ending; rather it spreads horizontally in all directions and across its' own self-created time and space. As an interpretive metaphor it is made up of a whole set of entities, such as human, nonhuman, and humanchines. It has no overarching governor or controller, it exhibits no intentionality, although domains within the rhizome, or *plateaus*, may grow, expanding ubiquitously and changing its constituents, shape, and trajectory in accordance with inherent conventions, often localized rules, and self-imposed constraints, acting within external contingencies in the wider rhizome. The rhizome is guided in its growth and spread by a set of self-governing, inherent principles and parameters both necessary to and a condition of its dynamics and *autopoietic persistence* (Maturana and Varela 1980). Self-delineating plateaus form within the rhizome through processes of what Deleuze and Guattari identify as *territorialization*. These, in turn, deconstruct through the processes of deterritorialization when constituents separate and disengage from each other, only to form anew with other constituents through processes of reterritorialization. Territorialization is the dynamic by which the rhizome grows, incorporates, ubiquitously spreads, and mutates:

> A multiplicity has neither subject nor object, only determinations, magnitudes, and dimensions that cannot increase in number without the Rhizomic multiplicity also changing in nature (the laws of combination therefore increase in number as the multiplicity grows) (Deleuze and Guattari 1988, p. 8).

Seen as a landscape, the rhizome is traversed by *nomads*. These are dynamic, highly mobile territorializations lacking in any form of structure, "bodies without organs"

(Deleuze and Guattari 1996; Deleuze and Guattari 2004). Nomads are dynamic localized autopoietic territorializations within the rhizome. Their migrations are achieved through processes of mapping and mutual transcription. The nomads traverse the rhizome in the way a mini-whirlpool skitters across flowing water, incorporating and ejecting water molecules as it goes, constituting bodies without organs in that they have form, made up of a locally territorialized multiplicity of constituents, without predetermined structures. Nomads are created by constituents spontaneously territorializing and deterritorializing to facilitate it traversing across the rhizome, and over time creating pathways across it.

The mechanism of territorialization, deterritorialization, and reterritorialization that create the dynamics and substance of the rhizome govern the coming together of the heterogeneity of its human and/or nonhuman constituents, their agency and their subsequent splitting up. These territorializing processes, in turn, constitute the dynamic network growth and chart its trajectory across time and space.

In contrast, and in opposition to the rhizome, is the tree, the vertical arboreal structure of the state, the establishment institutions, for example, the clinical royal colleges, the academe Française, multinational corporations. Deleuze and Guattari see these as vertical hierarchical arboreal structures. Such tree-like structures seek to exert control over all that they territorialize within their own hierarchical autopoiesis (Maturana and Varela 1980), including the rhizome from which they grow. The state can be seen as arboreal structures, domains in which arboreal political parties war against each other across the wider rhizomic polity. A tree, for Deleuze and Guattari (2004), is a site of power and domination. It is controlled from the top down to the bottom. Those at the top look down on and manipulate those below, who in turn look up to them. Communication in the tree is, vertical, from top to bottom, in contrast to the rhizome, in which communication is horizontally and universally ubiquitous. From branch tip to root, a tree is a control structure that grows from a single seed. It punctures the rhizome and disfigures it, seeking to territorialize those constituents who constitute the rhizome for its ends, creating striped, striated, patterns within its smooth surface. The tree's roots spread out and further distort the smoothness of the surrounding rhizome. Trees, and those at the top of its hierarchy, also seek to territorialize those who constitute the rhizome for their own power and instrumental ends. It forms a unity of which Deleuze and Guattari say, "The notion of unity appears only when there is a power takeover in the... [Rhizomic] multiplicity...or a corresponding subjectification proceeding" (1988, pp. 8-9).

The tree, for Deleuze and Guattari, is both a sign and a signifier, representing status, command, and domination, the "tree articulates and hierarchies." They cite Rosenstiehl and Petitot's (1974) critique of the arboreal as a representation for hierarchy and repressive power. The arboreal invokes "the imagery of command trees," identifying them as centered systems or hierarchical structures, and that "accepting the primacy of hierarchical structures amounts to giving arborescent structures privileged status." For them, grand science (e.g., Physics, Chemistry), even establishment "high art," constitutes arboreal structures that are aligned with, and patronized by the state and economic elites. Equally so are the multinational corporations, which also become territorialized into government, military, and industrial arboreal complexes that emerge vertically out of the rhizome that is a country's, and indeed the world's polity. The roots of these vertical hierarchies generate striated domains within the, otherwise smooth, free flowing, rhizome. The state epitomizes the arboreal, the tree of command, but so do multinational

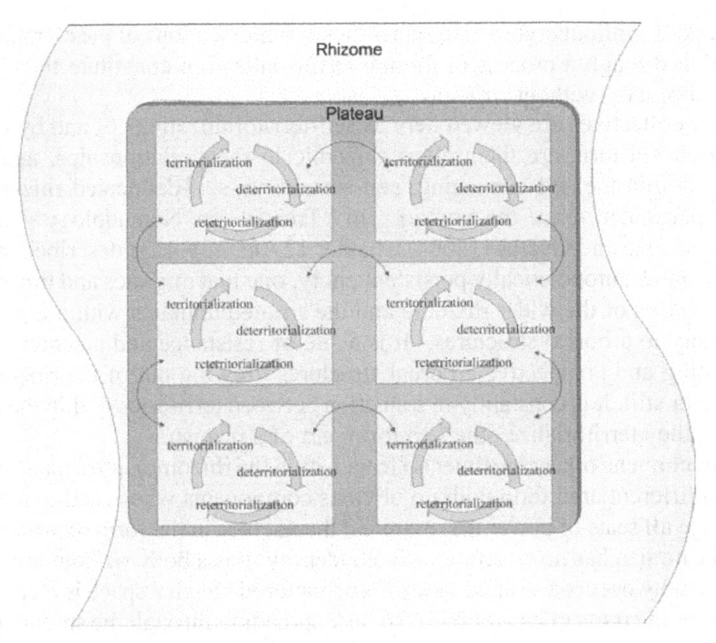

Figure 2. Rhizome and Plateau Representation

companies and, indeed, all hierarchical structures. Arboreal systems epitomize cybernetic (Ashby 1956; Wiener 1948) hierarchies, in which information, power, and control flow vertically rather than horizontally across the rhizome, centered in its plateaus. There is no one governing language within the rhizome; rather, there is a multiplicity of dialects, languages and idioms. The rhizome is truly heterogeneous in nature.

The rhizome is constituted, over time, out of an incessant myriad of instances of territorialization, deterritorialization, and reterritorialization. If many constituents territorialize together out of common interests, mutuality, interdependency, and, in some instances, the need for collaborative agency, then multiplicities of them may form a semi-permanent, heterogeneous plateau within the wider rhizome (see Figure 2). Such rhizomic plateaus not only embody and exhibit the principles of the rhizome set out above, but are also delineated as persistent territorializations, forming localized assemblages of affinity within the wider rhizome. A plateau is

> a continuous, self-vibrating region of intensities whose development avoids any orientation toward a culmination point or external end....[It is also] any multiplicity connected to other multiplicities by superficial underground stems in such a way as to form or extend a Rhizome (Deleuze and Guattari 1988, p. 22).

Here, persistent self-territorialized networks of humans and nonhuman constituents form plateaus (Deleuze and Guattari 1996, 2004; see Table 1).

There are no singularities within the rhizome; all are pluralities. The subjects within the rhizome are best served by being thought of as collective individuals—what we call *humanchines*. However, following Deleuze and Guattari, they constitute at their most simplistic a body without organs. Humanchines are intersections of the human and non-human, which through a process of mutual territorialization constitute the contingent body without organs, without structure.

All such collectives are viewed here as self-territorializations of and by their constituents; these, in turn, are themselves rhizomic in their relationships, actions, and behaviors. Within the self-convening, self-sustaining, self-delineated rhizome, there emerges a particular form, the nomad. In "Treatise on Nomadology – The War Machine," Deleuze and Guattari (2004, Chapter 12, pp. 351-423) describe a nomad as an auto-convened, autopoietically persistent entity, one that migrates and traverses both the smooth spaces of the wider rhizome and the striated domains within it induced by their proximity to arboreal structures. It is a site of resistance and a center for forces against existing and prospective arboreal structures growing out of the rhizome. The nomad is never still, it is constantly in transition between territories within the rhizome, until, that is, they territorialize others to form part of a plateau.

Nomads convene other constituents (from within the rhizome) into a plane of consistency, an undifferentiated form with no obvious components whose self-actuating role is to challenge all seats of power that manifest themselves in the form of arboreal structures. As the nomad has no structure or fixed identity, it is a body without organs. This challenge initially occurs a striated space is encountered. Such a space is created within the rhizome by the roots of an arboreal structure spreading through the smooth rhizome; the tree is a simile for the arboreal hierarchy that is the state, but could also be an enterprise. On meeting an arboreal structure, such as an arm of the state or enterprise, the nomad, autonomically, reterritorializes itself into a war machine, whose role is to challenge not only behaviors of that arboreal state, but its very existence. The war machine is a ubiquitous feature of the rhizome's thousand plateaus.

Deleuze and Guattari decry, "The tree is already the image of the world, or the root the image of the world-tree. The tree articulates and hierarchizes tracings; tracings are like the leaves of a tree." Further, almost in exasperation, they go on, "We're tired of trees. We should stop believing in trees, roots, and radicles. They've made us suffer too much." The rhizome and the tree are in a relationship of struggle over space and across time. Nomads form within the rhizome. They are rather like rebel guerrilla bands, localized territorializations of constituents within the rhizome formed to constitute war machines. Their role is to challenge the threats and incursions of the arboreal state into the rhizome. The nomad does this by cutting off the trees roots, preventing the tree from creating distorting striped patterns within the rhizome and territorializing constituents into its vertiginous trunk. As a manifestation of this struggle, rhizomes can be observed growing out of the tree or undermining its roots, deterritorializing and reterritorializing the tree's constituents back into its plateaus. The relationship between the state tree and the rhizomic is one of incessant struggle over the territorializing, deterritorializing, and reterritorializing of their constituents

"The war machine is the invention of the nomads (insofar as it is exterior to the State apparatus and distinct from the military institution) and exists outside of the State apparatus" (Deleuze and Guattari 2004, p. 419). The smooth (nonhierarchical) spaces that nomads traverse (as revealed in Deleuze and Guattari 1996) are in conflict with the striated spaces formed within the rhizome by the arboreal state and the corporation. A rhizomic nonstriated space is formed by the rhizome, one traveled by the nomad, constitutes a *plane of consistency*. As such, the rhizome

> knows nothing of differences of level, orders of magnitude, or distances. It knows nothing of the difference between the artificial and the natural. It knows nothing of the distinction between contents and expressions, or that between forms and formed substances (Deleuze and Guattari 2004, p. 77).

The relationship between the rhizome and the arboreal are not mutually exclusive, they intermesh, as Deleuze and Guattari point out:

> a tree branch or root division may begin to burgeon into a Rhizome. The coordinates are determined not by theoretical analyses implying universals but by a pragmatics composing multiplicities or aggregates of intensities. A new Rhizome may form in the heart of a tree, the hollow of a root, the crook of a branch. Or else it is a microscopic element of the root-tree, a radicle that gets Rhizome production going (2004, p. 15).

The capacity of rhizomatics as an extended metaphor and theoretical framework through which an informatics case study is explored in the following interpretation of the case of Ivy University.

3 INFORMATION SYSTEMS AND THE RHIZOME: THE CASE OF IVY UNIVERSITY

This exploration of rhizomatics is as an interpretive framework with which to enhance understanding and inform informatics interpretive research and practice (Klein and Myers 1999; Walsham 1993). This paper will employ the vocabulary of rhizomatics as an extended metaphor (Lakoff and Johnson 1980; Ricoeur 1978) throughout a narrative reporting of the case of Ivy University (Scott and Wagner 2003; Wagner and Newell 2004) as a means of both enriching and illuminating it. Following that, a brief reflection and evaluation of rhizomatics and its potential as an ignoble science is discussed. The case presented and rhizomatically analyzed here was constructed from reviewing the published papers of the studies carried out into the case of Ivy University by Wagner and Newell (2004). Direct correspondence and discourses with Wagner were also used to ensure the efficacy of this interpretation.

3.1 A Rhizomatic Interpretation of Ivy University

An informatics example of arboreal territorialization struggles within a rhizomic setting will be illustrated here, based on the richly exemplified exploration of the introduction of an enterprise resource planning application into a major university within the United States, which was given the aptly rhizomatic *nom de guerre* of Ivy (Scott and Wagner 2003; Wagner and Newell 2004). The case is set within a complex socio-technical rhizome, a complex heterogeneity in the form of Ivy University.

Table 1. Major Rhizomatic Concepts

Rhizomatic Concepts	Explanation	Example (Ivy University Case)
Arboreal	In contrast to, and opposing the thizome, is the tree, the vertical arboreal structure of the state.	The president of Ivy University, accompanied by an important academic provost together with a vice president for Finance and Administration and their associated central administrative infrastructure. Controlling or even replacing departmental IS, territorializing them and their departments within and through an ERP that was arboreally linked directly to the executive, computerized IS as being a manifestation of the arboreal command tree.
Body without organs (BwO)	A reflexive in-folding of the social world which enables the emergence of subjectivity. While the BwO is the site of domination, it is also the site of resistance and refusal, and can be understood also as a limit constituted in this opposition of power and resistance. Within this dynamic struggle between domination and resistance, the BwO is constructed and reconstructed continually (Fox 1999).	Resulting from the dynamic struggle between the powerful president (the executive) and the resourceful local departments and their administrators.
Deterritorialization	To take the control and order away from a land or place (territory) that is already established. It is to undo what has been done; the freeing of labor-power from specific means of production.	ERP had been deterritorialized from the executive and reterritorialized into the various plateaus of the self-territorializing departments, for their own use, rather than the executive's. Only in this way was there departmental adoption.
Nomad	Nomads may follow customary paths, but the points along the way possess no intrinsic significance for them. They do not mark out territory to be distributed among people (as with sedentary cultures); rather, people are distributed in an open space without borders or enclosures. Nomad space is smooth, without features, and in that sense the nomad traverses without movement, the land ceases to be other than support (Fox 1999); dynamic highly mobile lacking in any form of structure, bodies-without-organs.	Resourceful administrators exported data from the bolt-on and imported it into Microsoft Excel enabling them to recreate silos of activity, answer faculty questions about funding availability, and maintain their academic temporal zones. These local shadow systems were shared among academic departments and resulted in the mobilization of a grassroots network that fused Ivy's historical working rhythms with ERP generated data.

Rhizomatic Concepts	Explanation	Example (Ivy University Case)
Nomadology	Multiplics narratives; creating an uninterrupted flow of deterritorialization which establishes a line of flight away from territories, grand designs and monolithic institutions, not something that is achieved once and for all, there is always another deterritorialization (Fox 1999).	Essentially the whole of the Ivy University case, from the president's attempt to introduce the ERP, to the subversive action by the departments and the subsequent withdrawal of the ERP and the revised scenario.
Plateau	"[A plateau] is always in the middle, not at the beginning or the end…a continuous, self-vibrating region of intensities whose development avoids any orientation toward a culmination point or external end" (Deleuze and Guattari 1988, p. 22).	University departments.
Reterritorialization	The restructuring of a place or territory that has experienced deterritorialization; it is the design of the new power.	The department administrators went about reterritorializing the data the ERP captured and collated and used it for themselves. They then reconfigured it and used it to resist the attempted arboreal territorialization of the departmental plateaus by the executive. When buy-in to the ERP was eventually achieved (2003), it was on the terms of the departmental administrative plateaus rather than in response to demands of the arboreal central executive.
Rhizome	"[A rhizome] connects any point to any other point... it brings into play very different regimes of signs, and even nonsign states....It is composed not of units but of dimensions....It has neither beginning nor end, but always a middle (milieu) from which it grows and which it overspills....A rhizome is made of plateaus" (Deleuze and Guattari 1988, p. 21).	University rhizomic heterogeneity was constituted out of academic and non-academic staff, buildings and resources, support staff, students and researchers, rules and regulations, research and teaching staff and teams, administration at the central, faculty and departmental levels territorializing into a multiplicity of departmental plateaus, colleges, schools, research centers, research projects and teams, and undergraduates and graduates undergoing a myriad of programs.
Territorialization	The outcome of dynamic relations between physical and/or psycho-social forces. Territorialization is an active process, whose agent may be human, animate, inanimate, or abstracted.	Ivy wanted to implement an integrated financial infrastructure that would unify its silos of administration. There was a machination of arboreal territorialization. The attempted implementation of the ERP was a machination of arboreal territorialization directed by a paternalistic VP toward achieving control over all the departments and their research budgets.

Rhizomatic Concepts	Explanation	Example (Ivy University Case)
War machine	Challenge to the behavior and existence of the arboreal state.	The departmental administrators, in league with their in-house applications and other data sources, had become self-territorialized war machines that carried on as before and, where possible, aligned the University's corporate information from the new ERP with their interests. Their purpose was to resist and undermine the striped patterns forming in the University rhizome.

Since the concepts involved in rhizomatics are unlikely to be that familiar, it was thought important to provide some explanation of the main concepts and examples (drawn from the Ivy University case) to illustrate them. However, it should be noted that these concepts are tightly intertwined and have been pulled apart only for analytical purposes. Still, the links are so close that some of them rely on each other for understanding.

Ivy is a large prestigious U.S. research-driven university governed by a provost and vice president along with a board of directors. The administration is centrally controlled but relatively decentralized in terms of the degree of autonomy experienced by academic departments and research teams. As such, Ivy has historically struggled with duplication and consistency of administrative activities done locally and then reported on centrally. "A decision was made in 1996 to modernize all administrative information systems through ERP software" (Wagner and Newell 2005a, p. 9).

This university rhizomic heterogeneity was constituted out of academic and non-academic staff, buildings and resources, support staff, students and researchers, rules and regulations, research and teaching staff and teams, administration at the central, faculty, and departmental levels territorializing into a multiplicity of departmental plateaus, colleges, schools, research centers, research projects, teams, and undergraduates and graduates undergoing a myriad of programs. Collectively they constituted the rhizome that was Ivy University. Through the center of Ivy University grows the extremely powerful arboreal president, who is accompanied by an important academic provost together with a vice president for Finance and Administration and their associated central administrative infrastructure. The University itself is overseen by a "corporation" that has a similar function to a traditional board of directors.

A detailed analysis of the introduction of the ERP into Ivy University is provided by Scott and Wagner (2003) and Wagner and Newell (2004). The goal of introducing the ERP was "the achievement of order…within the University around the applications implementation and alignment of faculty and department." The intent was to explore the forms of social ordering needed to create a workable ERP.

> In summer 1996, Ivy created an alliance with Vision Corporation to become their "showcase customer"…in order to develop and implement flagship technology. Ivy wanted to implement an integrated financial infrastructure that would unify its silos of administration.

There was a machination of arboreal territorialization.

Wagner and Newell report that "the vendor's desire to work with Ivy to convert its government/public sector package into a higher education 'solution' was attractive to Ivy who wanted to be seen as an administrative leader of higher education institutions" (Wagner and Newell 2005b, p. 450). The system vendor's reason for working with Ivy in this high profile university environment was to develop and thoroughly road test an extension of their existing ERP application as a way into the United States, and potentially global, university sector. A project manager was appointed at the project inception and, in 1997, a technical director was hired to take control of the technical progress of the Ivy project. The project itself was

> formed and functional teams were created with cobusiness and -technical leaders. The teams were comprised mostly of Ivy middle managers from central administration whose permanent positions had been backfilled for the duration of the project. Although an experienced ERP project manager had been hired, the real authority lay with the teams who communicated directly with Ivy's newly appointed VP for finance and administration (Wagner and Newell 2006, p. 46).

However, this centrally driven ERP initiative resulted in considerable resistance from the University's multiple departmental plateaus, including not only the departmental administrators but also their staff and their in-house developed, or procured, administrative information systems. The new ERP information technology was viewed as an arboreal machination of control and incorporation, an attempt to bring the faculty and departments under the control of the vice president. This managerial arborealization by the budget director, acting on behalf of Ivy' administration necessitated that the "day-to-day leadership of the initiative was driven by the VP's core group of middle managers, the majority of whom were considered functional business experts with only a cursory understanding of ERP technology" (Wagner and Scott 2001, p. 10).

The ERP was, from a rhizomic perspective, a panoptical device for attaining informational and organizational arborealization, part of the creation of a managerial command tree whose controlling roots, if allowed, would span the University rhizome. It would, as Scott and Wagner report, encompass not only the technology but centralize control over the departmental plateaus along with their faculty and support staff so that they could be managed centrally (i.e., arboreally) by the administration within a corporate hierarchical tree formation. As the budget director, acting on behalf of the administration's interests, says,

> mentality that we've had…for managing is primitive to say the best and it's very old-fashioned…the corporate world left it many years ago….Many faculty…think of things fundamentally wrong…we want to move people towards a management model where we're going to ask [them] to put together a time-phased business plan (Scott and Wagner 2003, p. 305).

This executive and managerial call for a business plan from each department was a further machination for asserting the University administration's corporate arboreal control over the rhizomic University plateau of departments out of which it had grown and over which it sought dominion. The departmental business plans, their subsidiary status and intended integration within the University executive's overall business plan were a further attempt at the arborealization of the University departments. In a similar way, the attempted implementation of the ERP was a machination of arboreal territorialization directed by a paternalistic vice president toward achieving control over all the departments and their research budgets that had been territorialized locally within the departments through a principle of mutuality to form the wider rhizome that was Ivy University.

The intended role for the ERP actor was being created by the supplier on behalf of the vice president. It was one of controlling or even replacing departmental IS, territorializing them and the departments within and through an ERP that was arboreally linked directly to the executive. The process was one in which ERP information resonated with Deleuze and Guattari's (2004) view of computerized IS being a manifestation of the arboreal command tree. The ERP being implemented here was a machination by the executive of departmental arboreal territorialization with little reciprocity.

However, by 1998, the project was in crisis. There was considerable resistance to it from the departments and their administrators and the vice president challenged Vision on this matter. A new ERP module specifically for the university market was being developed to address this issue at the end of 1998. It was intended to be a means, by the executive, of territorialization, subsuming local departmental control over research project financial information and management, the applications on which they ran, and its data. The move by the executive to territorialize and control research budgets was a means of arboreally centralizing control over them by the administrative trunk of Ivy. This entailed the ERP provider, who had never encountered such academic environments before, in developing new modules to meet the requirements of what until now had been an untapped market. It was this opportunity that attracted them to the Ivy project in the first place. The project afforded the company with an opportunity to develop and road test modifications within the academic market to their existing ERP product and services. The core of this was to develop a new module for their ERP specifically to meet the needs of the university market.

The ERP (see Figure 3) went live and, by July 1999, it was collating and delivering data to the departments. However, it was not successful at being a machination of arboreal territorialization. The data it collected and collated, while timely and available, was not of the categories and qualities that were meaningful to the departments and their administrators under the new conditions within Ivy. The ERP had, therefore, failed to become part of Ivy's departments. As a machination of arborealization it was a failure. The departments, in the summer of 1999, revolted at what they saw as an attempted arboreal territorialization of them and their information by the arboreal executive.

> In this legacy environment, a great deal of power remained within departmental units whose administrators used shadow systems to help them translate between academic/programmatic needs and institutional reporting requirements, whereas in the ERP-enabled environment these administrators were required to work within a system designed to meet central administrative needs. What was previously a relatively straightforward accounting system was made complex, non-intuitive and difficult from a departmental perspective. This lead to a prominent controversy at Ivy that nearly stalled the use of the ERP by the academic constituency because they felt their needs were not considered in its design (Wagner and Newell, 2005a, pp. 22-23).

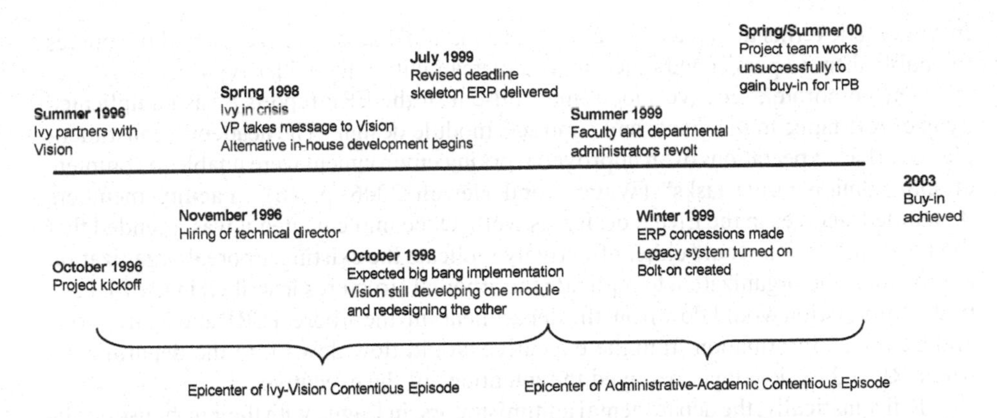

Figure 3. Timeline of Ivy ERP Project Highlighting Contentious Episodes (after Wagner and Newell 2005b)

As this attempt at arborealization, by trying to bring the departments and their IS into the executives arboreal plan, was thwarted concessions were made and the new module was never fully finalized or implemented. The departments and their local administrators resisted the vice president's machination of arborealization through the ERP. This was intended to be achieved by imprinting a hierarchical arboreal model into the ERP and onto the departmental plateaus formed within the wider University rhizome, over which the central administration would, via the ERP, have dominion. Instead, the departmental plateaus retained the territorialization of their own IS for themselves. In doing so, they maintained their own autonomy, as plateaus, within the wider rhizome of the University's collegiate plateau. The ERP providers, in 1999, failed to develop fully a university "bolt-on" module that would have overtly territorialized the various departmental plateaus' own IS into the central administration's ERP. This in turn afforded them with a means of managing the departments' operations, finances, and personnel related to research grants and projects by the central, arboreal, administration.

The department administrators went about reterritorializing the data the ERP captured and collated and used it for themselves. They then reconfigured it and used it to resist the attempted arboreal territorialization of the departmental plateaus by the executive. As Scott and Wagner reported from Spring/Summer 2000,

> The "bolt on" remained unused by administrators who conscripted actors from their past and present (e.g., legacy reports, spreadsheet tools, faculty advocates) in order to create an alternative to the ERP system. In this way they were trying to actively redefine the future of Ivy's administrative environment. Resourceful administrators exported data from the bolt-on and imported it into Microsoft Excel enabling them to recreate silos of activity, answer faculty questions about funding availability, and maintain their academic temporal zones. These local shadow systems were shared among academic departments and resulted in the mobilization of a grassroots network that fused Ivy's historical working rhythms with ERP generated data (Scott and Wagner 2003, p. 306).

In rhizomatic terms they reterritorialized informational actors from a myriad of sources to enable them to persist and function as a plateau within the wider Ivy rhizome.

The corporate executive suboptimal roll-out of the ERP reports "was complicated by user resistance to the grants and contracts module design. The academic constituencies who had expectations of an improved working environment were unable to complete crucial administrative tasks" (Wagner and Newell 2006, p. 46). Faculty members demanded changes in the ERP's design as well. Once implemented it was intended that the ERP application would either effectively replicate the existing arboreal organization and/or force the organization to replicate the arboreal structures inscribed in the ERP, or both. Information would flow from the departments up the arboreal ERP and instructions and corporate information from the executive would flow down it to the departments. The ERP and its developer betrayed the intentions of the executive.

Rhizomatically, the departmental administrators, in league with their in-house applications and other data sources, had become self-territorialized war machines that carried on as before and where possible aligned the University's corporate information from the new ERP with their interests. Their purpose was to resist and undermine the striped patterns forming in the University rhizome and its departmental plateaus as result of the arboreal ambitions of the executive and its vice president. The local administrators and their departmental plateaus had, by 2002, turned the tables on the executive's arboreal ERP. The administrators territorialized the ERP academic module for themselves. The departments also territorialized and used the data the ERP produced, not for what was intended by the executive, but for their own purposes, and integrated it with the data from their own information systems and sources and carried on independently. In this sense there was, by 2003, some buy in but it was on these departmental administrators' own terms. The executive, it appears, had put themselves "up a tree" with the ERP. It could be said the executive was mistaken in introducing the ERP into the University plateau in the first place.

When buy-in to the ERP was eventually achieved (2003), it was on the terms of the departmental administrative plateaus rather than in response to demands of the arboreal central executive; as Wagner and Newell report, "While it was possible for the project team to move forward during implementation by using hierarchical power to impose a particular decision, this did not lead to a successful ERP system in-use (2005a, p. 23). As a machination of arborealization by the executive of the departmental plateaus of the Ivy University's rhizome, the ERP failed. Rhizomatically what was happening in Ivy between the centered departmental plateaus and their nomadic administrator/local data systems and the arboreal University executive and its ERP was, as Deleuze and Guattari succinctly capture was an attempt at the

> arborification of multiplicities....[which] occurs when the stems form segments that striate space in all directions, rendering it comparable, divisible, homogeneous...conversely, and without symmetry, the stems of the Rhizome are always taking leave of the trees, the masses and flows are constantly escaping... a whole smoothing of space, which in turn reacts back upon striated space (1988, p. 557).

By using their own data sources to initially react to and resist central arborealization and then demanding on their own terms, and not the executive's, a new ERP departmental administration module was developed to support what the administrators did in their departmental plateaus, allowing the departmental administrators to territorialize the ERP on their own terms. Buy-in, by the administrative war machines, was achieved in 2003, but this was at a price of it being on the departmental terms and not those of the vice president and the centralized administration.

The overall result was that the vice president moved to another university with a few of his core team remaining as full-time project employees. The vice president saw the ERP become a naturalized object, part of the furniture (i.e., territorialized). But, it is argued here, only as part of an idiosyncratic community that was the University rhizome and its departmental plateaus. In rhizomatic terms the ERP had been deterritorialized from the executive and reterritorialized into the various plateaus of the self-territorializing departments, for their own use, rather than the executive's. Only in this way was there departmental adoption. The University had once again reasserted its rhizomic nature. It had resisted the arboreal territorialize of the power alliance of the executive and the ERP and restored itself to its true character as a self sustaining plurality punctuated by an arboreal administration. In Deleuze and Guattari's terms, it became a multidepartmental, self-territorializing series of plateau, a rhizomatic body without organs: "The body without organs is not a dead body, but a living body all the more alive and teeming once it has blown apart the organism and its organization....Lice hopping on the beach. Skin colonies....The full body without organs is a body populated by multiplicities" (2004, p. 34) and, it might be said here, informated too.

Rhizomatics has been deployed here as an extended metaphor to provide interpretive insight into the Ivy narrative. In so doing, the paper has tested out rhizomatics as an interpretive framework for appreciating the nature and complexities of information systems. It also challenges Deleuze and Guattari's opinion that information systems and associated technologies can only be arboreal and centralist in nature.

The next section examines the set of principles that underpin rhizomatics. It explores the utility of these rhizomatic principles in providing a further interpretive, and more structured, tool with which to explore instances of real world informatics cases, in this instance, of Ivy.

3.2 Interpreting Rhizomatic Principles in the Context of Information Systems

Rhizomatic has been used, until this juncture, as an extended metaphor with which to interpretively explore and enrich the narrative of the Ivy University. Deleuze and Guattari, however, also set down a series of rhizomatic principles, essentially, properties of the rhizome. These principles can generate a series of rhizomatic questions with the potential to explore further about Ivy and its complex relationship with IS. We will considering each rhizomatic principle in turn.

3.2.1 Principles 1 and 2: Connection and Heterogeneity

Connectivity is seen as "the capacity to aggregate by making connections at any point on and within itself"; heterogeneity is seen as "the capacity to connect anything with anything other, the linking of unlike elements" (http://capitalismandschizophrenia.org/index.php/Rhizome).

Can IS be seen as network heterogeneities, integrations of humans and nonhumans and as a result rhizomic in nature? In the Ivy case, the technological IS was implemented as an attempt by the executive at arborealization, to territorialize the departmental plateaus in line with their corporate interests. However the wider Ivy rhizome, with its heterogeneity of human and nonhuman actors, their connections, and self installed in-house IS, resisted this to the point where they reasserted their autonomy.

Can, IS alternatively be viewed as a machination of territorialization, linking other human and nonhuman constituents across the rhizome? The executive and the introduction of the ERP was a corporate machination of territorialization. The arboreal roots of an executive sought to bring central corporate governance to the rhizomic Ivy University. Buy-in to the ERP application only occurred when there was localization and departmental territorialization of the IS across the University, a suboptimal use of the new ERP.

3.2.2 Principle 3: Multiplicity

Multiplicity consists of "multiple singularities synthesized into a 'whole' by relations of exteriority" (http://capitalismandschizophrenia.org/index.php/Rhizome).

What are the ramifications of the rhizomic concept of multiplicity and heterogeneity for the development and creation of IS and the many forms they take? From the case of Ivy, it can be seen that the ramifications of the rhizomic concept of multiplicity and heterogeneity were ignored initially by the corporate executive and this was replicated within the arboreal nature of the ERP they procured and its suppliers provided. The multiplicity concept may be ignored in command/control tree corporate environments, where data flows up and down its trunk. However in contexts such as Ivy, where there are multiple stakeholders, individuals, groups, and organizations forming self-organizing plateaus, with localized power, multiplicity has to be included both when designing the technology and implementing the system. An ERP has an arboreal image inscribed in it, while the administrators and the IS of Ivy's multiple departments were localized territorializations. It was only when localized modules and functionality, which could accommodate to the multiplicity of departmental plateaus, and the ERP became more rhizomic and less arboreal that it was finally adopted.

3.2.3 Principle 4: Asignifying Rupture

Asignifying rupture refers to "not becoming any less of a Rhizome when being severely ruptured. It is the ability to allow a system to function and even flourish despite local 'breakdowns,' thanks to deterritorialising and reterritorialising processes" (http://capitalismandschizophrenia.org/index.php/Rhizome).

Do human social and/or informated socio-technical systems exhibit rhizomic asignifying properties? In the case of Ivy, the introduction of the corporate ERP was an attempt to arboreally punctuate and control the Ivy University rhizome from the center. The aim of the executive's introduction of the ERP was to prevent Ivy University departmental plateaus from territorializing the research funding and operational resources so as to control them. However, the departments and their administrators simply drew on their own localized shadow accounting systems to circumvent the arboreal ambitions of the executive. The University rhizome, wherever there was an attempt by the executive to punctuate it, reasserted itself and fought on.

3.2.4 Principles 5 and 6: Cartography and Decalcomania

Cartography can be described as "the method of mapping for orientation from any point of entry within a 'whole,' rather than by the method of tracing that re-presents an a priori path, base structure or genetic axis"; decalcomania forms "through continuous negotiation with its context, constantly adapting by experimentation, thus performing a non-symmetrical active resistance against rigid organization and restriction" (http://capitalismandschizophrenia.org/index.php/Rhizome).

Can information technologies/systems and their human partners be considered as nomads capable of creating informational pathways traversing across the wider socio-technical rhizome? The administrators of the local departmental plateaus, along with their accounting shadow systems, clearly found ways of nomadically traversing Ivy's rhizome. This they did this to territorialize the information (from a range of sources) to maintain their independence. In addition, the departmental plateaus could manage their fiscal resources, while resisting the attempt of the centre to arborialize them. This was achieved without the ERP system's attempted arboreal territorialization. They became war machines convened out of the Ivy rhizome to challenge the legitimacy of the arboreal executive to arboreally territorialize them.

Can IS design, development or procurement, and implementation into the social or organisational rhizome be viewed as instances of cartography and decalcomania? The introduction of the ERP into Ivy University can be seen as attempted decalcomania. It was one of taking the arboreal image of the organization, which was cartographically inscribed in the ERP during its implementation processes, and imprinting it onto Ivy University and its departmental plateaus. Unfortunately, the arboreal cartography built into the ERP application, and even its special bolt on for Universities, did not map onto the Ivy rhizomatic network. As a process of decalcomania, of inscribing the arboreal image of the organization built into the ERP onto Ivy University, it failed. Some attempts at cartographical adjustment, in the form of the bolt-on module, did provide better results. However, the individual departmental administrators had far better maps with which to traverse the Ivy rhizome and create pathways in it to access the information they needed to circumvent and subvert central executive's attempt at an arboreal take-over of the whole rhizome.

4 CONCLUSION

In these explorations of rhizomatic informatics it has been posited that the work of Deleuze and Guattari has the potential to be an interpretive framework for the understanding of the nature of information systems and its associated practices as they occur in the real world. The concept of the rhizome and its arboreal foe, it is argued, provides a means not only of conceptualizing IS within social and organizational settings but offers a set of questioning principles on which the concept of a rhizomatic informatics may be based and IS case studies interpreted.

In juxtaposing the totalizing notion of the arboreal power tree to that of the heterogeneity of the ehizome, with its multiplicity of plateaus and trajectories, rhizomatics offers an image of information systems that are not solely linked to corporate needs. Rather, as argued and illustrated here through the Ivy University case, it constitutes a potential with which to underpin a more open and accessible form of IS interpretive research and participatory practices, under the ubiquitous concept of rhizomatic informatics. Specifically, the organic nature of some IS lend themselves to be seen as rhizomatic in nature (e.g., the Internet). In addition, the dynamic element inherent in the theory can also be key in understanding the developments in specific IS—not just where the IS starts, and what intentions are instantiated during the design and development, but how it develops through use over time and what trajectory is seen (cf. the technology-in-practice concept, Orlikowski 2000).

Looking at the specific issues raised by the rhizomatic principles, we can say

- IS can be seen as rhizomatic in nature (i.e., as network heterogeneities, integrating humans and nonhumans).
- IS can also be viewed as a machination of territorialization, linking other human and nonhuman constituents across the rhizome.
- Multiplicity has to be included both when designing the technology and implementing the system, or the rhizomatic properties of the organization may work against the effectiveness of the IS.
- Human social and informated socio-technical systems do exhibit rhizomatic asignifying properties, in that they are endlessly flexible and often able to reassert themselves in the face of attack and disruption.
- Information systems and technologies and their human partners can be considered as nomads capable of creating informational pathways traversing across the wider socio-technical rhizome. Again, this needs to be considered during design and development.
- IS design, development or procurement, and implementation into the organizational rhizome can be viewed as instances of cartography and decalcomania. If the new or revised information system is seen as attacking the existing rhizomatic stability (shaky at best anyway), then the war machine swings into operation and the guerilla resistance might begin. The implications for change management are significant, with sufficient time and energy needing to be allocated to this process.

These ideas also question Deleuze and Guattari's own view of computerized information systems as characterized as being purely instruments of corporate arboreal control. Rather, information technologies and data sources can be territorialized equally into, and become a constituent within, the Rhizome and the plateaus that emerge from it.

The rhizomatic principles of Deleuze and Guattari have been used here in two ways. First, they have been used as an extended metaphor, a way to enrich the case narrative of Ivy University and its ERP implementation. Second, the formal principles of rhizomatics and the questions they pose for and reveal about the nature of information systems have also been explored, through their application to the Ivy case study. Both suggest that rhizomatics offers an interpretive framework for the understanding of the nature and complexities of information-rich real-world contexts and a means of informing practice.

Acknowledgments

The authors wish to acknowledge the insightful and critically constructive comments of Dr. Erica Wagner with respect to the whole of this paper and in particular to those sections relating to the case study of Ivy University. We also greatly appreciate her giving us specific consent to use the case of Ivy University reported here.

References

Ashby, W. R. 1956. *An Introduction to Cybernetics,* London: Chapman & Hall.

Callon, M. 1986. "Some Elements of a Sociology of Translation: Domestication of the Scallops and the Fishermen of Saint Brieuc Bay," in *Power Action and Belief: A New Sociology of KNowlege?*, J. Law (ed.), London: Routledge and Kegan Paul, pp. 196-233.

Deleuze, G., and Guattari, F. 1983. *Anti-Oedipus: Ccapitalism and Schizophrenia,* London: Athlone.

Deleuze, G., and Guattari, F. 1988. *A Thousand Plateaus: Capitalism and Schizophrenia,* London: Athlone.

Deleuze, G., and Guattari, F. 1996. *Nomadology: The war machine,* New York: Semiotexte.

Deleuze, G., and Guattari, F. 2004. *A Thousand Plateaus: Capitalism and Schizophrenia,* London: Continuum.

Fox, N. 1999. "Nomadology," *Nick on the Web* (http://www.wisdomnet.co.uk/nick/nomad.html).

Giddens, A. 1979. *Central Problems in Social Theory,* Basingstoke, UK: Macmillan.

Giddens, A. 1984. *The Constitution of Society: Extracts and Annotations,* Berkeley, CA: University of California Press.

Jones, M. 1999. "Structuration Theory," in *Rethinking Management Information Systems,* W. L. Currie and R. Galliers (eds.), Oxford, UK: Oxford University Press.

Jones, M. R., and Karsten, H. 2008. "Giddens's Structuration Theory and Information Systems Research," *MIS Quarterly* (32:1), pp. 127-157.

Klein, H. K., and Myers, M. D. 1999. "A Set of Principles for Conducting and Evaluating Interpretive Field Studies in Information Systems," *MIS Quarterly* (23:1), pp. 67-93.

Lakoff, G., and Johnson, M. 1980. *Metaphors We Live By,* Chicago: University of Chicago Press.

Land, F. 1979. "Organisational Problems of Implementing Distributed Systems, in Infotech State of the Art 1979 Report," *Distributed Systems* (2), pp. 113-121.

Latour, B. 2005. *Reassembling the Social: An Introduction to Actor–Network Theory,* New York: Oxford University Press.

Law, J. 1987. "Technology and Heterogeneous Engineering: The Case of the Portuguese Expansion," in *The Social Construction of Technical Systems: New Directions in the Sociology and History of Technology*, W. E. Bijker, T. P., Hughes, and T. Pinch (eds.), Cambridge, MA: MIT Press, pp. 111-134.

Law, J. 1992. "Notes on the Theory of the Actor–Network: Ordering, Strategy and Heterogeneity." *Systems Practice* (5:4), pp. 379-393.

Lee, A. S. Zmud, R., Robey, D., Watson, R., Zigurs, I., Wei, K. K., Myers, M., Sambamurthy, V., Webster, J., and Agarwal, R. 2001. "Research in Information Systems: What We Haven't Learned," *MIS Quarterly* (25:4), pp. v-xv.

Maturana, H. R., and Varela, F. J. 1980. *Autopoiesis and Cognition: The Realization of the Living*, Dordrecht, Holland: D. Reidel Publishing Company.

Mumford, E. 1995. *Effective Systems Design and Requirements Analysis: The ETHICS Approach,* Basingstoke, UK: Macmillan.
Orlikowski, W. 2000. "Using Technology and Constituting Structures: A Practice Lens for Studying Technology in Organizations," *Organization Science* (11:4), pp. 404-428.
Ricoeur, P. 1978. *The Rule of Metaphor: The Creation of Meaning in Language,* London: Routledge.
Rosenthiehl, P., and Petitot, J. 1974. "Automate a social et systèmes acentrés," *Communications* (22), pp. 45-62.
Scott, S. V., and Wagner, E. L. 2003. "Networks, Negotiations, and New Times: The Implementation of Enterprise Resource Planning into an Academic Administration," *Information and Organization* (13:4), pp. 285-313.
Seto, I. 2006. "Organization of Knowledge and the Hyperlink: Eco's *The Name of the Rose* and Borges' *The Library of Babel,*" *Library Student Journal*, November (http://www.librarystudentjournal.org/index.php/lsj/article/view/34/36).
Wagner, E. L., and Newell, S. 2004. "'Best' for Whom?: The Tension between 'Best Practice' ERP Packages and the Diverse Epistemic Cultures of Diverse University Context," *Journal of Strategic Information Systems* (13:4), pp. 305-328.
Wagner, E. L., and Newell, S. 2005a. "Making Software Work: Problem Solving a Troubled ERP Implementation," CHR Working Paper Series Center for Hospitality Research, Cornell University.
Wagner, E. L., and Newell, S. 2005b. "Making Software Work: Producing Social Order Via Problem Solving in a Troubled ERP Implementation," in *Proceedings of the 26th International Conference on Information Systems*, D. Avison, D. Galletta, and J. I. DeGross (eds.), Las Vegas, December 11-14, pp. 447-458.
Wagner, E. L., and Newell, S. 2006. "Repairing ERP: Producing Social Order to Create a Working Information System," *Journal of Applied Behavioral Science* (42:1), pp. 40-57.
Wagner, E. L., and Scott, S. V. 2001. "Unfolding New Times: The Implementation of Enterprise Resource Planning into an Academic Administration," Working Paper Series #98, Department of Information Systems, London School of Economics.
Walsham, G. 1993. *Interpreting Information Systems in Organizations,* Chichester, UK: Wiley.
Wiener, N. 1948. *Cybernetics: or Control and Communication in the Animal and the Machine* (2nd ed.), Cambridge, MA: MIT Press.

About the Authors

Chris Atkinson is a senior lecturer in Information Systems in the Manchester Business School. Until recently he was in the UMIST School of Informatics and prior to that with Brunel University's Department of Information Systems and Computing. Originally a civil engineer, he undertook an M.Sc. and read for a Ph.D. at Lancaster University under Professor Peter Checkland in soft systems with a particular focus on systemic metaphor and its role in organizational problem solving. He has worked as an academic, consultant, and practitioner, focusing on how to integrate information systems development and organizational change, especially within healthcare settings. To that end, he has evolved and deployed the Soft Information Systems and Technologies Methodology (SISTeM). Callon and Latour's actor-network theory, along with other theories such as Deleuze and Guattari's rhizomatics and Bakhtine's carnival have recently emerged as important frameworks for his IS research, practice, and methodological development. Integrating structuration theory with ANT to create the theoretical hybrid StructurANTion has also proved fruitful as a further area for research and development. His field of study and practice has centered on working with multi-professional teams, clinicians, managers, and information systems practitioners in effecting integrated organizational development. Chris may be contacted at Christopher.Atkinson@manchester.ac.uk.

Laurence Brooks is a lecturer in the Department of Information Systems and Computing at Brunel University, UK. He previously was a lecturer in the Department of Computer Science at the University of York and before that held a research post at the Judge Business School, University of Cambridge. He gained a Ph.D. in Industrial Management from the University of Liverpool and a B.Sc. in Psychology from the University of Bristol. His research interests focus on how IS and organizations can better fit together and create more positive outcomes. To that end, he has focused on the role that social theory (such as structuration theory, actor network theory, StructurANTion theory) might play in contributing to our understanding of information systems in areas such as health information systems and collaborative work support systems. Other research interests include early requirements engineering, the use of cognitive mapping for creating insights into complex scenarios, and exploring the role of culture in IS development and use. He is the current President of the UK Academy for Information Systems (UKAIS) and has co-chaired their last two annual conferences. Laurence can be reached at Laurence.Brooks@brunel.ac.uk.

7 THE INFLUENCE OF SUBGROUP DYNAMICS ON KNOWLEDGE COORDINATION IN DISTRIBUTED SOFTWARE DEVELOPMENT TEAMS: A Transactive Memory System and Group Faultline Perspective

Yide Shen
Michael Gallivan
Robinson College of Business
Georgia State University
Atlanta, GA U.S.A.

Abstract *With the globalization of the software industry, distributed software teams (DSTs) have become increasingly common. Among the various social aspects that are essential to the success of distributed software projects, the focus of this research is the impact of inter-subgroup dynamics on knowledge coordination. To address this research question, we extend and apply theory from two primary sources: transactive memory systems theory and the faultline model. We describe a field survey study that is in progress. The findings from this study will inform managers on how DSTs develop capabilities to perform successfully across temporal, geographic and cultural boundaries.*

Keywords Distributed software team, inter-subgroup dynamics, knowledge coordination, transactive memory systems, faultline

1 INTRODUCTION

With the globalization of the software industry, distributed software teams (DSTs) have become increasingly common (Carmel and Agarwal 2002; Herbsleb and Mockus

Please use the following format when citing this chapter:

Shen, Y., and Gallivan, M., 2008, in IFIP International Federation for Information Processing, Volume 267, Information Technology in the Service Economy: Challenges and Possibilities for the 21st Century, eds. Barrett, M., Davidson, E., Middleton, C., and DeGross, J. (Boston: Springer), pp. 103-116.

2003; Kotlarsky and Oshri 2005; Sarker and Sahay 2004). For distributed software development projects to be successful, managers need to focus not only on technical aspects but also social factors (e.g., trust, social ties, formal and informal communication, etc.) that are crucial to these projects (Kotlarsky and Oshri 2005; Orlikowski 2002). Among the various social aspects that are essential to the success of dispersed software projects, the focus of this research is the impact of *inter-subgroup dynamics on knowledge coordination*. Specifically, we investigate

RQ1: Do subgroup dynamics affect team members' knowledge coordination?
RQ2: Does knowledge coordination benefit team performance and member satisfaction?

There are four reasons why this is an important area of research. First, software development is an excellent example of service work that is both highly paid and responsible for rapid growth in the economies of many developing countries. Second, software project failures occur due to coordination problems (Bohem 1981; Kraut and Streeter 1995), especially when projects are large (Brooks 1995) and members are geographically distributed (Herbsleb and Grinter 1999). Coordination is a crucial process in software development that ensures knowledge is properly acquired, shared, and integrated among various members, teams, and organizations. Third, while DSTs utilize a wide range of communication tools such as groupware and codified KMS, coordination breakdowns still occur (Kotlarsky and Oshri 2005). This suggests that teams need to develop *distributed organizing*[1] capabilities to complement existing technical solutions, in order to deal effectively with knowledge coordination challenges in dispersed environments (Orlikowski 2002). Finally, inter-subgroup dynamics—the relationships among subgroups within the overall project team—affect distributed teams' ability to share knowledge, because subgroups often emerge within larger groups (i.e., the notion of group faultlines), which can have negative consequences on member trust, information sharing, and overall coordination. This has been widely shown in laboratory experiments of members charged with performing a task requiring overall group coordination (Li and Hambrick 2005).

2 THEORY BUILDING AND HYPOTHESES DEVELOPMENT

2.1 Knowledge Coordination in Distributed Environments

Prior studies have identified many factors contributing to collaborative work, such as social ties, formal and informal communication, trust, and rapport (Kotlarsky and Oshri 2005). Among them, research on traditional colocated software teams has found that expertise coordination plays a significant role in software teams' performance, above

---

[1]Distributed organizing is defined as "the capability of operating effectively across the temporal, geographic, political, and cultural boundaries routinely encountered in global operations" (Orlikowski 2002, p. 249).

and beyond the mere presence of expertise, professional experience, and software methods employed (Faraj and Sproull 2000). However, knowledge coordination is never an easy task, especially when teams operate across temporal, geographic, and cultural boundaries (Herbsleb and Moitra 2001; Kotlarsky and Oshri 2005; Orlikowski 2002). Among other challenges, the problem of *where to locate project knowledge when needed* is a major challenge (Herbsleb and Mockus 2003). When knowledge is distributed among various stakeholders (Curtis et al. 1988), each member needs to know where to look for information before he or she is able to find and apply that knowledge. Field studies found that problems and questions that require timely solution often occur in software teams (Paasivaara and Lassenius 2003), and are particularly common in distributed projects (Carmel and Agarwal 2001). But few projects are proactive in planning for this kind of knowledge-sharing ahead of time, causing project members to spend much time just trying to find someone with the necessary knowledge, wasting both time and energy (Paasivaara and Lassenius 2003). In DSTs, geographical distance, time-zone differences, and organizational or national culture differences make it even more challenging to coordinate knowledge. Research shows that software developers find it much more difficult to identify distant colleagues with needed expertise and to communicate with them effectively, compared to when all members are local (Herbsleb and Mockus 2003). Herbsleb et al. (2000, p.3) described how one global project team faced this challenge of identifying who knows what, so that "difficulties of knowing who to contact about what, of initiating contact, and of communicating effectively across sites, led to a number of serious coordination problems."

2.2 Transactive Memory and TMS in DSTs

This social aspect of "knowing who knows what" is also labeled *transactive memory*, the knowledge that a person has about what *another* person knows. A transactive memory system (TMS) is a group-level concept, referring to "the operation of the memory systems of the individuals and the processes of communication that occur within the group" (Wegner 1987, p. 191). It describes the active use of members' transactive memories to complete a group task cooperatively. According to TMS literature, researchers generally agree on three facets that reflect the presence of TMS (Lewis 2003; Liang et al. 1995; Moreland and Myaskovsky 2000): *specialization* (the existence of specialized team knowledge), *credibility* (members' trust and reliance on each other's knowledge), and *coordination* (coordinated task processes). By convention, TMS researchers also agree that the higher the levels of these three facets of TMS, the more developed is the group's TMS—and the more value this TMS has for effective knowledge coordination.

Both laboratory and field studies have been conducted to specify the antecedents and consequences of TMS. Table 1 lists the antecedents of TMS development both in traditional, colocated teams and in distributed teams, while Table 2 summarizes research on the consequences of TMS.

Of the studies listed in these tables, only one quantitatively examined the levels of TMS and their antecedents in distributed teams (Kanawattanachai and Yoo 2007). Results suggest that the three TMS dimensions have different effects on distributed team performance. *Specialization* and *credibility* (these dimensions are labeled as *expertise*

Table 1. Antecedents of TMS Development

Factors Facililitate (+) or Hinder (–) TMS Development	Prior Literature
In Traditional Colocated Teams	
Group training (+)	Liang et al. 1995; Moreland 1999; Moreland et al. 1996, 1998; Moreland and Myaskovsky 2000
Performance feedback about one anothers' training performance (+)	Moreland and Myaskovsky 2000
Membership change (–)	Lewis et al. 2007; Moreland and Argote 2003
Distributed expertise (also moderated by *member familiarity*) (during project planning phase) (+)	Lewis 2004
Face-to-face communication (during project implementation phase) (+)	Lewis 2004
Non-face-to-face communicatoin (moderated by TMS developed in planning phase) (during project implementation phase) (+)	Lewis 2004
In Distributed Teams	
Task-oriented communication (e-mail, message, etc.) (early project stage) (+)	Kanawattanachai and Yoo 2007

Table 2. Consequences of TMS in Teams

TMS's Impact on Teams	Prior Literature
Individual-level learning	Lewis et al. 2003
Team-level learning	Lewis et al. 2003
Viability	Austin 2003; lewis 2004; Liang et al. 1995; Moreland and Myaskovsky 2000; Yoo and Kanawattanachai 2001
Team performance	Austin 2003; Kanawattanachai and Yoo 2007; Lewis 2004, 2005; Liang et al. 1995; Moreland 1999; Moreland et al. 1996, 1998; Moreland and Myaskovsky 2000
Successful collaboration	Kotlarsky and Oshri 2005

location and *cognition-based trust* in this research) have no direct impact on performance; instead, their effect is mediated by *coordination*. In addition, the latter effect occurred only in the final stages of the project, once members had learned to work together (Kanawattanachai and Yoo 2007).

Another quantitative study by Faraj and Sproull (2000) did not explicitly mention TMS; however, it introduced the concept of *expertise coordination*, which overlaps with

TMS to a large extent.[2] These authors studied traditional, colocated software teams, finding that *expertise coordination* had a positive effect on team performance. Members' ability to coordinate expertise exerted a strong effect on performance, above and beyond the mere *presence* of member expertise. A case study of two globally distributed system projects by Kotlarsky and Oshri (2005) found that TMS is a key contributor to successful collaboration. Based on these prior results, we propose that

> *Hypothesis 1*: The level of a team's TMS will be positively related to performance in DSTs; however, the only direct effect is through the *coordination* dimension of TMS.

2.3 Gaps in Transactive Memory Systems Literature

2.3.1 Additional Outcome Variable for TMS's Impact on Team

In addition to the conventional outcome variables that have been studied in TMS research (team performance, individual- and team-level learning, etc.), research on groups suggests that team members' psychological well-being is also an important outcome to consider. Thus, we include *member satisfaction* with the team as a second outcome variable. Studies of virtual teams suggest that teams who overcome coordination barriers are more likely to be satisfied with each other (Maznevski and Chudoba 2000; Piccoli et al. 2004). Thus, we posit that

> *Hypothesis 2*: TMS will be positively related to member satisfaction in DSTs; however, the only direct effect on satisfaction is through the coordination dimension of TMS.

2.3.2 Lack of Research on Subgroup Dynamics' Influences on TMS

While providing insightful perspectives of TMS development in teams, the prior TMS studies have the limitation that they focus on teams (either face-to-face teams or virtual teams) that treat each member as an "independent actor" (Li and Hambrick 2005) contributing his/her profile to the overall team diversity. From a group diversity perspective, prior TMS research has examined member heterogeneity and homogeneity among *individuals*, but has not addressed how patterns of difference between *subgroups within a larger group* might influence TMS development. In DSTs, however, dissimilarity between subgroups based on location, culture, and possibly language will have stronger negative effects on knowledge sharing and member behavior than overall member heterogeneity (Li and Hambrick 2005).

---

[2]Compared to the three facets of TMS (specialization, credibility, and coordination), two dimensions of Faraj and Sproull's expertise coordination construct (knowing the location of expertise and recognizing the need for expertise) map to the specialization facet of TMS; their third expertise coordination dimension (bringing expertise to bear) maps to the coordination facet of TMS. Faraj and Sproull do not consider any construct analogous to credibility.

For example, it is not uncommon to see a student project team splitting into one subgroup with international students versus another comprised of domestic students. In a distributed team environment, location differences become salient as the team engages in its task. In turn, these salient location differences can affect team dynamics such as trust, conflict, and communication patterns. For instance, one field study found that hybrid teams composed of two or three subgroups of colocated members experienced more conflict and less trust than fully-distributed teams.[3] Polzer et al. (2006) compared hybrid teams where some members were colocated to fully distributed teams. While the latter had to rely on listservs for information sharing, those teams with some colocated members were able to substitute face-to-face communication for some messages that would otherwise have been sent via listserv to all members. While this had obvious benefits (i.e., in terms of saving time), it also had the undesirable consequence that co-located members started behaving as a faction, making statements such as "the three of us would like to" and "we at [Australian university] decided to take some action" (Polzer et al. 2006, p. 688). This led to a reduction in the level of information sharing across all team members; ultimately, communication was reduced to communication between the subgroups. Panteli and Davison (2005) observed this phenomenon in their virtual student teams where the volume of communication that reached *all* member was lowest in the teams with strong, colocated subgroups.

Thus, when obvious subgroups emerge within a given team (due to locational or other factors, such as gender, race or ethnicity), this has negative effects on team performance and other outcomes (Li and Hambrick 2005). While this effect has been labeled group *faultlines* and widely studied in the groups literature, the notion of sub-group dynamics and group faultlines has not explicitly been linked to the research stream on TMS. To make this link explicit, and to explore its implications for DSTs, we leverage the notion of faultlines to theorize about how such faultlines that emerge in distributed teams can impair overall team performance and satisfaction via reductions in the level of TMS.

2.4 Group Faultlines: Subgroup Dynamics in DSTs

2.4.1 Concept of Group Faultlines

Faultlines are "hypothetical dividing lines that may split a group into subgroups based on one or more attributes" (Lau and Murnighan 1998, p. 328). In organization studies research, the focus has usually been on demographic factors that distinguish team members from each other. For example, a group comprised of two Asians in their 20s and two Caucasians in their 50s (Group 1 in Figure 1) has the potential to split into sub-groups consisting of young Asian versus mid-age Caucasian members. Group 1 is defined as a group where strong faultlines occur; in contrast, faultlines are weaker in completely heterogeneous groups (Group 4) or in completely homogeneous groups (Group 3). By definition, the latter groups have weak faultlines. The stronger the fault-lines, the more likely the team will split into factions, leading to the potential for inter-

---

[3]Polzer et al. (2006) defined fully distributed teams as those where each individual member worked in a separate location.

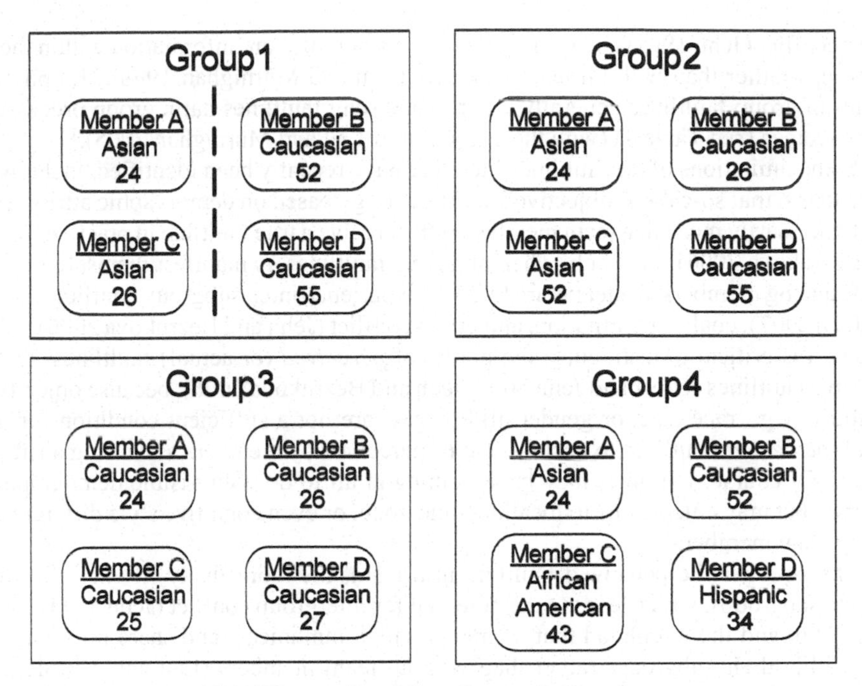

Figure 1. Groups with Different Faultline Strength Levels

Table 3. Direct Impacts of Strong Faultlines

Prior Literature	Faultline Base†		Direct Effect of Strong Faultines (+ = positive impact; – = negative impact)
Earley & Moskowski 2000 (study 2)‡	Nationality	–	Worse proceses (team identity, group efficacy, role expectations, intrateam communication) Worse outcomes (team performance, satisfaction with team's performance
Lau & Murnighan 2005	Ethnicity and sex	+	Less relationship conflict Better group outcomes (psychological safety, group satisfaction)
Molleman 2005	Gender, age, and having a part-time job	–	Lower group cohesion Higher team conflict
Li & Hambrick 2005	Age, tenure, gender, and ethnicity	–	Higher emotional conflict Higher task conflict
Polzer et al. 2006	Geographic location	–	Higher conflict Lower trust
Rico et al. 2007	Educational background and conscientiousness	–	Worse performance Lower level of social integration

†We define the attribute or set of attributes based on which group faultlines are formed as the faultline base.
‡This paper reported results from three studies that examined the relationship between team and nationality heterogeneity and effectiveness performance. The faultline concept was reflected well only in Study 2's operationalization of heterogeneity.

group conflict (Jehn 1995) and the risk that members will share information within their subgroups rather than with all team members (Lau and Murnighan 1998). Empirical studies of group faultlines generally report that strong faultlines harm group processes and outcomes (see Table 3) (with the exception of Lau and Murnighan 2005).[4]

Some limitations of the faultlines literature have recently been identified, including the criticism that so-called "objective" faultlines (e.g., based on demographic attributes) don't necessarily mean that team members will perceive a true faultline in practice (Jehn and Bezrukova 2006). It is only when objective faultlines *do* manifest themselves as a divide among members that teams are likely to experience inter-subgroup conflict (Greer and Jehn 2007), coalition formation, and group conflict (Jehn and Bezrukova 2006). We believe it is critical to distinguish the notion of *perceived* (or actual) faultlines[5] from *objective* faultlines (Greer and Jehn 2007; Jehn and Bezrukova 2006), because objective faultlines (e.g., race, age, or gender differences) are not a sufficient condition for an actual, *perceived* faultlines to occur—that is, perceived by members as causing a rift or divide. Of course, faultlines may have nothing at all to do with visible demographic factors—but may emerge due to location, time-zone, or even cognitive style differences among team members.

The stronger the perceived faultline in a team, the more likely it will split into discrete subgroups, which leads to the potential for intergroup conflict (Jehn and Bezrukova 2006) and the likelihood that members will communicate and share information only within their subgroups rather than with all team members (Lau and Murnighan 1998).

2.4.2 Impacts of Perceived Faultlines on DSTs

In distributed environments, location differences become salient as the team engages in its task. Such location differences may, in turn, shape team dynamics and outcomes, including trust, conflict, and communication patterns. In field settings, researchers also observed that subgroups tend to withhold information from each other (Cramton 2001) or share knowledge only within their subgroups, with rare collaboration with other subgroups (Gratton et al. 2007). These studies suggest that strong perceived faultlines cause the team to disintegrate into subgroups, with members communicating and sharing knowledge only within their subgroups (Cramton 2001). This leads to the existence of uniquely held information among one or a few members of the team. Such uniquely held information is less likely to be salient to other members, causing knowledge gaps and misunderstanding (Stasser and Titus 1985).

Thus, failure of information exchange due to perceived faultlines has two negative consequences: first, it contributes to the existence of uniquely held information in distributed teams, which becomes a source of confusion and misunderstanding; second,

---

[4]Some authors reject a simple linear relationship between faultlines and group outcomes. For example, Gibson and Vermeulen (2003) and Thatcher et al. (2003) report a curvilinear relationship between the faultline strength and various outcomes.

[5]Jehn and Bezrukova (2006, p. 6) define perceived faultlines in groups "when members actually perceive these divisions and the group behaviorally splits into two subgroups based on the alignment of two or more demographic attributes."

even when all members of a team have the same information regarding certain areas of the project, the problem of salience may still occur due to lack of information in other project aspects. In other words, people may be aware of the existence of project-related knowledge distributed in the team, but they do not realize the usefulness or importance of this knowledge because of their different schema (resulting from uniquely held information), thus they don't use that information or they use it in a way that creates misunderstanding or confusion. For example, research on product development teams found that members who lack a shared understanding of their domain activities often fail to take advantage of each others' knowledge due to differences in skills and experience (Dougherty 1992). Due to these differences, members often lack cues that can help them judge the credibility and quality of knowledge from their remote colleagues, which in turn leads them to ignore or misunderstand that knowledge (Carlile 2002; Dougherty 1992).

The knowledge management literature recognizes that effective coordination through electronic media depends on having a common understanding about the problems at hand, clear norms of behavior, and a context for interpreting knowledge (Davenport and Prusak 1997; Dougherty 1992; Krauss and Fussell 1990). When members of distributed teams have different information, they are more likely to filter out or misconstrue information held by others. Based on this logic, we anticipate a negative relationship between perceived fautlines and the coordination dimension of TMS:

Hypothesis 3: Perceived fautlines will be inversely related to the coordination dimension of TMS in DSTs.

Perceived faultlines may also damage members' attitudes toward each other. Groups with strong faultlines have higher levels of conflict, which causes members to avoid communicating and sharing information with other subgroup members. According to attribution theory, when people lack situational information due to failure of information exchange, they tend to explain others' behavior as resulting from individual disposition, rather than due to the situation (Nisbett et al. 1973). This causes people to reach negative conclusions about others, particularly members of other subgroups. Based on this logic, we suspect that groups with strong faultlines will be less likely to have high levels of member credibility, compared to teams with no perceived faultlines (or only weak ones).

Hypothesis 4: Perceived fautlines will be inversely related to the credibility dimension of TMS in DSTs.

Figure 2 shows our overall research model.

3 RESEARCH METHOD

We plan to survey of DSTs in multiple organizations. Data will be collected from team members and aggregated to the project team level. Where possible, validated measures from prior studies will be adapted. Table 4 summarizes measures that we are currently pilot testing with distributed student teams.

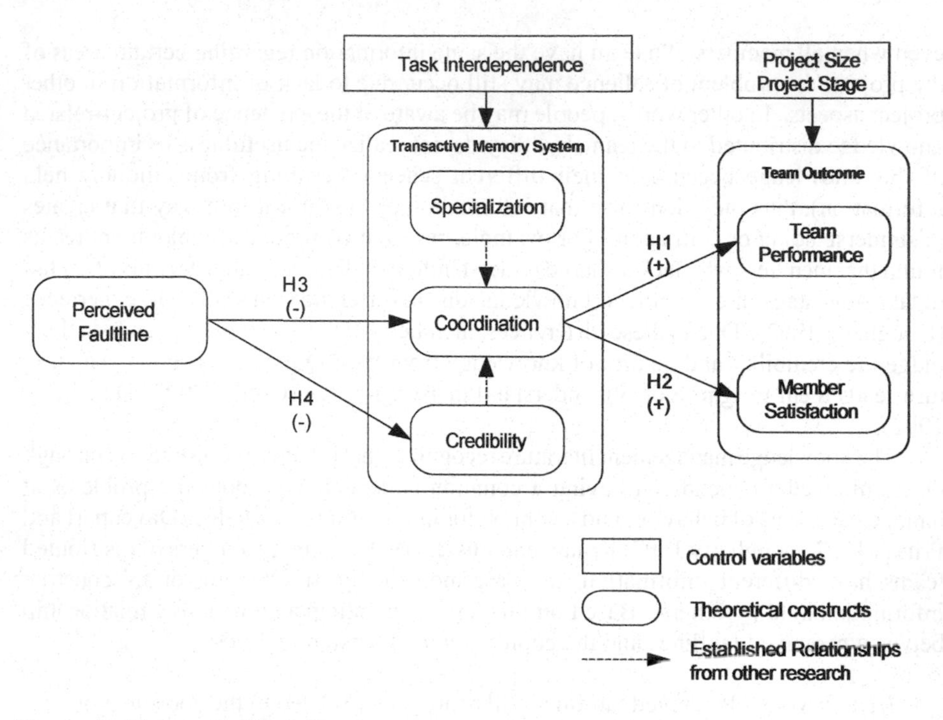

Figure 2. Research Model

Table 4. Measurement Items Used for Each Construct

Constructs and Measurement Items	References
Model Variables	
Perceived Faultlines	Original measures
Transactive Memory System	Lewis 2003
Team Performance	Henderson and Lee 1992
Member Satisfaction	Piccoli, Powell, and Ives 2004
Control Variables	
Project Stage	Original measures
Team size	Piccoli, Powell, and Ives 2004
Work Interdependence	Pearce and Gregersen 1991

4 EXPECTED CONTRIBUTION

4.1 Theoretical Contributions

Our study contributes to theory by examining how groups develop a TMS, taking into account subgroup dynamics based on faultlines triggered by demographic, location, or other differences. Moreover, we will focus on those faultlines that are *perceived* by group members (Greer and Jehn 2007; Jehn and Bezrukova 2006), rather than just the presence of objective faultlines (i.e., those assumed to occur based on demographic attributes). Prior research suggests that groups with strong faultlines have higher levels of conflict, which make members *intentionally* withhold information from other members. Our study will emphasize other by-products of faultlines (i.e., lack of effective coordination, reduced performance, and member satisfaction). Second, by measuring TMS as a consequence of group faultlines, we will provide insights into why such problems in developing an effective TMS occur, thus opening the "black box" that prior researchers have posited between objective faultlines and performance. By measuring the level of group TMS as a downstream result of actual, perceived faultlines, we hope to show how perceived faultlines impair team performance and member satisfaction.

4.2 Contribution to Practitioners

Our results will emphasize that managers should pay attention to subgroup dynamics that emerge within their teams. Especially in DSTs, faultlines can easily emerge due to salient location or culture differences. Second, the results from our study may encourage managers to take steps to increase the level of TMS among team members, especially members from subgroups who differ in terms of cultural or locational attributes.

References

Austin, J. R. 2003. "Transactive Memory in Organizational Groups: The Effects of Content, Consensus, Specialization, and Accuracy on Group Performance," *Journal of Applied Psychology* (88:5), pp. 866-878.

Bohem, B. W. 1981. *Software Engineering Economics*, Englewood Cliffs, NJ: Prentice-Hall.

Brooks, F. 1995. *The Mythical Man-Month: Essays on Software Engineering*, Reading, MA: Addison-Wesley.

Carlile, P. R. 2002. "A Pragmatic View of Knowledge and Boundaries: Boundary Objects in New Product Development," *Organization Science* (13:4), pp. 442-455.

Carmel, E., and Agarwal, R. 2001. "Tactical Approaches for Alleviating Distance in Global Software Development," *IEEE Software* (18:2), pp. 22-29.

Carmel, E., and Agarwal, R. 2002. "The Maturation of Offshore Sourcing of Information Technology Work," *MIS Quarterly Executive* (1:2), pp. 65-77.

Cramton, C. D. 2001. "The Mutual Knowledge Problem and its Consequences for Dispersed Collaboration," *Organization Science* (12:3), pp. 346-371.

Curtis, W., Krasner, H., and Iscoe, N. 1988. "A Field Study of the Software Design Process for Large Systems," *Communications of the ACM* (31:11), pp. 1268-1287.

Davenport, T. H., and Prusak, L. 1997. *Working Knowledge: How Organizations Manage What They Know*, Boston: Harvard Business School Press.

Dougherty, D. 1992. "Interpretive Barriers to Successful Product Innovation in Large Firms," *Organization Science* (3:2), pp. 179-202.

Earley, P. C., and Mosakowski, E. 2000. "Creating Hybrid Team Culture: An Empirical Test of Translational Team Functioning," *Academy of Management Journal* (43), pp. 26-49.

Faraj, S., and Sproull, L. 2000. "Coordinating Expertise in Software Development Teams," *Management Science* (46:12), pp. 1554-1568.

Gibson, C. B., and Vermeulen, F. 2003. "A Healthy Divide: Subgroups as a Stimulus for Team Learning Behavior," *Administrative Science Quarterly* (48), pp. 202-239.

Gratton, L., Voigt, A., and Erickson, T. 2007. "Bridging Faultlines in Diverse Teams," *MIT Sloan Management Review* (48:4), Summer, pp. 22-29.

Greer, L. L., and Jehn, K. A. 2007. "Where Perception Meets Reality: The Effects of Different Types of Faultline Perceptions, Asymmetries, and Realities on Intersubgroup Conflict and Workgroup Outcomes," paper presented at the Academy of Management Conference, Philadelphia, PA.

Henderson, J. C., and Lee, S. 1992. Managing I/S Design Teams: A Control Theories Perspective," *Management Science* (38), 1992, pp. 757-777.

Herbsleb, J. D., and Grinter, R. E. 1999. "Architectures, Coordination, and Distance: Conway's Law and Beyond," *IEEE Software* (16:5), pp. 63-70.

Herbsleb, J. D., and Mockus, A. 2003. "An Empirical Study of Speed and Communication in Globally-Distributed Software Development," *IEEE Transactions on Software Engineering* (29:3), pp. 1-14.

Herbsleb, J. D., Mockus, A., Finholt, T., and Grinter, R. E. 2000. "Distance, Dependencies, and Delay in Global Collaboration," in *Proceedings of the 2000 ACM Conference on Computer Supported Cooperative Work*, New York: ACM Press, pp. 319-329.

Herbsleb, J. D., and Moitra, D. 2001. "Global Software Development," *IEEE Software* (18:2), pp. 16-20.

Jehn, K. A. 1995. "A Multimethod Examination of the Benefits and Detriments of Intragroup Conflict," *Administrative Science Quarterly* (40), pp. 256-282.

Jehn, K. A., and Bezrukova, K. 2006. "The Effects of Faultline Activation on Coalition Formation, Conflict, and Group Outcomes," working Paper, Leiden University.

Kanawattanachai, P., and Yoo, Y. 2007. "The Impact of Knowledge Coordination on Virtual Team Performance Over Time," *MIS Quarterly* (31:4), pp. 783-808.

Kotlarsky, J., and Oshri, I. 2005. "Social Ties, Knowledge Sharing and Successful Collaboration in Globally Distributed System Development Projects," *European Journal of Information Systems* (14:1), pp. 37-48.

Krauss, R., and Fussell, S. 1990. "Mutual Knowledge and Communicative Effectiveness," in *Intellectual Teamwork: The Social and Technological Bases of Cooperative Work*, J. Galegher, R. E. Kraut, and C. Egido (eds.), Hillsdale, NJ: Lawrence Erlbaum, pp. 111-144.

Kraut, R. E., and Streeter, L. A. 1995. "Coordination in Software Development," *Communications of the ACM* (38:5), pp. 69-81.

Lau, D. C., and Murnighan, J. K. 1998. "Demographic Diversity and Faultlines: The Compositional Dynamics of Organizational Groups," *The Academy of Management Review* (23:2), pp. 325-340.

Lau, D. C., and Murnighan, J. K. 2005. "Interactions Within Groups and Subgroups: The Effects of Demographic Faultlines," *Academy of Management Journal* (48:4), pp. 645-659.

Lewis, K. 2003. "Measuring Transactive Memory Systems in the Field: Scale Development and Validation," *Journal of Applied Psychology* (88:4), pp. 587-604.

Lewis, K. 2004. "Knowledge and Performance in Knowledge-Worker Teams: A Longitudinal Study of Transactive Memory Systems," *Management Science* (50:11), pp. 1519-1533.

Lewis, K. 2005. "Transactive Memory Systems, Learning and Learning Transfer," *Organization Science* (16:6), pp. 581-598.

Lewis, K., Belliveau, M., Herndon, B., and Keller, J. 2007. "Group Cognitions, Membership Change and Performance: Investigating the Benefits and Detriments of Collective Knowledge," *Organization Behavior and Human Decision Processes* (103:2), pp. 159-178.

Lewis, K., Gillis, L., and Lange, D. 2003. "Who Says You Can't Take it with You? Transferring Transactive Memory Systems Across Tasks," *Academy of Management Best Paper Proceedings*, pp. A1-A6.

Li, J., and Hambrick, D. C. 2005. "Factional Groups: A New Vantage on Demographic Faultlines, Conflict, and Disintegration in Work Teams," *Academy of Management Journal* (48:5), pp. 794-813.

Liang, D. W., Moreland, R. L., and Argote, L. 1995. "Group Versus Individual Training and Group Performance: The Mediating Role of Transactive Memory," *Personality and Social Psychology Bulletin* (21), pp. 384-393.

Maznevski, M. L., and Chudoba, K. M. 2000. "Bridging Space over Time: Global Virtual Team Dynamics and Effectiveness," *Organization Science* (11:5), pp. 473-492.

Molleman, E. 2005. "Diversity in Demographic Characteristics, Abilities and Personality Traits: Do Faultlines Affect Team Functioning," *Group Decision & Negotiation* (14:3), pp. 173-193.

Moreland, R. L. 1999. "Transactive Memory: Learning Who Knows What in Work Groups and Organizations," in *Shared Cognition in Organizations: The Management of Knowledge,* L. Thompson, D. Messick, and J. Levine (eds.), Mahwah, NJ: Lawrence Erlbaum, pp. 3-31.

Moreland, R. L., and Argote, L. 2003. "Transactive Memory in Dynamic Organizations," in *Learning and Managing People in the Dynamic Organization,* R. S. Peterson and E. A.Mannix (eds.), Mahwah, NJ: Lawrence Erlbaum, pp. 135-162.

Moreland, R.L., Argote, L., and Krishnan, R. "Socially Shared Cognition at Work: Transactive Memory and Group Performance," in: *What's Social About Social Cognition? Research on Socially Shared Cognition in Small Groups,* J.L. Nye and A.M. Brower (eds.), Sage, Thousand Oaks, CA, 1996, pp. 57-84.

Moreland, R. L., Argote, L., and Krishnan, R. 1998. "Training People to Work in Groups," in *Theory and Research on Small Groups,* R. S. Tindale, L. Heath, J. Edwards, E. J. Posavac, F. B. Bryant, Y. Suarez-Balcazar, E. Henderson-King, and J. Myers (eds.), New York: Plenum, pp. 37-60.

Moreland, R. L., and Myaskovsky, L. 2000. "Exploring the Performance Benefits of Group Training: Transactive Memory or Improved Communication?," *Organization Behavior Group Decision Processes* (82:1), pp. 117-133.

Nisbett, R., Caputo, C., Legant, P., and Marecek, J. 1973. "Behavior as Seen by the Actor and as Seen by the Observer," *Journal of Personality and Social Psychology* (27:2), pp. 154-164.

Orlikowski, W. J. 2002. "Knowing in Practice: Enacting a Collective Capability in Distributed Organizing," *Organization Science* (13:3), pp. 249-273.

Paasivaara, M., and Lassenius, C. 2003. "Collaboration Practices in Global Inter-Organizational Software Development Projects " *Software Process: Improvement and Practice* (8:4), pp. 183-199.

Panteli, N., and Davison, R. M. 2005. "The Role of Subgroups in the Communication Patterns of Global Virtual Teams," *IEEE Transactions on Professional Communication* (48:2), pp. 191-200.

Pearce, J. L., and Gregersen, H. B. 1991. "Task Interdependence and Extrarole Behavior: A Test of the Mediating Effects of Felt Responsibility," *Journal o Applied Psychology* (76:6), pp. 838-844.

Piccoli, G., Powell, A., and Ives, B. 2004. "Virtual Teams: Team Control Structure, Work Processes and Team Effectiveness," *Information Technology & People* (17:4), pp. 359-379.

Polzer, J. T., Crisp, C. B., Jarvenpaa, S. L., and Kim, J. W. 2006. "Extending the Faultline Model to Geographically Dispersed Teams: How Colocated Subgroups Can Impair Group Functioning," *Academy of Management Journal* (49:4), pp. 679-692.

Rico, R., Molleman, E., Sánchez-Manzanares, M., and Van der Vegt, G. S. 2007. "The Effects of Diversity Faultlines and Team Task Autonomy on Decision Quality and Social Integration," *Journal of Management* (33:1), 2007, pp. 111-132.

Sarker, S., and Sahay, S. 2004. "Implications of Space and Time for Distributed Work: An Interpretive Study of US-Norwegian System Development Teams," *European Journal of Information Systems* (13:1), pp. 3-20.

Stasser, G., and Titus, W. 1985. "Pooling of Unshared Information in Group Decision Making: Biased Information Sampling During Discuss," *Journal of Personality and Social Psychology* (48), pp. 1467-1478.

Thatcher, S. M. B., Jehn, K. A., and Zanutto, E. 2003. "Cracks in Diversity Research: The Effects of Diversity Faultlines on Conflict and Performance," *Group Decision Negotiation* (12), pp. 217-241.

Wegner, D. M. 1987. "Transactive Memory: A Contemporary Analysis of the Group Mind," in *Theories of Group Behavior,* B. Mullen and G. R. Goethals (eds.), New York: Springer-Verlag, pp. 185-208.

Yoo, Y., and Kanawattanachai, P. 2001. "Development of Transactive Memory Systems and Collective Mind in Virtual Teams," *The International Journal of Organizational Analysis* (9:2), pp. 187-208.

About the Authors

Yide Shen is a Ph.D. candidate in the Department of Computer Information Systems at Georgia State University. She holds a Master's degree in MIS from University of Nebraska at Omaha and a Bachelor's degree in Accounting from Wuhan University, People's Republic of China. Her research focuses on knowledge coordination issues in distributed software development teams. Her work has been published in *Decision Support Systems*, in the proceedings of the America's Conference on Information Systems, Hawaii International Conference on System Sciences, and in *Managing Global Information Technology: Strategies and Challenges* (P. Palvia, ed., Ivy League Publishing). Yide can be reached at yshen@cis.gsu.edu.

Michael Gallivan is an associate professor in Georgia State University's Computer Information Systems Department. He holds a Ph.D. from MIT, an MBA and MHA from the University of California, Berkeley, and a BA from Harvard University. Mike studies how organizations adapt to technological innovations, how they develop competitive advantage through outsourcing IT, and how technical workers learn in their jobs. His work has been published in journals such as *MIS Quarterly, Journal of Management Information Systems*, *European Journal of Information Systems*, *Information & Organization, Information Systems Journal*, *The Data Base for Advances in Information Systems*, *IEEE Transactions on Professional Communication*, and *Information & Management*. Mike can be reached at mgallivan@gsu.edu.

8 THE SERVICE BEHIND THE SERVICE: Sensegiving in the Service Economy

Neil C. Ramiller
School of Business Administration
Portland State University
Portland, OR U.S.A.

Mike Chiasson
Department of Management Science
Lancaster University
Lancaster, U.K.

Abstract *In this modest essay, we reflect on the crucial role of sensegiving, and hence sensemaking, in the creation of IT-enabled service encounters. "Creation" has two meanings here: (1) the* ***design*** *of repeatable (reproducible) IT-enabled services, and (2) the on-going coproduction of IT-enabled service* ***events*** *by service providers and recipients. Indeed, we will argue that the sense given by diverse and role-differentiated actors constitutes in its own way a crucial and pervasive service that enables services in the more familiar sense. Sense-giving, as a "service behind the service," is of particular salience when it comes to novel IT-enabled services, because of the challenges posed by their innovative character. As a practical matter, we are especially interested in how failures in the delivery of innovative services can be caused by shortfalls in sensemaking and sensegiving, and how the difference between successful and failed service outcomes commonly turns on choices made during the design of IT-enabled service systems. These designs either recognize and embrace, or marginalize and ignore, the required and novel sensemaking and sensegiving of employees and customers. We also recognize that system designs are rarely determinative (as constraining as they might prove to be), and that service outcomes will still depend on the variable appropriation of information technology in real situations of practice. We conclude our essay by outlining some research directions in IT-enabled service delivery, arising from these issues.*

Please use the following format when citing this chapter:

Ramiller, N. C., and Chiasson, M., 2008, in IFIP International Federation for Information Processing, Volume 267, Information Technology in the Service Economy: Challenges and Possibilities for the 21st Century, eds. Barrett, M., Davidson, E., Middleton, C., and DeGross, J. (Boston: Springer), pp. 117-126.

Keywords Sense-making, sense-giving, service delivery, coproduction, information systems design

1 INTRODUCTION

The concept of service has enjoyed increasing prominence in discourse about the emerging global economy. Indeed, as the call for papers for this working conference observes, growth in the service economy has been profound and global in its scope. Services are of particular interest to our research community because of the role that information technologies are playing in the transformation of the service sector, through their support of innovative business models, work practices, conversions of products into services, and amplifications of traditional products and services with information.

In innovations of these kinds, there is always much to be learned and decided about the role that information technology can and ought to play. In light of the learning involved, we reflect in this essay on the crucial place that participants' sensemaking and sensegiving hold in the creation of new and reconstituted service encounters. We will argue that sensegiving, in particular, is usefully conceptualized as a service in its own right, which enables the design and delivery of services in the more familiar sense. While sensegiving has always played such a role in services, we will suggest that it is particularly salient where novel IT-enabled services are concerned. This is the case in part because of the demands that innovation makes on sensemaking (Swanson and Ramiller 2004). But it is also true because the potential of IT to transform services presents management (especially) with choices concerning how much to take the sense-making and sensegiving of participants involved in service delivery into account during technology design and operation.

Analytically, we will locate sensegiving, as this "service behind the service," in two interrelated domains. Sensegiving is necessarily an activity in the organizational *design* of repeatable and reproducible IT-enabled services. It is also an aspect of the communicative exchange that takes place between service provider and service recipient during their *coproduction* of actual service *events*. What is of interest and importance, for both practice and research, is that sensegiving can be more or less effective during service delivery. Effectiveness, in-turn, can depend on the choices made during the design of information technology in recognizing and allowing, or marginalizing certain forms of sensemaking and sensegiving between employees and customers. Technological designs also influence the amount of "play" that the participants in service encounters can find to improvise, whether by means of, or in spite of, the technical facilities that are provided.

Our discussion proceeds as follows. First, we consider in further depth how sense-making and sensegiving fit in a fundamental way into the constitution of services. We then explore the alternative fates of sensegiving, relative to its role in the kinds of service transformations that are appearing and helping to define the service economy. We follow up, then, by considering the relationship between sensegiving and *effective* service delivery. While conceding the complex and problematic relationship between tech-nological constraint and allowance, and service effectiveness, we also note that its overlapping and sometimes conflicting meanings help illuminate the important part that power, politics, and participation play in service design (as process) and service designs (as outcomes). We conclude our essay by reflecting, in a preliminary way, on how academic research might help illuminate sensemaking and sensegiving in various IT-enabled services.

2 SENSEMAKING AND SENSEGIVING AS SERVICE

As the call for papers observes, service activities entail "a negotiated and often co-generated exchange between a provider and a...customer in the provision of largely intangible assets." The provider and the recipient, acting in concert with and by means of other actants which may include information technologies of one kind or another (Callon 1986), coproduce a service within institutionalized and communicative rules and norms. These rules help make the encounter stable, efficient, understandable, and predictable to the participants—an important part of service delivery. They also make it possible for us to recognize more or less discrete categories of services (e.g., software support, catalog ordering, healthcare provision, hospitality services, distance education, and so on) which are also important in shaping provider and customers expectations. With increasing levels of standardization, often at the initiative of the provider using information technology, the service encounter can become even more predictable. But even in the strictest of institutional and standardized domains, every service event entails some degree of sensemaking and sensegiving on the participants' part, even if this is undertaken mainly to apply, and signal compliance with, the established norms. Especially in those cases where there is innovation in a service, or at least a lack of familiarity, which for a service recipient can amount to much the same thing as innovation (Rogers 2003), service sensemaking and sensegiving naturally become much more significant and demanding.

Sensemaking is as it sounds: making sense of things, where understanding is as yet lacking (Weick 1995). Sensemaking should not be construed as passive, but as integrally joined with action in cycles of enactment-and-sensemaking that shape as well as interpret reality. In the organizational context, then, sensemaking is "activity that talks events and organizations into existence" (Weick et al. 2005, p. 413), a fact that further "suggests that patterns of organizing are located in the actions and conversations that occur on behalf of the presumed organization and in the texts of those activities that are preserved in social structure" (ibid).

The notion that services are coproduced by provider and recipient further evokes an important aspect of sensemaking: that one party's interpretations are built on the interpretations of others, to a large degree. In this regard, the call for papers notes, "Services exchange often involves many complex combinations of both explicit and tacit knowledge as providers and customers attempt to collectively coordinate and integrate their knowledge in service delivery." Such combinations, coordination, and integration evoke the sensemaking and sensegiving that are integral in the live constitution of the service encounter, as each participant must interpret context, objectives, tasks, means, and the evidence (from speech and action) concerning others' interpretations, while signaling to others their own interpretations.

Sensemaking is, therefore, fundamentally social in character. In describing the actions of an experienced nurse who is trying to find the language that will trigger a doctor's decision making in a critical medical event, Weick and his colleagues remark,

> The...nurse absorbs the complexity of the situation...by holding both a nurse's and doctor's perspectives of the situation while identifying an account of the situation that would align the two. What is especially interesting is that she tries to make sense of how other people make sense of things, a complex determination that is routine in organizational life (Weick et al. 2005, p. 413).

The enactment side of sensemaking in a social context is often accomplished by sensegiving, which is "undertaken to create meanings for a target audience...the content of sensegiving (present versus future image) and the target (insider versus outsider) affect how people interpret the actions they confront" (Weick et al. 2005, p. 416).[1]

In the singular service event, the participants typically will, to the extent possible, construct the emergent social reality using familiar means. However, where novelty is relatively high, the sensemaking and sensegiving that are required may increase substantially. This is true in the constitution of the particular service event, but can also return to the general invention or reinvention of the service practice in the first place. Figure 1 suggests the added complexity we must take into account.

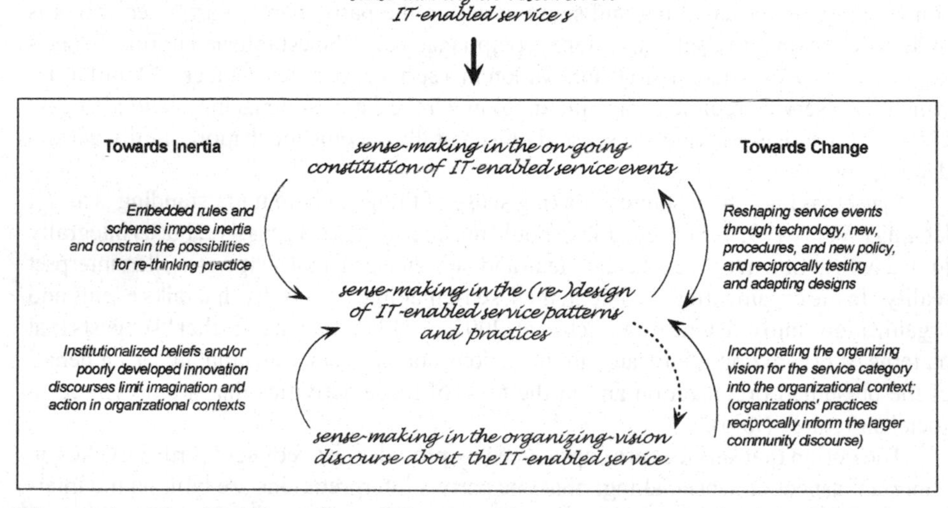

Figure 1. The Stratigraphy of Sensemaking (and Sensegiving) in IT-Enabled Services

---

[1]With sensegiving, the loop must of course be closed by the recipient being open to incorporate that gift of sense into their own sensemaking. Boland and Tenkasi have noted this in connection with their concept of *perspective-taking*. They state, "producing knowledge to create innovative products and processes...requires the ability to make strong perspectives within a community, as well as the ability to take the perspective of another into account" (Boland and Tenkasi 1995, p. 350).

Here we call attention to the middle layer in Figure 1. As a service-behind-the-service undertaken in the cause of new designs (and redesigns) in some service domain, sensemaking (the production aspect) and sensegiving (the delivery aspect) together can change how the organization and its members see that domain. The goal is to make the change more sensible and legitimate, and hence as a practical matter, more tractable. Of course, in doing this, sensemaking and sensegiving also prepare those actors for action that help them shape the material reality during actual service encounters.

Note that the creation of an information system to enable one or more new or transformed services is itself a kind of service rendered to others by the systems development project team. And many a project success or failure has hinged on the quality of the coproduction of sense between the largely "technical" project team members and the people in the business. Of course, this is far from a matter of making sense only of the technology's role. It is also a matter of constructing or reconstructing the cognitive and behavioral patterns, and the deeper structures of meaning and legitimacy which enable service experts to map customer requests and needs into cognitive and procedural templates that deliver the correct service quickly, and in ways that are consistent with agreed patterns of identity and ethical appropriateness.

What we have just said might be read as a claim for a top-down relationship from service design to service event, but this is not our intention. Note the reciprocal arrows in Figure 1. Yes, new service designs, in the form of technology and organizational policies and procedures, surely affect the production of the events. But these designs, as we have already suggested, are of a different nature from the corresponding service events themselves. Accordingly, the coproduction of the service by provider and receiver on the ground is always, in part, an improvisational act. Moreover, in the right circumstances, reflection on completed service events and their outcomes can lead to changes in policies and procedures that may even bring about an adaptive redesign of the technology. More subtly and subversively, reflective action may cause the technology's selective and "unfaithful" appropriation (DeSanctis and Poole 1994), through, for example, work-arounds and other activities which are invoked in order to provide a desirable and needed service to clients.

Figure 1 includes yet another layer in which sensemaking and sensegiving matter to the IT design and service delivery, and this is wider public discourse on IT-enabled services. This is where the *organizing vision* for the IT-enabled service arises, as Swanson and Ramiller (1997) would label it. The public discourse is not idle chatter, nor simply a playground for pundits and journalists, but rather helps to define how people and their organizations think substantively about the possibilities, while also lending normative force to the service innovation (Swanson and Ramiller 1997, 2004). In this way, the public discourse can have an institutionalizing effect that helps to move the innovation toward a taken-for-granted status, even as it serves as a resource in an organization's local sensemaking.[2] The universal usefulness of the public discourse, however, is not a given. It may be immature, or poorly developed for a particular application or industry, or it may be poorly developed for a particular organization's goals and existing practices

[2]Institutionalization occurs "where societal expectations of appropriate organizational form and behavior come to take on rule-like status in social thought and action" (Covaleski and Dirsmith 1988, p. 562).

(Willis and Chiasson 2007), or it may encourage unmindful innovation (Swanson and Ramiller 2004). The upshot, then, is that the public discourse will sometimes fail to adequately inform and support an organization's local efforts in service design.

In summary, sensemaking and sensegiving together constitute an indispensable service in the (re)production of actual service events and in the transformation of service patterns or designs. Sensemaking and sensegiving are also integral to the broader institutional discourse about IT-enabled services and various service categories. In terms of a subject matter of concern to a particular academic community, attention to this service-behind-the-service can reasonably focus on the sense that can, will, and might be made about the information technology that helps (or restricts) the delivery of specific service events, and the designs, patterns, and concepts of service.

3 THE FATE OF SENSEGIVING IN CONTEMPORARY SERVICE TRANSFORMATIONS

So, in focusing our attention in this way, what do we see? While this may be largely a matter for empirical research to decide, we would like to identify what we believe are some general and possible tendencies. We propose that, more commonly than not, the emphasis in information technology design within service industries has been on the restriction of individual and collective sensemaking, in order to rationalize and freeze past or to reengineer future practices in order to define fixed, reliable, and measurable services.[3] We have all personally experienced, in many of our own service encounters, how IT may also decrease and hamper the provider's capability and flexibility in responding to the true complexity of the recipients' needs. We believe that management's first (and sometimes, worst) inclination has been to use technology to eliminate the numerous and "idiosyncratic" approaches that employees have cultivated "haphazardly" like weeds in an untended garden (see Head 2003). In short, we view the trend in IT-enabled service delivery as being closely akin to management's historical choices in machine automation (Noble 1984).

Moving beyond this tendency to naturalize this mechanistic strategy through a rhetoric of technological determinism,[4] alternatives are available that proceed from a different set of assumptions about the range of sensemaking and sensegiving allowed and supported during service events. One possibility is a shift in system design away from a closing-down of sensemaking toward the development of systems that actively support it. This has both *process* and *design* aspects.

When it comes to the processes of specifying, designing and configuring IT to enable service production, a system that is intended to support a more active, on-going and future sensemaking will require deeper and broader participation in the sensemaking that

---

[3]In the first case, this can be likened to the popular expression "paving the cowpaths," and in the latter case to "building a highway to the future." In both cases, the image is one of linear, predictable, and determinative process.

[4]"[N]aturalization gives to particular ideological representations the status of common sense, and thereby makes them…no longer visible as ideologies" (Fairclough 1995, p. 42).

occurs during IT design itself. That is, if empowerment in sensemaking and sensegiving during actual service situations is the objective, then it stands to reason that a similar empowerment will be needed in the creation of the system that will help to enable such outcomes. This may point toward more "agile" socio-technical strategies that can support the meaningful and extended engagement of the future system users (Cockburn 2002), and leverage their insights about practice in a way that fosters flexibility in design (Brown and Duguid 2000). This is in contrast to more traditional structured approaches and to development strategies that are commonly reflected in ERP and business process reengineering, which sequester and limit user participation.

When it comes to the content of technology designs, our basic position is that the particulars of design will typically best be left to those who know the work. Nevertheless, we have some general suggestions. Technological and system designs which allow and foster technical change in order to explore and support new service sensegiving and sensemaking practices would be increasingly desired. Often in sharp contrast, today's enterprise resource planning systems, despite considerable diversity in modularity and implementation options, appear to largely close-down and lock into place particular service practices (Head 2003). This is especially true where technology designers consider service primarily to be about the legitimized handling and entry of data across an assembly-line of personnel, instead of as a source of information and organizational memory that could help in supporting the development and delivery of a successful service outcome.

In addition to supporting both current and future service delivery possibilities, information technology infrastructures need to be open to change and modification, perhaps by the individuals and groups exploring and creating new service patterns from novel sensegiving and sensemaking events. Design will need to focus less on the specific information and programs to support and freeze current service delivery, but on an infrastructure that will allow a reshaping of specific data structures, content, and user access to support revised service delivery (Chiasson and Green 2007).

One way to consider this theoretically is to consider system development as a task to design and build *boundary objects* (Star and Griesemer 1989), which allow value-added exchange across the heterogeneous interests of producers and suppliers. Here, boundary objects would support not only support-specific service events, which always involve some form of sensemaking and giving between producer and client, but also facilitate sensemaking and sensegiving toward new service encounters and service designs which support these more systematically.

In making the claims about the common restriction of sensemaking during current system design, and having spoken broadly in favor of empowerment and against automation and for sensemaking and sensegiving, we will certainly also concede that there is much to be said for efficiency and consistency in service, and for the standardization that is often required to realize it. The challenge, then, is that the various approaches to IT design in the support of service *effectiveness* may not appear commensurate. Does effectiveness mean the efficiency (speed, resource consumption) and consistency with which services are rendered? Does it mean the range of problems that can be solved, including the numerous exceptional cases? How does resiliency in service delivery figure into effectiveness? Do feelings of empowerment and satisfaction on the provider's and/or the recipient's part matter to the service outcome?

These overlapping and apparently conflicting meanings point to the fact that service effectiveness is, to a degree, a question of consistency and change—where particular service patterns at some level and in certain circumstances can be dealt with quickly and efficiently, to the satisfaction of everyone. Effective service delivery in these circumstances, nevertheless, still depends on the active engagement and sensing of the producer and consumer in order to identify which efficient service requires delivery. In other new and novel circumstances, however, expanded sensemaking and giving needs to be possible in order to deal with unique situations, and the required novelty needs to be technologically supported in order to allow for this mindful deviation.

At the same time, it must be recognized that the supported services and the extent of mindful deviation also depends on whose interests are being served, and, to some extent, whose ox may be gored (which sadly, more often than not, is the customer's). There is, in short, an element of power and politics involved in the design of technology in the delivery of services. The prerogatives of sensegiving at the design stage may have an important effect on the sensemaking and sensegiving that are subsequently granted to others, in great part because of the way that "power is expressed in acts that shape what people accept, take for granted, and reject" (Weick et al. 2005, p. 418).

4 TOWARD RESEARCH: MAKING SENSE OF SENSEGIVING AS SERVICE

We have argued here that the task of designing and applying information technology in the delivery services is, in great part, a means to support service sensemaking and sensegiving. These design activities can themselves be considered as a kind of service-behind-the-service, where sense-making and sensegiving about the tools and techniques to support many actual service encounters are considered. Moreover, the quality of this service-behind-the-service in any particular case will depend upon interactions among the organization's service design, participants' situational appropriation of that design, and the larger public discourse on the possibilities for service innovations (Figure 1 summarizes these interactions).

Figure 1 also identifies a place for academic research as its own kind of service in sensemaking and sensegiving. The following are preliminary ideas, but we would suggest that future research might usefully address any of the layers shown in the figure. For example, researchers might study close-up the sensemaking work of participants as they cocreate on-going service events using IT, or explore the sensemaking of diverse project members in the course of initiatives to redesign service categories and the creation of new service patterns. Researchers could also experiment with alternative modes of organizing and allocating opportunities for sensegiving in design projects, or study the organizational uptake and utilization of public discourses on service innovations in local sensemaking and legitimation activities. Finally, researchers could be involved in identifying and evaluating the implications for sensemaking and sensegiving of various technical design innovations, or in identifying alternative appropriations in practice.

In all of these research areas, making sense of the participants' own production of sense can be the foundation for meaningful service that the researchers themselves can provide. The trick, then, will be to create opportunities for academic sensegiving that can help produce improvements in IT-enabled practice, specifically through a guided concern for the sensegiving of the practitioners themselves. Again, this is not to claim that sense-making and sensegiving have only just lately become important in the service domain. Rather, we hope to be convincing that these very old social processes deserve our community's refreshed attention, especially if we are to move beyond simplistic and breathless enthusiasm for the service economy, and to arrive instead at a critical understanding that embraces the variable realities of participation, power, design, and outcomes associated with the IT-enabled services.

References

Boland, R. J., Jr., and Tenkasi, R. V. 1995. "Perspective Making and Perspective Taking in Communities of Knowing," *Organization Science* (6:4), pp. 350-372.

Brown. J. S., and Duguid, P. 2000. "Balancing Act: How to Capture Knowledge Without Killing It," *Harvard Business Review* (78:3), May-June, pp. 73-80.

Callon, M. 1986. "Some Elements of the Sociology of Translation: Domestication of the Scallops and the Fishermen of St. Brieuc Bay," in *Power, Action and Belief: A New Sociology of Knowledge?*, J. Law (ed.), London: Routledge, pp. 196-223.

Chiasson, M., and Green, L. W. 2007. "Questioning the IT Artefact: User Practices That Can, Could, and Cannot Be Supported in Packaged-Software Designs," *European Journal of Information Systems* (16), pp. 542-554.

Cockburn, A. 2002. *Agile Software Development*, Boston: Addison Wesley.

Covaleski, M. A., and Dirsmith, M. W. 1988. "An Institutional Perspective on the Rise, Social Transformation, and Fall of a University Budget Category," *Administrative Science Quarterly* (33), pp. 562-587.

DeSanctis, G. R., and Poole, M. S. 1994. "Capturing the Complexity in Advanced Technology Using Adaptive Structuration Theory," *Organization Science* (5:2), pp. 121-147.

Fairclough, N. 1995. *Critical Discourse Analysis: The Critical Study of Language*, London: Longman.

Head, S. 2003. *The New Ruthless Economy: Work and Power in the Digital Age*, Oxford, UK: Oxford University Press.

Noble, D. F. 1984. *Forces of Production: A Social History of Industrial Automation*, New York: Knopf.

Rogers, E. M. 2003. *Diffusion of Innovations* (5th ed.), New York: Free Press.

Star, S. L., and Griesemer, J. R. 1989. "Institutional Ecology, 'Translations' and Boundary Objects: Amateurs and Professionals in Berkeley's Museum of Vertebrate Zoology, 1907-1939," *Social Studies of Science* (19:3), pp. 387-420.

Swanson, E. B. and Ramiller, N. C. 1997. "The Organizing Vision in Information Systems Innovation," *Organization Science* (8:5), pp. 458-473.

Swanson, E. B. and Ramiller, N. C. 2004. "Innovating Mindfully with Information Technology," *MIS Quarterly* (28:4), pp. 553-583.

Weick, K. E. 1995. *Sensemaking in Organizations*, Thousand Oaks, CA: Sage Publications.

Weick, K. E., Sutcliffe, K. M., and Obstfeld, D. 2005. "Organizing and the Process of Sensemaking," *Organization Science* (16:4), pp. 409-421.

Willis, B., and Chiasson, M. 2007. "Do the Ends Justify the Means? A Gramscian Critique of the Processes of Consent During an ERP Implementation," *Information Technology and People* (20:3), pp. 212-234.

About the Authors

Neil Ramiller is the Ahlbrandt Professor in the Management of Innovation & Technology at Portland State University's School of Business Administration. Neil holds a Ph.D. from UCLA's Anderson School of Management. His primary research activities address the management of information-technology innovations, with a particular focus on the role that rhetoric, narrative, and discourse play in shaping innovation processes and negotiating multi-party interests within organizations and across interorganizational fields. He also conducts work on the social construction of information technology scholarship, and the implementation of the "linguistic turn" in information technology studies. Neil has presented his work at a variety of national and international conferences, and published articles in a number of journals, including *Journal of the Association for Information Systems*, *MIS Quarterly*, *Information & Organization*, *Information Technology & People*, *Organization Science*, *Journal of Management Information Systems*, and *Information Systems Research*. Neil can be reached at neilr@sba.pdx.edu.

Mike Chiasson is currently an AIM (Advanced Institute of Management) Innovation Fellow and a Senior Lecturer at Lancaster University's Management School, in the Department of Management Science. Before joining Lancaster University, he was an associate professor in the Haskayne School of Business, University of Calgary, and a postdoctoral fellow at the Institute for Health Promotion Research at the University of British Columbia. His research examines how social context affects IS development and implementation, using a range of social theories (actor network theory, structuration theory, critical social theory, ethnomethodology, communicative action, power-knowledge, deconstruction, and institutional theory). In studying these questions, he has examined various development and implementation issues (privacy, user involvement, diffusion, outsourcing, cyber-crime, and system development conflict) within medical, legal, engineering, entrepreneurial, and governmental settings. Most of his work has been qualitative in nature, with a strong emphasis on participant observation. Mike can be reached at m.chiasson@lancaster.ac.uk.

Part 2:

IT-Enabled Services in Industry Settings

9 POSSIBILITIES AND CHALLENGES OF TRANSITION TO AMBULANT HEALTH SERVICE DELIVERY WITH ICT SUPPORT IN PSYCHIATRY

Synnøve Thomassen Andersen
Finnmark University College
Alta, Norway

Margunn Aanestad
University of Oslo
Oslo, Norway

Abstract *Transformations of established institutional orders are to be expected alongside the appearance of novel ICT-enabled models of service delivery. Such transformations are neither simple nor short-term, but involve complex and fundamental changes in normative, regulative, and technical aspects. In this paper we describe the initial stages of a project redesigning psychiatric services for children and adolescents. New collaboration models, supported by new ICT applications, were introduced into the ordinary structures of health care services in Finnmark, the northernmost county of Norway. Our aim is to contribute to the understanding of how the preexisting technical and organizational systems and models, the installed base, impact radical change. We address the potential for user-driven innovations, and focus our analysis on the quality of generativity in the existing information infrastructure. The salient challenges in our case were related to the existing technical information infrastructure in the health sector. The new solution was pushed outside the established information infrastructure. While this may be only a temporary situation, we argue that, as a result of this process, the new solution emerged with a greater potential for future generativity than it would have had it been more linked to the existing information infrastructure.*

Keywords Health care, psychiatry, mobile technologies, installed base, generativity

Please use the following format when citing this chapter:

Andersen, S. T., and Aanestad, M., 2008, in IFIP International Federation for Information Processing, Volume 267, Information Technology in the Service Economy: Challenges and Possibilities for the 21st Century, eds. Barrett, M., Davidson, E., Middleton, C., and DeGross, J. (Boston: Springer), pp. 129-141.

1 INTRODUCTION

The shift to a service economy is pervasive in many sectors, not the least the health care sector. The shape of this trend in many Western countries is to emphasize decentralization and patient empowerment, motivated by the realization that future care models must change in order to be economically feasible and sustainable. In particular, patients with chronic diseases should be allowed to manage their diseases themselves as far as possible. This draws strength from the findings that the mobilization of a patient's own resources, as well as family and community resources, contributes significantly to the healing process (Ball and Lillis 2001; Brennan and Safran 2003). The trend also coincides with current management practice, oriented towards achieving quality and efficiency improvements through market mechanisms, with a client or customer focus. Information and communication technology systems are expected to be crucial elements in such a redesigned health care sector, enabling a shift from a provider-centric model of service delivery to a more flexible configuration of services.

However, introducing novel collaboration forms and technologies into existing structures is challenging. A domain such as the health care sector is comprised of well-established institutional orders supported by regulation, administrative procedures, as well as large-scale, complex networks of communication technologies. When new ICT-enabled models of service delivery are introduced, they have to be accompanied by transformations in the normative, regulative, and technical aspects. The research questions we address in this paper relate to this tension between radical change and inertia: How do such ICT-triggered transition processes play out in practice? In particular, how much room is there for user involvement in these processes, and what are the effects of this involvement?

In order to conceptualize this, we draw on the theoretical perspectives associated with studies of information infrastructures: large scale, complex, and networked information and communication technologies (Hanseth and Lyytinen 2004). We focus on the notion of an *installed base* that emphasizes the historical evolution of an information infrastructure from some preexisting and inherited socio-technical legacy system. Moreover, we bring in the notion of *assemblages* (Lanzara 2008; Sassen 2006), which can help us to emphasize the normative and institutionalized dimensions associated with information infrastructures. Since our focus is on the potential for users' contribution in change processes, we complement this literature with the concept of *generativity* (Zittrain 2006), which presents some aspects that make a technological system more or less open to third-party innovation. In the following section we present the theoretical framework for our study. The research methods are then presented, followed by the case study. In the analysis section, we focus on the limited generativity of the previous assemblage, and the effect as well as the potential of this on the change process discussed in this paper. The conclusion is presented in the final section.

2 THEORETICAL FRAMEWORK

In order to explore the possibilities for intervention and change in large-scale, socio-technical networks, we need theoretical resources. Some useful concepts such as installed base, assemblage, and generativity are briefly presented in the following sections.

2.1 Information Infrastructures, Assemblages, and Installed Base

Information infrastructures denote large and complex networks of information systems and communication technologies (Monteiro and Hanseth 1995). The concept of information infrastructure should not be taken to indicate only the technical components, but to also include social and organizational elements (e.g., work practices, routines, regulations) (Star and Ruhleder 1996). Information infrastructures have attributes that make them different from traditional, standalone IT systems. As an information infrastructure grows, it may acquire momentum (Hughes 1983) and become increasingly entrenched and irreversible. Concepts from the field of network economy used to explain this include self-reinforcing effects, path dependence, and lock-in (Arthur 1989; David 1985). Due to its emphasis on historical development, the notion of an installed base is central within this perspective. Information infrastructures evolve from and are conditioned by what is already in place. For instance, legacy systems as well as sociotechnical institutions, regulations, and procedures can significantly shape the evolving information infrastructure. The installed base is a central resource for any change attempt; it is what facilitates and enables evolution through extension or substitution. On the other hand it can be an obstacle, resisting change due to lock-in. One has to design both "with" and "against" the installed base at the same time.

This approach conveys an understanding that the traditional control-oriented strategies have their limits and calls for different approaches to design (Hanseth et al. 2001). Incremental, evolutionary, and bottom-up approaches are seen as more appropriate approaches, as they are more sensitive to the installed base, exploit the self-reinforcing mechanisms (such as bootstrapping; see Hanseth and Aanestad 2003), and are geared toward avoiding lock-in situations (Hanseth and Lyytinen 2004; Monteiro 1998). However, in emphasizing the limitations of traditional design approaches, this literature may have downplayed other aspects. A recent study of innovation in the mobile sector stated that "existing conceptualization portrays information infrastructures as autonomous and under-theorizes the multiplicity, the agency and the interrelations between information infrastructure builders and their institutional context in building processes" (Nielsen 2006, p. vii). Thus we argue that we also need to conceptualize the institutional context in which information infrastructures develop.

On the one hand, existing solutions, prevailing routines and social institutions may be entrenched and cannot be changed in an instant. On the other hand, the evolving information infrastructure may enable entirely new constellations and arrangements of agencies. The notion of *assemblages* has been used to denote how technical communication networks and political, institutional, and organizational entities are imbricated (Lanzara 2008; Sassen 2006). In public sector administrative practices, institutional configurations and ICT are becoming intimately intertwined. The notion of assemblage points to the resulting loosely structured institutional ecology of multiple heterogeneous actors, such as political authorities, bureaucratic organizations, ICT providers, and technical agencies. Each of these act according to their own logic, which may not overlap. Change derives from and is conditioned by older capabilities, material orders, institutional conditions, and technological infrastructures.

2.2 Making Room for Users: Generative Technologies

We also want to focus on the need for user or other third-party participation in the design process. There is a popular belief that radical change in the way health care is offered is more likely to come from a demand- or needs-driven development (i.e., from the clients or patients) than from reforms to health care services "from within." These arguments have their parallel in information systems. For instance, Ciborra (1994) argues that grassroots-driven change is more likely to lead to radical innovation than top-down driven projects. Here, the "establishment" on the top is eager to preserve the status quo, while more marginalized actors are in a position where they have less to loose and more to gain. Without fully buying into these arguments, our goal is to examine the space available for users to be involved in infrastructural change activities.

Studies show that different technologies and different development models facilitate this to varying degrees. Abbate's (1994) comparison of the Internet model with the telecom model showed that different configurations allowed for different usages and different paths of evolution. The Internet turned out to be an enormously flexible platform for multiparty innovation due to its simple network protocols, end-to-end intelligence, and open development processes. In describing the Internet's unprecedented growth and innovation (and the backlash against this), Zittrain (2006) employs the notion of *generativity*. He suggests that rather than talking about *open* versus *proprietary* technologies, we should consider *generative* versus *closed* configurations of technologies. Technologies classified as generative (such as the Internet) are technologies that remain open to third-party (user) innovations. Rather than going into his argument about the future of the Internet, we will utilize his definition of generativity as a function of four aspects. A technology is generative if it has (1) capacity to leverage a set of possible tasks, (2) adaptability to a range of different tasks, (3) ease of mastery, and (4) accessibility. The first aspect, *leverage of tasks*, is related to whether the technology makes the task easier. A very generative technology will allow a lot of effort to be saved, and/or allow effort to be saved on a lot of different tasks, not just one. The second aspect, *adaptability*, may relate both to *use flexibility*, whether the technology can be used as it is for different things, and to *change flexibility*, whether it allows changes to be made. The third aspect, *ease of mastery*, denotes the ability of the technology to avail inexperienced users to take advantage of it, at least partly. *Accessibility*, as the fourth aspect, signifies the degree to which a person can come to use and control technology. This includes the ease with which the necessary knowledge can be obtained. Accessibility can be influenced by financial, legal, secrecy, or scarcity barriers. According to Zittrain, the generativity of technologies increases the possibility that users can generate new and valuable usage patterns that become sources for further innovations.

3 RESEARCH METHODS

We present the findings of a study of the redesign of psychiatric services for children and adolescents, and describe the challenges and opportunities encountered during the initial project period. In this section, we elaborate on the research strategies used to gather empirical material.

3.1 Background Information about the Research Site

This study was conducted in Finnmark, the northernmost county in Norway, with a population of around 73,000 inhabitants. Out of these, approximately 20,000 are under the age of 20. In 2007, there were 950 children under the age of 18 receiving daily treatment in clinics for children and youth psychiatry (CYP). CYPs in Finnmark employ 50 or 60 persons. The county is 48,649 square kilometers, slightly larger than Denmark, and it is sparsely populated, as only 40 square kilometers (0.8 percent of the area) is inhabited. Consequently there are long distances between communities and, for most inhabitants, long distances to the nearest hospital or medical expert. The ICT infrastructure in the county is uneven, with some areas well covered by both broadband networks and telephone networks, other almost without any coverage at all.

The project described in this paper is called "Come Here! – ambulant teams and technology."[1] It was established on January 1, 2006, following a decision in the county's health authority to close down the only psychiatric hospital on July 1, 2005. The hospital was to be replaced by a decentralized model, where ambulant teams would conduct and support home-based treatment for both families and children (between 6 and 12 years). The project was started in order to develop and implement suitable technologies to support this new decentralized care model, with the hope that this might reduce waiting time and the need for hospitalization. The project will end December 31, 2008, after which the model will become the standard for psychiatric care for children and adolescent youth. Four communities were selected as pilots. At the time of writing, the training of users had been completed and deployment of the prototype solution was starting.

3.2 The Action Research Project

The first author has been the project manager in this project since its start and, as such, she has been directly involved in the work to develop and implement mobile technology. At the same time, she is pursuing a Ph.D. focusing on the change processes around this project. For this reason the action research method has been deployed in the project. In information systems, action research is a well-known methodology (Baskerville 1999; Checkland 1991; Susman and Evered 1978). A central premise in action research is to include people in the research domain, with the aim of changing the field in ways that both correspond with and challenge tradition. Even though this paper does not emphasize the action research process, this approach strongly influenced the way empirical material was gathered (see section 3.3) and comes with certain challenges that we will briefly discuss.

Action research gives unique access to the field as it allows intimate participation in core processes during the entire project. In addition, the researcher's role is more engaged as compared to a traditional observer. This implies the need to go beyond merely observing and describing from the researcher's own standpoint, to being able to understand, incorporate, represent, and negotiate the other participants' standpoints as

---

[1]The Norwegian title, "Kom hIT – ambulante team og teknologi," includes a pun through capitalizing IT in *hit* (meaning *here*).

well. As such this approach comes with an in-built validity check. The action research process is challenging, as it involves more work than an observational study, including supporting the process, keeping it on track, and activities to facilitate shared learning. Moreover, it entails challenges in balancing research interests with the practical needs of the project. There may be divergent interests between the role as a researcher and as a change agent, and it may be difficult to be fundamentally critical if one believes in and fights for the project's goals. In this paper, we are not assessing the solution or the change process as such, as these matters would have necessitated a longer discussion. We provide an account of concrete and publicly known events in the project's history, and believe that this makes the study less prone to biased representations.

3.3 Empirical Material Used in This Paper

The orientation of the action research project (to stimulate and facilitate user innovation) thus has defined the focus of this paper. The approach has also shaped the methods for collecting data, to which we now turn.

In the role as project leader, the first author has participated in multiple project activities which have generated the empirical material for this paper. A significant source of data is observations during formal meetings. There have been a total of 11 project team meetings, 9 steering group meetings, 5 meetings with contractors, and 4 meetings in the smaller "techno-group." The discussions and decisions from the meetings were documented by the researcher/project leader through extensive note-taking. The participants were aware of the project leader's research plans, and did not appear uncomfortable by her note-taking. The notes were then written up, shared, and discussed with the administrative participants, the project team, and the techno-group. The notes were examined by the meeting participants and were mostly accepted; only twice were revisions necessary. Another important data source was other documents produced during the project period, for instance, the project description (in several versions), summons, reports, letters, and contracts. Around 70 formal project documents and literally hundreds of e-mail messages have been produced during the project work so far. In addition, the first author has conducted three courses in using the mobile solution for the ambulant teams, the CYP specialists, as well as teachers, parents, and children. Moreover, during the initial phase (2006), four interviews with two officials from the CYP service and two representatives from the user organization, Mental Health, were conducted, focusing on gaining an understanding of the current situation.

4 CASE: "COME HERE! – AMBULANT TEAMS AND TECHNOLOGY"

Health Finnmark (the county's public health authority) established the project January 1, 2006, in order to develop and implement suitable technologies to support a new decentralized care model, and analyze if this model might reduce waiting time and the need for hospitalization. Thus the project was well anchored at the management level of authority. The health authority appointed the leader of the Clinic for Psychiatric

Health and Inebriation to collaborate with the project leader. During the first half of 2006, three meetings were held with the employees in the clinic and the ambulant teams. Tentative project descriptions were sketched and revisions discussed during this period. The final project team was established October 5, 2006, and consisted of the project leader, one member from each of the two ambulant teams, one member from the IT department in the county's health authority, and two members from the user organization, Mental Health. The project group decided to have a strong focus on user participation. The term *users* in this case means health care workers (CYP team members and CYP specialists) as well as parents, adolescents, and children.

When the project started, two important decisions had to be made: selecting the professional method and selecting the technology. The professional team decided to use the treatment method called the Parent Management Training – Oregon model. The PMT-O is an outpatient treatment model for parents with children that are difficult to raise. The method presupposes structured and intensive work to change the child's behavior, focusing on the behavior and situation in the family, at school, and in the community. The method aims at training parents to cope with and raise their child in a better way. During the initial meeting, the team, parents, and child define and prioritize the goals to be reached. Then they negotiate specific behaviors that should be encouraged or discouraged (for instance in relation to behavior during mealtime or when going to bed). The child's rewards in relation to these action points are defined, as well as how many scoring points certain behaviors should entail. Based on this, a reporting form is created. Parents in cooperation with the child are supposed to register the behavior and assign a score (between zero and five points) frequently, for example, during every meal or every evening when going to bed. These reports are the basis for the interaction between the family and the CYP team. Between the visits by the ambulant team, the parents should register the behavior. When this model was implemented, some parents would fax the forms to the team, but others kept the forms until the next time they interacted with the team. It was felt that this collaboration would benefit from more frequent reporting as well as enable easier and more frequent interaction. The project aimed at changing these communication patterns by introducing technology that let parents report on and register behavior immediately, allowing the CYP team to monitor progress on an ongoing basis.

Choosing the technology was the other component necessary in order to reach the project's goals. The health authority contacted the Norwegian Center for Telemedicine (NST) for advice on choosing suitable technology. While the NST proposed using video-conferencing technology on PCs, the steering group and the project team felt that third-generation mobile technology was more suitable. The arguments in favor of mobile technologies were multiple: A technical solution based on video-conferencing had high costs. The cost of purchasing mobile phones would be moderate as compared to the purchase of PCs, web cameras, document cameras, etc. Such a solution might also require costly upgrades of the different studios of the out-patient clinics. Moreover, the ambulant teams were traveling a lot, and mobile phones would therefore be more practical communication equipment than portable PCs. Coverage of the mobile network is also better than coverage of the broadband network in the northern part of Norway. The training required for children, families, and ambulant CYP teams in order to use the video-conferencing solution was comprehensive, while knowledge and use of mobile phones is widespread in most every age group and every social layer of the population.

The decision to go for mobile technologies was made early in the project period, at the second project meeting. At that time, the NST did not have competency on 3G technologies, and thus the project continued without their formal participation.

The decision to use mobile technology was followed by an unexpected reluctance from the IT department. One participant from the IT department was appointed to the project group, but was not active in providing information requested. The IT department formally withdrew from the project in January 2007, nearly a year after the project started. The argument for withdrawal was related to the reorganization from January 1, 2006, in the regional health authority (Northern Norway Regional Health Authority, which manages health care in Finnmark and two other counties). During this reorganization, the IT divisions of the health authorities for the counties were centralized into one regional IT department. As this project was initiated on a county level, the regional IT division required extra resources for participating and servicing the project. The project did not have these extra resources, and the effect of this was that the project team continued without participation from the IT department. Despite multiple requests, the project team also did not receive technical information about the existing information infrastructure, the Norwegian Health Network, which is a broadband network connecting all health institutions. With support from the project group, the project leader developed the requirement specifications for the project using mobile technology and a custom-developed application. However, this situation lead to the project being alone in choosing and shaping the solution. This had some important consequences. On one hand, it allowed a lot of room for adapting to the wishes and needs of the users in terms of communication patterns and functionality. On the other hand, the project explicitly emphasized building a flexible solution; this was due to the lack of information about the existing technical infrastructure.

An application that supports the PMT-O model was developed with user participation and is now implemented on mobile phones. The application (based on J2ME, version midp2.0) is called "Come Here! MOBILE" and is implemented on Nokia E65 phones that are distributed to project participants. This application is a replication of the paper forms used to register the results on specific action points regarding the child's problems. The application is general, but adaptable in order to give every child and family the possibility to adapt it to the individual case management plans (see Figure 1). The data sent from the phone (e.g., a list of scores or a report on behavior during meal time) is transmitted over the GPRS/UMTS network, using the HTTPS protocol (with 256 bits encryption). All data are stored in an MSsql database and an IIS server, using the .Net platform. There is no local caching, so no sensitive data is saved on the phones. A general assessment of security issues and a risk analysis was performed by a consultant from KITH (the Norwegian Competency Center for Health Care IT), who concluded that the solution complied with the legal requirements. The CYP workers have access to the information from the server (which is located at the technology vendor's facilities) through Internet and virtual private network (VPN) channels. As this is outside the secure healthcare network, there is no direct import into the main patient record application, but it is possible to cut and paste information from the application into the "CYP Data," which is the main patient record application.

The project's anchoring in the top level management of the Health Finnmark authority implied that necessary changes related to administrative routine were accom-

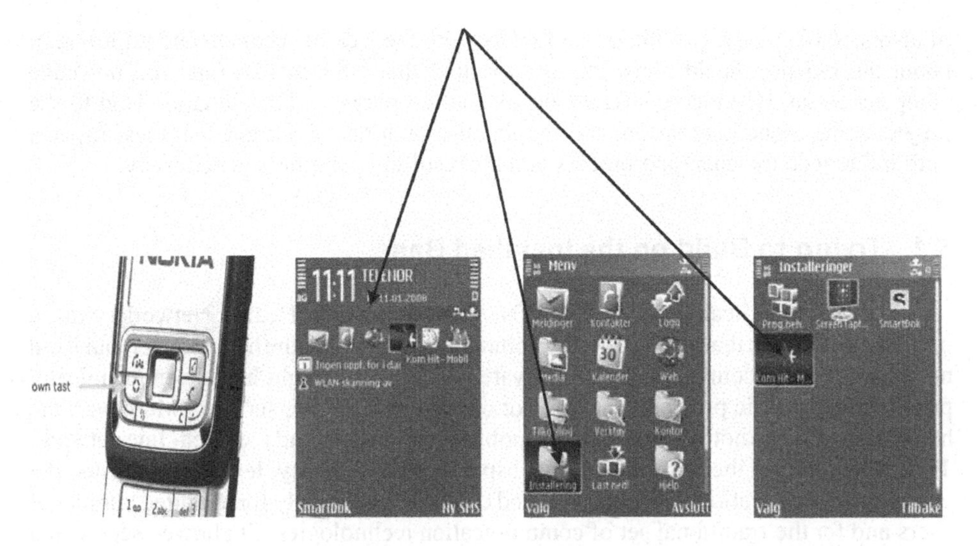

Figure 1. Three Ways to Start the Application (Dedicated Button, List of Programs, or Icon)

plished. These changes were not trivial, and related to defining new contract types, new models for purchase agreements, and new types of service models. For instance, the type of telephone subscription schemes and reimbursement models were discussed. However, these issues were resolved during the project period. The larger challenge remained, and decoupling from the IT department has still not been addressed satisfactorily.

In January 2008, the IT department responded with an assessment of the project's solution. Several critical comments were expressed, most of them relating to security and confidentiality issues. Specifically, the IT department commented that this solution presupposed direct IP-connectivity between the health network and the outside Internet; which was explicitly not permitted. Moreover, it allowed for a transaction being initiated outside the health network, while the current regulation says all transactions should originate within the secure health network (a pull rather than push-based communication pattern). The type of encryption selected also was not perceived to be sufficient. The IT department discussed the tradeoffs of a possible solution. If the solution was placed in a "demilitarized zone" outside the hospital, it would encounter legal challenges related to the storage of sensitive health data. Alternatively, a message-based solution could be placed inside the health network, but that would come with some lag in communications. While this description reflects the current situation at the time of writing, we expect that a permanent solution offering inclusion and coupling to the existing health network will be found before the project ends.

5 ANALYSIS AND DISCUSSION

Earlier, we presented the empirical material illustrating how the process of designing the new infrastructure was influenced by the contingencies and the resulting assemblage of actors. Concretely, two important factors were the lack of access to and information about the existing health network, and the fact that NST at that time did not have competence on 3G video-conferencing for mobile phones. This situation lead to the project being alone in choosing and shaping the solution. We argue that these aspects both influenced the changing process and the resulting solution's generativity.

5.1 Trying to Build on the Installed Base

The existing health network (NHN, the Norwegian Health Network) was a broadband network that allowed secure communication within the health sector, but it did not accommodate communication to private homes. This might have been technically possible for specific projects, but was not offered as a routine service. Moreover, the health network had not yet incorporated mobile technologies and the fixed- line network. Thus, according to the first of Zittrain's aspects of generativity, **leverage of tasks**, the existing communication network leveraged communication only for a limited number of users and for the traditional set of communication technologies. It allowed secure and fast communication, but only between those already on the network, using the technologies that were supported. This may work for the CYP offices but could not have included the families in a large-scale, routine solution. With regard to **adaptability**, the broadband network is a relative generic highway with lots of possibilities. However, its use is regulated, for instance, security considerations led to the requirement that initiative transactions had to be taken from inside (pull rather than push-based communication). The existing information infrastructure is, therefore, adaptable for the prevailing paradigm (i.e., the hospital-centered model). Moreover, the project history showed limited **accessibility** of the information infrastructure. While access to the broadband network might in principle have been possible, due to reorganization, changing responsibilities, and lack of resources, the county's IT department refused to support the project: It did not even release basic information about the infrastructure, which made the existing infrastructure less accessible, even for the employees in the health care sector. This effectively barred any external innovation from building directly upon it. It is difficult to assess to what degree the situation would have been different if the IT department had been able and willing to cooperate. For all practical purposes, we can regard the existing health network as closed for this innovation, and thus not particularly generative. The aspect of **ease of mastery** is also difficult to assess since there was no practical openness to interact with the infrastructure. While it would be available for use by the CYP specialists, it would not be available for technical modifications and tinkering. Even if it had been open for use, the adaptations required would have been technically complicated and would have required assistance from technically skilled personnel. Thus it was not easily mastered. The security aspects motivated a cautious progress in these kinds of use areas. When these factors are coupled with a stronger focus on regional and national standardization and centralization of ICT initiatives, we may see that there are internal barriers to innovation initiatives.

The resulting information infrastructure is not particularly generative information. Consequently, the new solution is not built directly upon the existing information infrastructure, but outside it, as a separately financed and technically independent communication path. Thus the non-generativity of the existing information infrastructure, due to the inertia of its installed base, actually spurred a more radical form for innovation, but in a decoupled way. The fact that the IT department relinquished control and withdrew left the space wide open for the users' participation and impact on shaping the solution. A different configuration of the assemblage would have resulted in a process that would have been conditioned in a different way, and probably would have been shaped more by the institutionalized communication patterns and solutions. This is evident from the *post hoc* comments to the solution from the regional IT department.

5.2 The New Solution: Will it Be Generative?

While in the last section we used the generativity concept to retrospectively analyze the reasons for the project's departure, here we want to employ the concept in a projective way. We will try to analyze the new care model and the technology and ask whether the solution may be generative for further innovation and use. To some degree, this will be an extrapolation from the existing system, rather than an empirically based analysis.

Leverage seemed to be important for the users who wanted mobile technology as a supplement in the therapy. They expected a tighter follow-up, with less travel and waiting time. The new mobile application allowed the user to accomplish tasks that couldn't otherwise be accomplished, such as immediate reporting and feedback. Also, the aspect of **adaptability** seemed relevant. The mobile technology has, in this case, been easy to modify for new purposes, to support psychiatric treatment through structured and nonstructured communication. The application allowed the modification of registration forms and report formats by the parents themselves. **Ease of mastery** defines how easy is it for broad audiences to both adopt and adapt a technology. The previous paper forms were readily mastered by the users, and they were also flexible and adaptable. The participant users in the group expressed clearly that they were familiar with the use of the mobile phone in general, and that participating in application development was meaningful. This application using mobile phones was supposed to have a lower entrance barrier than videoconferencing or PC-based systems. The users' participation in the development process, coupled with the absence of the established IT actors, resulted in the new application being adaptable and leveraging tasks. The resulting application has low technical complexity. Finally, **accessibility**, signifies the degree to which a person can come to control and use the technology. The mobile solution is available for the pilot users as a routine tool and allows the reporting forms to be adapted by the users themselves. The direct contact with the technology vendor who hosts the application also allows adaptations to be easily achieved.

The process of development led to the users having a significant role in defining the solution. For instance, the choice of mobile telephones rather than computers as the basic tool enabled user participation. This again led to a technically noncomplex application with intuitive usage patterns. Moreover, it is likely that the solution in general can be used in other geographical areas that may have a different organization of CYP services. It may also be adapted to serve other health care domains, especially those domains where self-monitoring and reporting are important. With respect to different diseases, information needs, information contents, and communication patterns may vary significantly. Generally in these usage settings the challenge is the need for security, which has priority, and is likely to influence the potential for unlimited innovation of usage patterns.

6 CONCLUDING REMARKS

We have described the process where the preexisting systems and models (the institutionalized assemblages) could not easily encompass user-driven innovations. Thus the solution was developed outside the existing healthcare network and outside the institutionalized locations for technology innovation. The controversies that emerged are illustrative of the types of challenges fundamental transformations of service models may encounter. It is telling to note that the most problematic aspects in this case were related to the inertia of the installed base of the technological infrastructure, rather than administrative or regulative aspects of the organization. We do not believe this is the end of the story; rather it is a snapshot during a particularly illustrative moment in the evolutionary history of the health sector's information infrastructure of Finnmark. The timing is crucial to be able to identify these controversies, which are often black-boxed when one studies only the resulting solution after the project has finished. Even though we believe that the involved health authorities (Health Finnmark and Northern Norway Regional Health Authority) will find an adequate solution in the end, we also believe that the generative characteristics of the new solution will be a lasting inheritance of this particular process. Since this new solution was developed in a somewhat marginal position, the stronger stakeholders did not have the opportunity to reproduce the institutionalized models and solutions. Thus, the initial stages of the project are interesting both for practical and theoretical reasons.

We have argued that because of the non-generativity of the existing installed base, the new solution emerged with larger potential for future generativity than it would have if the process had been different. The installed base is a central resource for any change attempt, and new solutions depart from and build on it in specific ways. This should constitute relevant topics for future research, both within this particular project and in other attempts to study the ongoing transformations of organizations and societies with information and communication technologies.

References

Abbate, J. 1994. "The Internet Challenge: Conflict and Compromise in Computer Networking," in *Changing Large Technical Systems,* J. Summerton (ed.), Boulder, CO: Westview Press, pp. 193-210.

Arthur, W. B. 1989. "Competing Technologies, Increasing Returns, and Lock-In by Historical Events," *The Economic Journal* (99:394), pp. 116-131.

Ball M. J, and Lillis J. 2001. "E-Health: Transforming the Physician/Patient Relationship," *International Journal of Medical Informatics* (61:1), pp. 1-10.

Brennan, P., and Safran, C. 2003. "Report of Conference Track 3: Patient Empowerment," *International Journal of Medical Informatics* (69), pp. 301-304.

Baskerville, R. 1999. "Investigating Information Systems with Action Research," *Communications of the AIS* (2:19), pp. 1-32.

Ciborra, C. 1994. "The Grassroots of IT and Strategy," in *Strategic Information Systems: A European Perspective*, C. Ciborra and T. Jelassi (eds.), Chichester, England: John Wiley and Sons, pp. 3-24.

Checkland, P. 1991. "From Framework through Experience to Learning: The Essential Nature of Action Research," in *Information Systems Research: Contemporary Approaches and*

Emergent Traditions, H-E. Nissen, H. K. Klein, and R. A. Hirschheim (eds.), Amsterdam: North-Holland, pp. 397-403.

David, P. A. 2005. "The Beginnings and Prospective Ending of 'End-to-End': An Evolutionary Perspective on the Internet's Architecture," Industrial Organization from Economics Working Paper Archive (EconWPA) (http://econpapers.repec.org/paper/wpawuwpio/0502012.htm).

Hanseth, O., and Aanestad, M. 2003: "Design as Bootstrapping: On the Evolution of ICT Networks in Health Care," *Methods of Information in Medicine* (42:4), pp. 384-391.

Hanseth, O., Ciborra, C., and Braa, K. 2001. "The Control Devolution: ERP and the Side Effects of Globalization," *SIGMIS Database* (32:4), pp. 34-46.

Hanseth, O., and Lyytinen, K. 2004. "Theorizing about the Design of Information Infrastructures: Design Kernel Theories and Principles," *Sprouts: Working Papers on Information Environments, Systems and Organizations* (4:4), pp. 207-241.

Hughes, T. P. 1983. *Networks of Power: Electrification in Western Society 1880–1930*, Baltimore, MD: The John Hopkins University Press.

Lanzara, G. F. 2008. "Mapping the Encounter Between ICT and Institutions: The Emergence of Techno-Institutional Assemblages," in *ICT and Innovation in the Public Sector: European Perspectives on the Making of e-Government*, F. Contini and G. F. Lanzara (eds.), Basingstoke, England: Palgrave MacMillan (forthcoming).

Monteiro, E. 1998. "Scaling Information Infrastructure: The Case of the Next Generation IP in the Internet," *The Information Society* (14:3), pp. 229-245.

Monteiro, E., and Hanseth, O. 1995. "Social Shaping of Information Infrastructure: On Being Specific about the Technology," in *Information Technology and Changes in Organizational Work*, W. J. Orlikowski, G. Walsham, M. R. Jones, and J. I. DeGross (eds.), London: Chapman & Hall, pp. 325-343.

Nielsen, P. 2006. *Conceptual Framework of Information Infrastructure Building: A Case Study of the Development of a Content Service Platform for Mobile Phones in Norway*, unpublished Ph.D. thesis, University of Oslo.

Sassen, S. 2006. *Territory, Authority, Rights: From Medieval to Global Assemblages*, Princeton, NJ: Princeton University Press.

Star, S. L., and Ruhleder, K 1996. "Steps Toward an Ecology of Infrastructure: Design and Access for Large Information Spaces," *Information Systems Research* (17:1), pp. 111-134.

Sussman, G. L., and Evered, R. D. 1978. "An Assessment of the Scientific Merits of Action Research," *Administrative Sciences Quarterly* (23), pp. 582-603.

Zittrain, J. 2006. "The Generative Internet," *Harward Law Review* (119:7), May , pp. 1974-2040 (http://eecs.harvard.edu/qr48/jzgenerativity.pdf).

About the Authors

Synnøve Thomassen Andersen is a Ph.D. student at the Department of Informatics, University of Oslo, Norway. She has worked as a Child Welfare Worker, and within informatics at Finnmark University College. Her research interests are related to global infrastructures, specifically in health care. Synnøve can be reached at synnovet@hifm.no.

Margunn Aanestad is a researcher at the Department of Informatics, University of Oslo, Norway. She worked in the health care and telecommunications fields before her doctoral study of surgical telemedicine. Her research interests are broadly related to large-scale information infrastructures, specifically in health care. Margunn can be reached at margunn@ifi.uio.no.

10 TRANSFORMING WORK PRACTICES IN A COMPLEX ENVIRONMENT

Riikka Vuokko
Helena Karsten
Åbo Akademi University
Turku, Finland

Abstract *Nursing work is intertwined with a number of technologies. This paper explores the work practices in a pediatric intensive care unit, and discusses some possible directions for introducing new technologies. Work in intensive care is approached as a set of complex and networked practices that are characterized by dynamism and reflexivity of situated action. We observed how, in the intensive care unit, the emerging issues and complexity of organizational action are anticipated with situational adaptability and self-ordering of action. Although the nurses are capable of adapting to rapidly changing situations, at the same time, the nursing practices are bounded by the situational rationalities, such as the information available on the patient. With new technologies, we see several opportunities for change in time-place arrangements, in coordination and communication practices, and in information sharing practices. The roles and tasks of the various actants may reformulate, and thereby possibly their skills and professional identities as well. All of this will take place when work practices, new technologies, and care processes are negotiated and made irreversible through the actions of the participants.*

Keywords Work practices, intensive care, nursing, actor-network theory, complexity theory

1 INTRODUCTION

This paper describes the work practices of nurses in a pediatric intensive care unit (ICU), which we wanted to record in order to understand how technologies and work practices relate to each other and in order to follow and document changes during and

Please use the following format when citing this chapter:

Vuokko, R., and Karsten, H., 2008, in IFIP International Federation for Information Processing, Volume 267, Information Technology in the Service Economy: Challenges and Possibilities for the 21st Century, eds. Barrett, M., Davidson, E., Middleton, C., and DeGross, J. (Boston: Springer), pp. 143-157.

after the adoption of new technologies. For example, electronic patient record systems are expected to enhance nursing documentation activities and thereby the quality and traceability of care (Berg 2001). As these systems enable rearranging work and reforming work practices, they also constrain work in unanticipated ways (Orlikowski 2002).

Continuous changes and diverse and often fast-paced work tasks contribute to the organizational complexity in hospitals, where a large number of interacting bodies act together to advance patients' health. Even though computers have been in use in this hospital since 1970s, there is no integrated system that would cover all areas of care. The patchy information technology infrastructure contributes to the complexity even when it is robust and reliable in parts. We can also assume that complexity in hospitals continues to grow as the hospitals become increasingly connected (Cohen 1999; Jacucci et al. 2006; Merali 2004) with, for example, municipal health centers, private clinics, government agencies (e.g., the national electronic patient record archive currently under construction), or pharmacies (electronic prescriptions will be used from 2009 on).

In this study, nursing work practices are explored as part of a socio-technical assemblage, where shared work practices are jointly formed and reformed (Latour 2005). Barnes (2001) argues that practices do not reduce to individuals, but that any social system is characterized by a set of shared practices generated in ongoing and self-reproducing action as "socially recognized forms of activity, done on the basis of what members learn from others, and capable of being done well or badly, correctly or incorrectly" (p. 19). Understanding such shared practices means an understanding of wider cultural phenomena in a given context. In this sense, shared practices are ways to distinguish one's own working community from others, from outsiders (Latour 2005). Practices form an invisible, taken-for-granted set of attitudes and reward structures (Haythornthwaite 2006) and in this way contribute to a transparent infrastructure that is inherent to a community (Star and Ruhleder 1996). Haythornthwaite (2006) sees practices as instantiated in the technologies used to accomplish work, that is, work practices are enacted through or around some kind of social or physical technology. Our aim is to observe the transparent infrastructure of the ICU nurses, also as manifested by the use of various technologies.

2 CONCEPTUALIZING INTENSIVE CARE WORKING

An intensive care unit is a small component in the overall complexity of a hospital. In itself, it is a complex system, as well. It involves many collaborating groups of people such as pediatricians, surgeons, assisting physicians, anesthesiologists, nurses specialized in intensive care, supporting staff, and multiple social, physical, or electronic technologies that all contribute to the complexity (Schneider and Wagner 1993; Wiener et al. 1979). However, there is a constant struggle to simplify work practices and create routines in order to resist the complexity and reduce the effort needed in work situations. This drive for simplicity includes not only emerging uses of various technologies but also embedded organizational roles, skills, and professional identities, hospital guidelines, and situational arrangements. What we see, however, is that much of the nurses' work relies on intuition (Effken 2001; Lauri and Salanterä 1998; Salanterä et al. 2003), especially when situations demand immediate action. The organizational members of a hospital unit form a community of self-organizing actors that interact in a dynamic, nonlinear fashion with a common history (Cilliers 1998) in this complex world.

In the classical systems paradigm, information systems are conceptualized as holistic, well-defined systems with clear boundaries, striving for stability. According to Merali (2004), such systems can be *complicated* in that they consist of many components. She contrasts these with *complex* systems that cannot be sufficiently understood through their components or subsystems. Merali describes technologies in organizations as facilitators of complex adaptive systems. For example, information technology can combine people, other technologies, and information in time and space in new ways. To understand how the role of various technologies is gradually formed in the pediatric ICU, we look at the intensive care unit as a complex system where changes may be nonlinear and emergent (Anderson 1999; Kim and Kaplan 2006).

Besides complexity theory, actor-network theory (ANT; Latour 1991, 2005) has been used to study and describe large and complex networks of technological innovation and change. Especially in recent theorizing, there has been considerable attention on complexity issues (Moser and Law 2006). In ANT, networks are constituted of a relevant social group of actors (Bijker 1995) that negotiate and interact with each other to solve a shared problem. While not underestimating the importance of human actors, we include also other relevant actors (or actants) such as technical artefacts, organizational rules, and scripts (Law and Callon 1995).

Howcroft, Mitev, and Wilson (2004) follow Latour (1999) by arguing that a new technology is conceived when a relatively stable heterogeneous network of aligned interests is created and maintained. Development and implementation of technologies involves building alliances between various actors, including individuals and groups, as well as other entities such as various technologies. As the actors are enrolled into a network and as the network evolves, the nature of the project and the identities and interests of the actors are themselves transformed (Law and Callon 1995). The results of the transformation process, the translation, are subsequently inscribed into technologies for immutability (Latour 2005).

A new system can have unanticipated outcomes. Latour (1999, 2005) describes how the process of translation is started when a problem of interpretation about the nature of the new technology emerges. The outcome or solution to such a problem can consist of years of development, as in Latour's (1999) example on Louis Pasteur's efforts in creating a first rabies vaccine. Equally well, the translation can be used as a metaphor to describe problem-solving on a smaller scale as, for example, when the practices of utilizing a new technology are negotiated and formed by organizational members.

When new technologies are taken into use and their utilization becomes routine, they disappear from the core of consciousness (Ciborra 2002; Orlikowski 2002), as their use no longer demands special effort and as they are embedded in the context. Latour (1999) describes the embedding of technology with the process of black boxing. Black boxing occurs when, for example, the organizational actors can use a solution or a tool although they cannot tell its inner functions. In a way, here, a complicated technical artifact or an information system can be simplified as a black box that can be utilized without special effort by the user. If the technology in use no longer meets the needs and goals of its users, the black box can be reopened and the translation process can start anew. Thus translation is a process that does not necessarily ever come to a definitive end (Latour 2005).

Similarly to ANT, in complexity theory the process of change is explained as continuous adaptation and adjusting to the changing environment. Both complexity theory and ANT deal with connected assemblages, that is, networks of interconnected nodes (Cilliers 1998; Latour 2005). According to Kim and Kaplan (2006) and Kaghan and Bowker (2001), both theories address the unexpectedness of changes influenced by the local or situational factors. The networks are relatively stable but in no way frozen in time or space. Complexity theory does not assume an end-point that can be reached. Also in ANT, the process of translation is never-ending and the multiple and complex "ordering" is never finished (Moser and Law 2006). Both theories acknowledge, however, that the networks or the actions of networks' members can be constrained by previous choices and that, in a sense, networks are constrained by a kind of path-dependency (Latour 2005; Merali 2004).

Network has been a prevailing metaphor for studies emphasizing connectedness in, for example, information science, organization science, and sociology (Castells 2000; Cilliers 1998; Latour 1999, 2005; Merali 2004). A network is dynamic, and has flexibility and adaptability to survive. In a network metaphor, interconnectivity has been described as negotiable, as voluntary or open-ended, or even as unpredictable. As such, the metaphor has fitted well to describe contemporary organizations and the changes in working life. In research, it means recognition of fragmentation and complexity (Knox et al. 2006). Still, the network metaphor has been criticized for lacking clear definitions, or for having multiple meanings (Cohen 1999; Doolin and Lowe 2002; Kaghan and Bowker 2001; Latour 2005). There is no agreement on the kinds of nodes and relations that can constitute a network. For example, power relations can be left undefined or even neglected when using the network metaphor. Moreover, Kaghan and Bowker (2001) criticize rationalist approaches in network theories as having tendencies of determinism, for example, when professionals or managers are portrayed as the "brains" that lead and regulate a change process. In this study, the actors in the intensive care unit are considered as "nodes" of their working network, and no single node is approached as the sole controller of complex interaction.

To explore the intertwined work practices and the use of technologies in intensive care, we adopt the conceptualization of both social and technical actors from ANT, as well as the process of translation that describes the co-construction or reforming of work practices and the technologies. From complexity theory, we adopt the definition of action as dynamic and nonlinear and as self-ordering or reflexive by nature. This includes the actor's adaptability to a changing situation or context while acknowledging also the path dependency on historical development and the bounded rationality of local action. Both ANT and complexity theory consider interaction and knowledge creation as emergent. The conceptual tools used in our study are discussed at length in Vuokko and Karsten (2007) and are summarized in Table 1.

3 RESEARCH METHODS

This study is part of a multidisciplinary research project, Louhi,[1] that aims to explore and build intelligent tools for natural language processing of medical records to support

---

[1]See http://www.med.utu.fi/hoitotiede/tutkimus/tutkimusprojektit/louhi/.

Table 1. Conceptual Tools for the Study

	Concepts for Studying Intensive Care Work
Actor-Network Theory	• Actor, actant (social and technical) • Network of shared interests • Transformation processes (translation, black-boxing) • Emergent interaction, emergent knowledge
Complexity Theory	• Action and interaction are dynamic and nonlinear • Self-organizing or reflexivity action • Adaptation to changing environment or situation • Path dependency on historical development • Bounded rationality of local action • Emergent interaction, emergent knowledge

nurses in an ICU. This study will form a background for evaluating the tools. We observed the nurses' work practices in a pediatric ICU in a university hospital, one of the five major teaching hospitals in Finland, with 953 beds. The pediatric ICU is one of the smallest wards, with readiness for 10 patients.

Two researchers (the first author and a research assistant) observed the daily life in the unit for 10 hours during April and May 2007. The observations concentrate on what kinds of actions took place in the unit and how various technologies participated in these (see Table 2). We carried out the observations in an environment that was new to us, and at first, it was difficult to "see" nurses' work practices. Also, we could not observe strictly according to our plans, as the situations at the ICU changed from day to day, and as we did not want to influence or hinder the work in any way. The parents of the children were usually present, and respecting their privacy led us not to do bedside observations when the patients were awake. Observing treatments to the patients, we admit, was beyond our courage.

Table 2. Planned Outline for Observations

Date Observer Others present
What is happening? What kind of action is visible? What consequences does the action have?
Who acts (roles)? Who participates?
What devices are used? How is technology used? How is technology a part of the action?
What are the contextual factors? What are the situational factors?
Special notes to consider in the future

The observation notes were systematically written down and, later, the field diaries were organized and shared by all researchers. In addition to observations, we collected a number of different forms. The nurses completed forms for a fictional patient to illustrate various details of their work practices. We also chatted with the nurses in their common room where they also held their reporting sessions and meetings.

The analysis of the data was interpretive. With the theoretical lenses provided (see Table 1), both observers coded and re-coded the data. In practice, this meant comparing our field notes, first, to find similarities in observations. We mapped out various actants and relations between them to describe relations between various staff members in and outside the ICU as well as interaction with machines and information. Physically, the ICU ward is clearly bounded, but on the level of interaction, the boundaries of the ward are more porous. For example, the ward works in close collaboration with the Intensive Care Nursery for premature infants located nearby and the adult ICU, which occasionally sends its overflow patients to the pediatric ICU.

Second, we looked through our data for all instances of adaptation or reorganizing of action to changing work situations. We discussed the possibilities and limitations for self-organizing in the ICU context, where the work tasks are regulated by diverse laws, guidelines, and standardized practices. One of our codes was exceptions: How are such situations handled in the ICU? To what extent are the nurses bound by the situation, the information, or other means they have at hand? To what extent can work in an ICU be planned in advance? In relation to this, we were interested in the nurses' means for simplifying the complex and the unpredictable.

The validity of analysis was upheld with both observers sharing similar experiences. The findings were also in line with our previous studies. We wrote a description of our observations in Finnish (Vuokko et al. 2007) in order to gather feedback from the ICU and from the Louhi project group. This resulted in a number of clarifications in details.

The resulting picture has some limitations that were caused by the relatively short observation time, by the observations being a first time experience for the researchers, and by the nature of intensive care working. Intensive care work practices showed up as being partly invisible and based on the intuition or practical experience of the nurses.

4 WORK PRACTICES IN THE PEDIATRIC ICU

The patients in the ICU come from the pediatric outpatient clinic, the first aid station, as transfers from other wards or other hospitals, or after demanding operations. Children in transit are admitted as well, for example, from passenger ships or airplanes. Many children are admitted during the night and many children are sent to other wards or hospitals during the afternoon. The beds are occupied 87.4 percent of the time. The patients stay on the average three days, but there is much variance. During our observations, we encountered both extremes of the patient situation: one day, all beds were occupied, and on another day, only one patient remained. During 2006, the ward cared for 638 patients. Of these, 16.3 percent required only observation, 63.1 percent required intensive care, and 20.6 percent demanded special intensive care.

Besides the pediatric nurses specialized in intensive care, the main actors we could identify included various kinds of medical specialists, especially in pediatrics, surgery, hematology, and anesthesiology. A key person was the ward secretary, acting as a broker for information. The secretary works 8:00 a.m. to 4:00 p.m. on weekdays. The unit has its own technician and a set of reserve machines in case one breaks down. Other important actors include the laboratory and imaging unit personnel, the cleaners, and the social workers in the pediatric clinic. The medical specialists have, in addition to the twice-daily rounds, many other duties in the teaching hospital and, therefore, they are seldom seen in the ward. There appeared also to be a number of nursing students present in the unit on a regular basis. Moreover, the child patients very often had at least one parent present.

Nonhuman actants in intensive care include, for example, respirators, ventilators, oxygen saturation meters, blood gas meters (Astrup for measuring the blood pH), and other technologies as well as information in, for example, patient records and hospital guidelines for specific procedures. These actants are immobile by nature, and there seems to be a tightly ordered positioning of supporting technologies and patient information within the pediatric ICU. For example, we witnessed a situation where a patient's daily nursing documentation was misplaced and no recording could be done before the documentation was found in the medicine room after some intense searching.

Physical environment. The pediatric ICU is physically a small unit with the patients' beds on two sides of the operations center: the youngest patients on one side (four beds) and older children on the other (four beds), with two separate rooms for isolation (two beds). The main part of the ward is a shared, open space with a multitude of technical equipment that all seem to give sound alarms. The open space was preferred by nurses as then they would have all the heavily medicated child-patients under their constant gaze. Moreover, the open space helps the nurses to communicate with each other. On the other hand, the open space in the unit created an atmosphere of restlessness and anxiousness, with the nurses constantly moving about, carrying and pushing around technical equipment, or the monitors for patients' vital signs raising alarms among all the other humming and bleeping devices. In contrast, we have found that the two ICUs for adults we visited tend to be quiet and peaceful, with the patients unconscious or sedated.

We spent much time in the operations center of the ICU. The operations center is the main station for documenting the nursing work, for sharing information about the patients, and for communicating with other facilities. On occasion, only the ward secretary is there, sitting at her computer, taking care of various administrative tasks such as attending to the telephone and the door bell, transcribing doctors' dictation, or ordering laboratory tests. There are two computers used by all; doctors seem more often to use the computer on the left and nurses read the lab results as they appear on the screen of the other. There are also monitors, phones and faxes, empty forms and folders, and maps full of hospital regulations. The patient folders are kept in a cart by the computers or, if work needs to be done on them, open on the table. Electronic mail is used to share information, for example, between units and work shifts. Orders for laboratory tests are done with one program, and for an x-ray with another one. Several other programs are in use, as well. Due to the high variety of child patients, the program used in adult wards for ordering food is not used here, but orders are written in a notebook. Thus the information environment (Lamb et al. 2003) is highly varied but centrally located.

Three shifts. The ICU has 31 nurses who work in three shifts. In the morning shift (7:00 a.m. – 2:00 p.m.), there are six to eight nurses working; in the evening shift (2:00 p.m. – 10:00 p.m.), there are five; and during the night shift (10:00 p.m. – 7:00 a.m.), there are four nurses. When one shift goes out and the other comes in, the outgoing nurses write their reports and the incoming shift holds a reporting session in the common room. This is carried out in this way to help all nurses in a shift know the basic information about all patients. To coordinate between shifts, there are also other practices that have been developed in the unit. For example, in the common room, there is a copy of the planned working hours on the notice board, and a notebook on the table. These are used, for example, to inform which nurses are coming to which shift, who is responsible for which patients and tasks, and who heads the shift. The ICU can share nurses and equipment with the nearby ward for premature infants, adding to flexibility in crisis situations.

The actual nursing work is structured according to several contextual and situational factors. The patients are assigned to nurses according to the nurse's experience and according to what kind nursing the patient needs. The parents help with immediate care and comfort, but still the work appears to be more hands-on than at the adult ICU. When there are more patients or when the patients are demanding, most of the nurses stay by the bedside of their patients and the operations center and the common room are deserted. There are also exceptions to the "one patient, one nurse" rule. Some patients, such as severe burn cases, may require two to three nurses. For a period, one nurse can care for several patients when their nurses are preparing medications in the "medicine room." Thus the nurses appear to take the role that is needed: nurse, secretary, specialist, teacher, laboratory assistant, or confidante.

To an observer, sometimes work in intensive care appears intense and requiring rapid action. At other times, very little seems to happen. For example, when there are few patients or when most patients require only light nursing, the nurses can immerse themselves in documenting, in finding more information, in planning the nursing work, or in other organizational supporting tasks such as ordering of supplies. While analyzing the observation data, we wondered how much of the actual work practices are invisible or hidden in the routines, self-evident to specialized nurses and impenetrable to us lay observers (cf. police work, Van Maanen 1988).

A notable practice we observed several times was that when something unexpected would come up, the nurses would together construct a suitable solution for the situation: "We need to cope no matter what happens." Often this was situational assistance for a colleague with a demanding patient (or a parent). Coping was extended to cover problems with technologies as well. Another typical feature in pediatric intensive care was the practice of double checking where, for example, a nurse would ask a colleague for reassurance that a medicine dose was correctly calculated. A third one was that some actions were not only recorded in writing but also said aloud: "I have now ordered the new tubes." The nurses spoke with each other even when attending to the patients. We even witnessed some questions being shouted to overcome the noise.

The nursing documenting practices were the most visible part of the activities. Much of the nursing tasks by the bedside might look self-evident or be unnoticeable. Without documentation, they would leave no trace and evaluating the care would be difficult. This is similar to other caring tasks: you only notice them when they are not done.

Doctors' rounds. The medical specialists do two daily rounds or inspections of the patients, one between 9:00 and 10:00 a.m. and the other one when the laboratory results have arrived, around 2:00 or 3:00 p.m. The operations center is full of people when the pediatricians, surgeons, haematologists, and anesthesiologists come to start rounds. The rounds structure the work into preparations, the actual visits at bedside, the planning of care in a separate office, doctors' dictations, and ordering of, for example, laboratory tests, made by the nurses or the ward secretary. Each nurse appears to participate in the rounds only when her "own" child is discussed, but the head nurse attends the whole time. If the ward secretary is available, she sits in on the planning sessions and does immediate recording and ordering.

Between the rounds, the nurses attend to updating patient records, which the physicians then use to get an overview of a patient's current status. In this sense, the physicians' inspections regulate the minimum intervals when nurses update the nursing documentation. In addition to the rounds, the repetitive tests, such as blood tests, regulate or order any working day. Data and information are constantly updated from various sources. The nurses need to extract the needed information and, with personal experience and previous information, process it to knowledge in use.

Documenting care is very seldom done by the bedside. The assisting nurse might write down on a piece of paper some items, but the nurse assigned to the patient does the actual recording in the operations center. The daily nursing records are penned on large paper sheets officially called control forms but in practice called just "the sheet." The vital signs, the fluids in and out, the measures taken, the medications, and the like are recorded during the day, starting at 6:00 in the morning. Several different methods of emphasis (colors, circles, triangles, arrows, and the like) are used, each with a specific meaning. The data points drawn will gradually turn into a graph, easy to interpret for a professional.

The data on a control form is used to get an overall view of a patient and even, on occasion, to predict trends of the patient's possible changes of status. For the nurses coming to the next shift, the data collected on the sheet provides quick and easy access to the current status of a patient, an overview (Robinson 1992). The control forms are not only for recording activities and test results but they also provide an important means for information exchange between nurses, doctors, and ward secretaries. Thus the control form is an example of black boxing when existing technology and routines are tied together in a way that, to us, seemed difficult to reopen or retranslate.

With the paper documents, it is also easy to place the documents side by side for comparison. This also has disadvantages, as small slips of paper, for example from the ultrasound or the Astrup machine, can easily float to the floor, and documents cover other documents. As physical pieces of paper of many sizes, they make an untidy bundle inside the folder.

Within the ICU, daily work takes place in a complex and interlaced network of interaction. Human actors are readily observable, and they can also explain the logic behind their actions. Still, other types of actants cause various actions and adaptations in ICU nursing work. These include the immaterial agency of information in forms, tables, notes, and other documents, and the technological actants, such as media for monitoring and test taking.

Preparing devices for a new patient. In the ICU, the pursuit of simplification can be observed in routine uses of technology. For example, when the nurses prepare to take a new patient into the unit, they start the preparations by collecting needed equipment and vitality monitoring devices by the bed for the new patient. The nurses know how to secure and attach the devices to the correct sensors and how to initialize the monitors according to the age of the patient. Only the breathing machine and the mechanical ventilator require adjustment by a physician; the nurses take care of the other machines.

In addition to patient monitoring, the nurses prepare and monitor the IV fluids. Other preparations are possible if the transfer information sheet implies such a need. Adjusting various machines is part of the routine, and the nurses are at most concerned with the right calculations but not on the functions of these machines. When more understanding about their inner function is needed, the nurses call for the support staff in charge of the technology.

Alarms. The vitality monitoring technology is so embedded in the routines that the nurses claim to be able to identify not only which device gives the alarm, but also why it alarms. In reality, as one device has one alarm sound, a nurse interprets from other situational features why the alarm goes off. For example, the nurse might have been already waiting for the alarm to sound so that the IV bag could be changed for a new one. On the other hand, when a patient is especially restless, such as a typical meningitis case, the nurse knows that the monitors give more alarms than needed. If the nurse becomes uncertain regarding the vitality monitors, a colleague is readily consulted and a solution is formed together. An instance of slight uncertainty occurred, for example, when a nurse came to work and started to prepare for a new patient. As one of the ducts in a vitality monitor had been replaced with a duct of a different color, she became uncertain of how the device should be connected to the sensors. The problem was quickly solved by asking the other nurses for information.

Transfer of a patient. As an example of the more complicated routines, we present here the transfer of a patient to another hospital as this was very visible in the total infrastructure with all the activities involved. The different work practices of the other hospital units would push in to the ICU, and a need to address and explicate carefully the details of the transfer situation emerged from these interactions. The transferring practices are coordinated by the transfer information sheet with a collection of patient information. The information is collected from various paper-based and electronic sources, and the sheet is filled in by hand. When the transfer time approaches, the last control tests are taken, and the patient is washed and otherwise prepared. The nursing documentation sheets concerning the patient are updated, and the patient's folder is checked when the physician's final dictation is attached to other documents. The patient's original diagnosis, all operations and major steps concerning the patient, detailed information of the patient's medication and the like are written on the transfer sheet. When a patient is transferred to another hospital, the nurse also fills in a form for travel expenses to be reimbursed for parents following the patient.

When the patient and all the needed documentation as well as the escorting nurse are ready, the patient is detached from the vitality monitors and the physical transfer takes place. The patient document folder remains in the ICU and only the transfer forms leave the unit at that time. When the escorting nurse comes back, she or he checks the patient as having moved out of the unit. The ward secretary makes sure that the status of the patient is changed also in the electronic hospital administration system. Even though the procedure for transfer is well described, the actual situation that occurred during the observations seemed hectic and almost chaotic, as two nurses were needed to prepare the patient. The child's own nurse, together with the ward secretary and helped by the head nurse, finalized the patient documentation and filled in the travel form for the accompanying parents. The preparations for transfer made small measures as well as nursing documenting practices visible. On top of that, the situation revealed how, in problematic situations, patient security and professional ethics guide what is done. The work practices in ICU were about sharing not only information but also responsibility for the patients.

5 DISCUSSION

In any action, when there is a multitude of actors and units of actors, the action itself is bound to add to the overall complexity of the situation. For example, in this paper we simplified the situation where work practices are carried out by listing only a few components of work practices, with actors, technologies, and information as influencing the situation. Changes in chosen areas influence other areas as well. We were interested in how new technologies might affect the work practices and division of actors or information. Such changes can be initiated also by minor issues such as varied interpretations of information, lack of technical support, or inflexibility of work practices. In the ICU, these emerging issues and the complexity of organizational action were anticipated with situational adaptability and self-ordering of action within the unit.

During the observations, we noted that working in the intensive care requires taking dynamic action as the work consists of adaptations to unexpected situations. There seemed to be constant changes as a variety of patients were transferred in and out of the unit. To an extent, the nurses were able to predict these changes by relying on their experience. Although the nurses are capable of adapting to rapidly changing situations, at the same time, the nursing practices are bounded by the situational rationalities, such as the information available on the patient and the time of day. In addition, selecting a specific technology meant emergence of path dependent features such as integration problems. The experience levels and organizational roles also set boundaries to decision making and action taking. For example, a nurse can be responsible for medicine orders, but the unit pediatrician is the one responsible for "ordering" or negotiating the patient transfers. Not only work practices are defined by organizational roles; documenting practices also are based on the task divisions.

In the pediatric ICU, transforming work practices and reinterpreting technology are anticipated to take place when, for example, the current work practices no longer respond to the technical tools in use. The need to change work arrangements can be a short-term situation, as in cases when some of the staff members of the pediatric ICU are called away from the unit to take care of a CPR alarm. A more stable change is needed when, for example, new features such as new information or new tools are introduced into the working environment. A new technology is embedded and fitted into the existing practices and routines by interpreting its beneficial usage in the given context. Through the adaptation to changing situations, not only is the technology being taken into use but at the same time, new work practices as well as new contextual knowledge may emerge.

The theoretical tools given by ANT appear to be well suited to describe and analyze a small, closed entity such as a hospital ward. Many other ANT studies focus on small entities, as well (Aanestad 2003; Latour 1991, 1999; Law and Callon 1995; Moser and Law 2006). Taking into consideration the whole hospital or the hospital district is more difficult. With this bracketing, it might be difficult to see whose interests are pulling a translation. New technologies might be introduced because of reasons not visible on the level of work practices. Even when use is mandated, enrollment is necessary for reducing problems with the change.

Also, tools given by complexity theory are well suited to study a small unit as they focus on networking systems with a shared set of practices, mediating information, and expertise. Merali (2004) argues that as complexity theory emphasizes dynamisms of interaction and is well suited for studying contemporary organizations. However, we noted some difficulties in implementing complexity theory concepts (see Anderson 1999; Desai 2005; Jacucci et al. 2006). First, some of the conceptualizations originated in positivist research in laboratory environments, and translating them for use in qualitative studies of social systems seems to require compromises. Second, complexity theory implies macro studies (Merali 2004) as it considers the environment of the system a crucial factor of development. Complexity theory emphasizes the actions of the agents within a network to the point where, at times, these agents become "heroes" in the survival of the fittest. Third, complexity theory is partly based in language studies (Cilliers 1998) and technologies emerge to the picture only through the meanings given by the organizational members.

When compared to ANT, complexity theory assumes that the networked relations imply power relations. Complexity theory implies that all organizations and systems are hierarchal, and that power issues are embodied in organizational relations (Merali 2004). Still, the organizational members have possibilities for reflective action and continued adaptation through feedback loops. In the ICU environment, such feedback is made possible in various meetings and gatherings as well as in formal documenting and evaluation of work. Through feedback and reflection among the organizational actors, possible changes in work practices.

Complexity theory also attempts to describe how local diversity emerges through various interpretations and reflective actions, whereas in ANT, the network attempts to achieve a shared translation. Room for possible variations was later given with interpretive flexibility. In complexity theory, the time line of the change or the study description is not always apparent. There seems to be no clear vision of whether social complexity studies should concentrate on describing past, present, or future action—or to combine all of these. In the ANT studies, historical development is often clearly described or the change is portrayed in time.

While the idea of studying hospital informatics as a complex—and not only complicated—system is still appealing, the limitations of complexity theory as understood in organizational studies are still considerable. The combination with ANT worked well, and this might be a direction to take in future studies, especially as ANT addresses technologies in a more sophisticated way. Seeing the ward as a part of a larger whole would seem to presuppose a societal level theory, such as structuration theory (Jones and Karsten 2008).

6 FUTURE RESEARCH

Working in the pediatric ICU is expected to change when the planned organizational implementation of an electronic patient record system takes place during the winter of 2007–2008. One goal with the new system, as stated in the hospital board decision, is that the nurses could prioritize more of the work time spent at the patient's bedside. This would be possible by receiving data from various vital monitoring systems directly into the new patient record system and by placing computers at the foot of every bed, like they are in the adult ICU.

In the future, it is anticipated, nursing documenting will demand less effort, and the nurses will be able to better concentrate on the practical treatments at the bedsides. The quality of the electronic patient records is expected to be improved with more accuracy and better readability (Bowles 1997). However, the quality might also deteriorate with the practice of cut and paste (e.g., Thielke et al. 2004) or with blind belief in the numbers given on the screen. There is still a long way to go for before electronic patient records can be shown to support decision-making in critical care (Bucknall and Thomas 1997).

To sum up, in the pediatric ICU we observed, we see several opportunities for change in time–place arrangements, in coordination and communication practices, and in information sharing practices. Documenting the care plans and the care given will change radically when direct connections between the monitors, the laboratory and imaging results, and the electronic patient record are established. The roles and tasks of the various actants will reformulate, and thereby possibly also their skills and professional identities.

Acknowledgments

This work was carried out within TEKES project *Louhi: Mining the Text in Patient Documentation* (grants 40435/05 and 40020/07). We thank the hospital for access and our research assistants Pekka Tetri and Joonas Peltola for support in the empirical data gathering.

References

Aanestad, M. 2003. "The Camera as an Actor: Design-in-Use of Telemedicine Infrastructure in Surgery," *Computer Supported Cooperative Work* (12:1), pp. 1-20.

Anderson, P. 1999. "Complexity Theory and Organization Science," *Organization Science* (10:3), pp. 216-232.

Barnes, B. 2001. "Practice as Collective Action," in *The Practice Turn in Contemporary Theory*, T. R. Schatzki, K. Knorr Cetina, and E. von Savigny (eds.), London: Routledge, pp. 17-27.

Berg, M. 2001. "Implementing Information Systems in Health Care Organizations: Myths and Challenges," *International Journal of Medical Informatics* (64:2/3), pp. 143-156.

Bijker, W. E. 1995. *Of Bicycles, Bakelite, and Bulbs: Toward a Theory of Socio-Technical Change*, Cambridge, MA: The MIT Press.

Bowles, K. 1997. "The Barriers and Benefits of Nursing Information Systems," *Computers in Nursing* (15:4), pp. 191-196.

Bucknall, T., and Thomas, S. 1997. "Nurses' Reflections on Problems Associated with Decision-Making in Critical Care Setting," *Journal of Advanced Nursing* (25:2), pp. 229-237.

Castells, M. 2000. "Toward a Sociology of the Network Society," *Contemporary Sociology* (29:5), pp. 693-699.

Ciborra, C. U. 2002. *The Labyrinths of Information: Challenging the Wisdom of System*, Oxford, UK: Oxford University Press.

Cilliers, P. 1998. *Complexity and Postmodernism: Understanding Complex Systems*, London: Routledge.

Cohen, M. 1999. "Commentary on the Organization Science Special Issue on Complexity," *Organization Science* (10:3), pp. 373-376.

Desai, A. 2005. "Adaptive Complex Enterprises," *Communications of the ACM* (48:5), pp. 33-35.

Doolin, B., and Lowe, A. 2002. "To Reveal Is to Critique: Actor-Network Theory and Critical Information Systems Research," *Journal of Information Technology* (17:2), pp. 69-78.

Effken, J. A. 2001. "Informational Basis for Expert Intuition," *Journal of Advanced Nursing* (34:2), pp. 246-255.

Haythornthwaite, C. 2006. "Articulating Divides in Distributed Knowledge Practice," *Information, Communication and Society* (9:6), pp. 761-780.

Howcroft, D., Mitev, N., and Wilson, M. 2004. "What We May Learn from the Social Shaping of Technology Approach," in *Social Theory and Philosophy for Information Systems*, J. Mingers and L. Willcocks (eds.), Chichester, UK: John Wiley & Sons, Ltd., pp. 329-371.

Jacucci, E., Hanseth, O., and Lyytinen, K. 2006. "Introduction: Taking Complexity Seriously in IS Research," *Information Technology & People* (19:1), pp. 5-11.

Jones, M., and Karsten, H. 2008. "Review: Giddens's Structuration Theory and Information Systems Research," *MIS Quarterly* (32:1), pp. 127-157 (plus Appendix A, available at http://www.misq.org/archivist/vol/no32/issue1/JonesAppendix.pdf).

Kaghan, W. N., and Bowker, G. C. 2001. "Out of Machine Age? Complexity, Sociotechnical Systems and Actor-Network Theory," *Journal of Engineering and Technology Management* (18:3/4), pp. 253-269.

Kim, R. M., and Kaplan, S. M. 2006. "Interpreting Socio-Technical Co-Evolution: Applying Complex Adaptive Systems to IS Engagement," *Information Technology & People* (19:1), pp. 35-54.

Knox, H., Savage, M., and Harvey, P. 2006. "Social Networks and the Study of Relations: Networks as Method, Metaphor and Form," *Economy and Society* (35:1), pp. 113-140.

Lamb, R., King, J. L., and Kling, R. 2003. "Informational Environments: Organizational Contexts of Online Information Use," *Journal of the American Society for Information Science and Technology* (54:2), pp. 97-114.

Latour, B. 1991. "Technology Is Society Made Durable," in *A Sociology of Monsters: Essays on Power, Technology and Domination*, J. Law (ed.), London: Routledge, pp. 103-131.

Latour, B. 1999. *Pandora's Hope: Essays on the Reality of Science Studies*, Cambridge, MA: Harvard University Press.

Latour, B. 2005. *Reassembling the Social. An Introduction to Actor-Network Theory*, Oxford, UK: Oxford University Press.

Lauri, S., and Salanterä, S. 1998. "Decision-Making Models in Different Fields of Nursing," *Research in Nursing and Health* (21:5), pp. 443-452.

Law, J., and Callon, M. 1995. "Engineering and Sociology in a Military Aircraft Project: A Network Analysis of Technological Change," in *Ecologies of Knowledge: Work and Politics in Science and Technology*, S. L. Star (ed.), Albany, NY: SUNY Press, pp. 281-301.

Merali, Y. 2004. "Complexity and Information Systems," in *Social Theory and Philosophy for Information Systems*, J. Mingers and L. Willcocks (eds.), Chichester, UK: John Wiley & Sons, Ltd., pp. 407-446.

Moser, I., and Law, J. 2006. "Fluids or Flows? Information and Qualculation in Medical Practice," *Information Technology & People* (19:1), pp. 55-73.

Orlikowski, W. J. 2002. "Knowing in Practice: Enacting a Collective Capability in Distributed Organizing," *Organization Science* (13:3), pp. 249-273.

Robinson, M. 1992. "Computer Supported Cooperative Work: Cases and Concepts," in *Readings in Groupware and Computer Supported Cooperative Work: Assisting Human–Human Collaboration*, R. Baecker (ed.), San Mateo, CA: Morgan Kauffmann, pp. 29-49.

Salanterä, S., Eriksson, E., Junnola, T., Salminen, E. K., and Lauri, S. 2003. "Clinical Judgement and Information Seeking by Nurses and Physicians Working with Cancer Patients," *Psycho-Oncology* (12:3), pp. 280-290.

Schneider, K., and Wagner, I. 1993. "Constructing the 'Dossier Representatif': Computer-Based Information Sharing in French Hospitals," *Computer Supported Cooperative Work* (1:4), pp. 229-254.

Star, S. L., and Ruhleder, K. 1996. "Steps Towards an Ecology of Infrastructure: Design and Access for Large Scale Information Spaces," *Information Systems Research* (7:1), pp. 111-134.

Thielke, S., Hammond, K., and Helbig, S. 2007. "Copying and Pasting of Examinations Within the Electronic Medical Record," *International Journal of Medical Informatics* (76:Supplement 1), pp. S122-S128.

Van Maanen, J. 1988. *Tales of the Field: On Writing Ethnography*, Chicago: University of Chicago Press.

Vuokko, R., and Karsten, H. 2007. "Working with Technology in Complex Networks of Interaction," in *Organizational Dynamics of Technology-Based Innovation: Diversifying the Research Agenda*, T. McMaster, D. Wastell, E. Ferneley and J. I. DeGross (eds.), Boston: Springer, pp. 11-22.

Vuokko, R., Tetri, P., Peltola, J., and Karsten, H. 2007. *Hoitajien työkäytänteet lasten teho-osastolla*, National Report 13, Turku, Finland: Turku Centre for Computer Science (in Finnish).

Wiener, C., Strauss, A., Fagerhaugh , S., and Suczek, B. 1979. "Trajectories, Biographies and the Evolving Medical Technology Scene: Labor and Delivery and the Intensive Care Nursery," *Sociology of Health and Illness* (1:3), pp. 261-283.

About the Authors

Riikka Vuokko is a research fellow in social and health informatics at the Åbo Akademi University in Finland and the Zeta Emerging Technologies Laboratory at Turku Centre for Computer Science (TUCS). With a Master's in Ethnology, Riikka is a trained ethnographer. Currently she is working on theories of practice, power, and identity, putting the finishing touches on her dissertation. Her publications have appeared in outlets including the conference proceedings of Hawaii International Conference on System Sciences, the America's Conference on Information Systems, the European Conference on Information Systems, Ethicomp, and IFIP Working Group 8.6.

Helena Karsten is a research director in Information Systems at the Åbo Akademi University in Finland and the head of the Zeta Emerging Technologies Laboratory at Turku Centre for Computer Science (TUCS). Her research interests include the interweaving of work and computers, the use of IT to support collaboration and communication, and social theories informing theorizing in information systems. She is an associate editor for *The Information Society* and an editorial board member for *Information Technology and People*. She can be reached at eija.karsten@abo.fi.

11 VIRTUALITY AND NON-VIRTUALITY IN REMOTE STOCK TRADING

Roger F. A. van Daalen Fuente
Mike W. Chiasson
Paul R. Devadoss
Lancaster University Management School
Lancaster, U.K.

Abstract *Advances in information technology allow for remote working, leading to suggestions that remote individuals can operate in virtual instead of face-to-face teams. This paper considers the continuation of face-to-face communication in a European group of stock traders, despite the capabilities of information technology to individuate the work. The case illustrates that traders prefer and need to work in face-to-face settings for various reasons. Short-term reasons arise from a need for instant and effortless communication in their manipulation of market prices and for instant knowledge sharing, leading to both higher individual and collective profits. Long-term reasons arise from a need for continuous learning by novices and experts, as stock markets and stock prices settle into behavioral patterns over longer periods of time. The implications for computing and work are discussed.*

Keywords Community of practice, remote working, electronic trading, knowledge work, information systems, virtuality

1 INTRODUCTION

Knowledge exchange appears to be a central issue in the so-called knowledge economy (e.g., Drucker 1993), and the growing importance of specialist knowledge has become of paramount importance (Blackler 1995). The implication is that specialists now own the means of production, resulting in human capital being of greater importance to the firm than any other form of capital (Starbuck 1992).

Please use the following format when citing this chapter:

Van Daalen Fuente, R. F. A., Chiasson, M. W., and Devadoss, P. R., 2008, in IFIP International Federation for Information Processing, Volume 267, Information Technology in the Service Economy: Challenges and Possibilities for the 21st Century, eds. Barrett, M., Davidson, E., Middleton, C., and DeGross, J. (Boston: Springer), pp. 159-172.

At the same time, information technology has "compressed time and space," allowing for global access to scarce distributed expertise (Ives and Jarvenpaa 1991, p. 33). This allows knowledge workers to work together without regular face-to-face contact through advances in information technology. Given this, it has been repeatedly suggested that effective virtual communities could be formed (Mowshowitz 2002) without (regular) face-to-face communication.

An industry that should be strongly influenced by these developments is the financial services industry. Some have suggested that the prime reason for stock exchanges and the development of this financial system was to lower transaction costs in exchanging capital (North 1991). By moving toward electronic trading, for example with the introduction of the electronic communication networks (ECNs) on the NASDAQ in 1997, stock transaction costs were significantly lowered. The system also allowed for faster trade execution, and provided more complete price information to traders (McAndrews and Stefanidis 2000).

This change allowed for the globalization of trading on financial markets, as access to the markets is now possible from any location. Electronic trading was to replace paper trading, so that traders could do business from anywhere. Barrett and Walsham (1999) argued that this change to remote trading (where traders do not need to meet in person) could radically change the way in which traders establish, continue, and enhance their relationships with each other.

In this paper, we argue that while this form of knowledge work has been influenced by advances in IT, our empirical work illustrates how and why the work is still done in face-to-face groups. Our conclusion is that the face-to-face group continues to act as a socio-technical support system, leading to substantial learning benefits and coordinated action.

The paper is structured as follows. First, a literature review is provided on how the move to electronic trading in financial markets could individuate work. Then, benefits of working in groups in a face-to-face context are put forward. This is followed by the methodology. The case study then explores how and why a case of European traders on the New York stock exchanges continues to work in face-to-face groups. We argue that the fusion of technical and social contact leads to coordinated trading and influence, and to member learning, which is made possible by the immediacy of direct communication. The implications for theory and practice are discussed in the final section.

2 ELECTRONIC TRADING: DETACHED, INDIVIDUALIZED, AND VIRTUAL?

The change from paper trading to electronic trading on financial markets is often viewed as an opportunity to decrease the need for human contact in the financial industry, which affects the cost and benefits of market reach. This can lead to changes in the nature and location of contact between individual traders. In the extreme case, trading work can become individualized, detached, and remote.

Barrett (1999) and Barrett and Walsham (1999) have shown how the digital transformations of trading work are often resisted by the traders themselves, for various human and social reasons. A member of the establishment commented that "[traders] feel they have to see the whites of their eyes and to see if their hands are trembling [in business transactions]" (Barrett and Walsham 1999, p. 13), signaling a continuing interest in maintaining rich and intense communications when trading by traders. This is contrasted with key players' views that individualization is possible and that traders' contacts can be severely limited. In summary, stock market leaders believed that "electronic trading support enables remote trading to develop with, perhaps, only occasional need for face-to-face interactions to establish or re-establish business relationships" (Barrett and Walsham 1999, p. 20).

Consistent with these possibilities, Knorr Cetina (2002a; 2002b; Knorr Cetina and Bruegger 2000, 2001, 2002) views the change in stock trading work by computer systems as transforming the work from being embodied by a dispersed network of trading partners into a "postsocial" world, wherein humans and objects have changed to relate in new ways. While traders sit next to each other on trading sites, she feels that the computerization of traders' work increasingly disengages them from the local setting. Her view also represents a belief that detachment and individualization is increasingly possible with technological advances in financial work.

Other research (Barrett and Scott 2000, 2004; Scott and Barrett 2005) indicates that the new work environment with IT will lead to a need for different skills. In a physical trading pit, physical cues lead to an embodied feeling for price movements, whereas in the virtual trading pit, intellectual skills will dominate. The informants in that research argue that this will lead to more of a calculative and individualized work environment.

In contrast, Millo et al. (2005) oppose the view that a move from buildings and paper to electronic systems will create a detachment in trading. They argue that the introduction of electronic markets transforms where face-to-face contact takes place, but does not eliminate it. IT does redistribute trading work, but has not lead to isolated individuals working in remote locations. This appears to contrast a common perception that financial markets will be virtualized and individuated by computerized technology.

Thus, while the detached, virtualized, and individualized nature of financial work is possible through the introduction of IT, perhaps motivated by the cost and benefits of globalized work, others argue that it will not and cannot completely do so. To address theoretically why, we turn to two areas of literature on knowledge exchange—dispersed team work and communities-of-practice—that consider the formation and modes of communication in effective groups. This will provide us with a series of theoretical viewpoints to analyze our trading case.

3 KNOWLEDGE EXCHANGE

Sapsed and Salter (2004), in reviewing the dispersed team work literature, suggest that knowledge is ordinarily described as locally embedded and difficult to transfer over distances. They also suggest that face-to-face interaction is critical in facilitating the transfer of complex knowledge, and for building trust, commitment, and social capital among participants. They argue that often the absence of face-to-face interaction produces distrust among distant partners, thereby inhibiting the sharing of knowledge.

In general, spatial proximity is considered to enhance organizational communication as it permits intense and ongoing face-to-face interactions. The reason for this is that it is the "richest" form of interaction, despite various (electronic) forms of communication such as instant messaging, groupware, videoconferencing, etc. (Daft and Lengel 1986). While it has also been found that experiences through other communication media can also be rich (Carlson and Zmud 1999), most studies find that colocation leads to better knowledge sharing and overall performance than dispersion (Kiesler and Cummings 2002).

As a result, while time and distance can be overcome to virtually connect distant coworkers to each other's knowledge (Finholt et al. 2002) and has been found to allow for collaboration across unprecedented geographic distances (Walsh and Maloney 2002), spatial proximity and face-to-face contact is, at least on occasion, important for the transfer of complex knowledge and the building of social capital (Sapsed and Salter 2004). It thus appears that more is at stake in than just the *ability* to communicate—being the development of trust, commitment, and social capital is important to create a *desire* for exchanging knowledge.

Building on this finding, community-of-practice theory deals specifically with how knowledge is shared and distributed in a work context. Communities-of-practice (COP) are groups of people who share a concern, a set of problems, or a passion about a topic, and who deepen their knowledge and expertise in an area by interacting on an ongoing basis (Wenger 1998; Wenger et al. 2002). Knowledge in COP theory is accumulated at the worksite in a situated sense, and is the result of group processes, and mutual engagement in a common action or idea, which leads to mutual accountability among participants, and a shared repertoire of the "way things are done." Furthermore, Brown and Duguid (1991) built on Orr's (1987, 1990) work to explain how knowledge pertains and is transferred in informal relationships, through shared insights and narratives.

Knowledge in a community of practice is accumulated at the work site during situated work, and is the result of group processes (Lave and Wenger 1991). Community-of-practice theory has been said to have originated in a wider tradition of learning, education, and cognitive theories (Fox 2000). Specifically, it has been addressed as a specific version of social learning theory, wherein individual members learn in the workplace (are situated) by participating in shared activity. Knowledge and practice as inherently intertwined was put forward to challenge the then-prevalent view of learning constituted as students "receiving" knowledge in a classroom, and allowing them to exercise that knowledge later on (Lave and Wenger 1991).

Others using this theory have considered enterprises or organizations as constituting a multitude of such communities (Fox 1997a, 1997b). Following this line of reasoning, individual members communally learn by participating in a shared activity in a particular place. In this view, knowledge is transferred within a group by participants, tying learning to ongoing activities in practice.

In short, community-of-practice theory argues that knowledge and practice are intertwined. In this view, learning is acquired through performance. When this is done within a group sharing common interests, knowledge is spread through various means, including socialization. This allows for new entrants to learn quickly about an area of work, through an embodied apprenticeship with others. Both novices and the more experienced members of the community benefit from each others' presence as new insights are shared.

This communal conception of learning in a work context is, in some ways, at odds with the belief that financial work can be performed individually through digital systems. This leads to a key question: Why do communities-of-practice continue to appear when work could be completely individuated?

Our findings show that face-to-face work still occurs in particular trading communities and for particular reasons, in order to build what the electronic systems cannot deliver. We explore this through the description of a daytrader community in Europe, which trades on various New York stock exchanges.

4 METHODOLOGY

Data was gathered during the second half of 2007, as an interpretive (Orlikowski and Baroudi 1991; Walsham 1993, 1995, 2006) in-depth case study (Yin 2003) in and around the work site at TradeCo (a pseudonym) in a major European city. Data collection was focused on the activities and learning of expert and novice daytraders within the community.

The objective was to perceive the understanding of social situations from the viewpoint of participants in a daytrader community at TradeCo. To explore social phenomena within groups and to interpret the meaning attributed to actions by those groups, data collection consisted of semi-structured interviews, participant observation, and secondary data using internal reports of the firm to broaden possible interview topics.

Four different groups of individuals were analyzed in the case. The *owner* of the company, who is not directly involved in the company's day to day operations, is a gatekeeper to the financial trading capital. The three other groups perform a similar job, with slight variations. *Traders* trade for TradeCo on the stock markets and receive a portion of their gross profits as compensation. *Trainee* traders and *managers* generally do this same work, but trainees do not receive a salary (or any other kind of compensation) until they have proven themselves as good traders. They can do this by "graduating" from trainee to trader by earning $2,000 in one month. Managers, in return for a small portion of all the other traders' profits, are responsible for day-to-day operations. Both managers of this daytrader organization, 7 of the 12 traders, 2 trainees, and the owner were interviewed. These 12 semi-structured interviews varied in duration, ranging from 30 minutes up to 2 hours.

In the following, the case data will be contextualized by a short history of the development of the stock market. Following this, the findings are provided, illustrating the activities and methods of learning in this community of traders. Finally, the reasons for their colocated work in both the short- and long-term are discussed.

5 CASE DATA

Since the stock market's conception in 1602, people have made a living on the quick trading of stocks, based on price fluctuations. Some traders specialize in very short-term investments, so that each working day represents a cash profit or loss, using only daily holdings in shares. Making a living based on such short-term stock investments is believed to have first been described in the book *The Day Trader's Bible* (Wyckoff 1919).

With the introduction of electronic communication networks (ECNs), daytraders could work from any location, using computers linked to the respective stock markets. Extending a line of innovations in remote trading which began with the telegraph in the mid-1800s, ECNs are an evolutionary broadening of the instant and remote work possibilities of trading.

Many of the reasons for these innovations are related to the dynamic nature of the stock market, where information increasingly loses relevance quickly. Any new information must be acted upon quickly. With the development of the ECNs on the NASDAQ in 1997, fast trade execution was possible from any location in the world, and complete price information was made available to traders (McAndrews and Stefanidis 2000). This change has allowed for the globalization of trading on such markets, as access to financial markets has become possible from any location. The owner of TradeCo commented on trading from a distance:

> *You know, I've been in this business for almost a decade now. But even though my company has traded 1.5 trillion stocks on the NASDAQ and NYSE, I've only seen a physical stock market once, and that only happened because I took a wrong turn.*

In 2004, TradeCo was established, thereby allowing daytraders to work remotely from a location in Europe, operating on the NASDAQ and NYSE. TradeCo is a branch of a global daytrader organization that was founded in 1997.

The daytraders at TradeCo are focused on minimal price differences (in terms of cents) to "shave" stock price differences between ECNs or expected minimal price fluctuations on a single ECN. This type of trading is only possible because of the minimal transaction costs involved in exchanging stocks on an ECN.

5.1 Work Characterization

Day trading work has been described as modern knowledge work (Royal and Althauser 2003), as daytraders are essentially investment analysts. The various computer screens used by a trader signal a voluminous amount of information, which is far more detailed and instantaneous than what the occasional investor sees. Traders make short-term investments in stocks, their contribution being an analysis—primarily based on experience and hunches—of the stock's price, and the buying and selling of investments resulting from quick price changes, sometimes two or three cents. The objective is to benefit from such price movements.

Quick and experienced interpretation of the information is crucial to good decisions. The speed and direction of a price fluctuation is only an estimate, and some estimates are better than others. Novices in this particular profession have a very hard time making any money at the outset. In the words of an early daytrader,

> Let anyone who thinks he can make money analyzing the stock market [attempt to trade in a simulated mode]....Put my name down as the opposing side of every trade and when done send me a cheque for what you have lost (Wyckoff 1919, p. 80).

On average, more experienced traders have higher payrolls, supposedly through tacit knowledge. This allows them to make correct predictions of where the stock price is headed more regularly.

In becoming a good trader, one must figure out both the technical terms being used and how to use the tools at the trader's disposal. Again in the words of an early day-trader,

> It seems to us, based on our experience, that Tape Reading is the defined science of determining from the tape the immediate trend of prices. It is a method of forecasting, from what appears on the tape now in the moment, what is likely to appear in the immediate future. Tape Reading is rapid-fire common sense (Wyckoff 1919, p. 7).

Replace *tape* with *electronic price information* and *tape reading* with *the art of analyzing price information* and this definition is essentially the same today.

5.2 TradeCo as a Community of Practice

Day trading is a skill, which can only be learned by actually doing the job. Novices at TradeCo learn by employing daytrader tools in its specific setting, by mistake and correction, through continuous learning. This is far from an individual affair, as the aim and speed of learning by a trainee is often affected by their colocated partners.

For example, sitting near someone with a successful technique can facilitate an immediate transfer of both short and long-term knowledge, so that elements of that technique can be implemented by the trainee. The same holds for more experienced traders, as learning is a perpetual work in progress, and insights are shared quickly between face-to-face workers. As price and stock market behaviors take unexpected turns, learning is a continuous necessity. The situated nature of learning is noticeable for both novices and experts.

Members of the community gain legitimacy as they become experts, often seen and expressed by their income. Legitimization is primarily measured in a trader's salary, which also denotes his rank in the collective. Further to this, the number of screens a trader's desk sports indicates how much money he is able to earn. Trainees and ordinary traders start out with two computer screens. Once they earn $10,000, an extra screen will be added ceremoniously and yet another one when a trader has once earned $20,000 gross in a single month. When problems arise, the better trader will always be helped first by management. For example, software problems leading to ambiguity in the data stream from the ECNs ordinarily require a phone call to the ECNs to determine how many shares a trader owns or owes, in which case the biggest traders are assisted first.

Because of this, the community of daytraders at TradeCo can be seen as a hierarchical, quick moving, and learning community of practice. This is expressed by the situatedness of learning, and how legitimacy is expressed.

5.3 Short-Term Communal Learning Benefits

Within this community of practice, there are a number of short-term learning benefits from colocated work. First, colocation leads to the ability to take short breaks, and this enhances the scope of the market. This is possible because of the community's watchful eye and the quick communication through direct physical presence. Second, knowledge is also shared in terms of price behavior estimation from person to community. As they act as a community, opinions on price behavior "stack," so that knowledge becomes a community's opinion.

When little is happening on the stock market, and there are few opportunities for profits, traders resort to social chitchat with their close neighbors, discussing what the stocks might do. Some even go outside for a smoke, or for a glass of mineral water. This is always potentially dangerous, as the market can make a sudden move, so the few traders left "on watch" will shout a loud warning should anything unexpected happen. Traders will then run back to their stations, make a quick evaluation, and resume trading.

Such sharing is not just limited to observation of general trends, but also extends to a communal watching of events: As traders feel cognitively limited to monitor six or seven stock prices at the same time, and nobody can monitor the entire market at the same time, all daytraders specialize in a favorite stock to trade and a few others, which they will vary from day to day. When one daytrader picks up an interesting movement on a stock, he signals this to the group, who will then switch their collective view to this one. All participants can then join in on an unexpected price shift in the market, leading to profits that probably would have been missed by an isolated individual.

Within this environment, success depends on people helping each other, while at the same time benefitting from their collective behavior. During conversations, traders exchange insights to achieve a common understanding of where prices are headed. "*Wow, the markets are in bad shape again!*" one of the best traders commented. This signals to the other traders that prices will fluctuate more aggressively than usual. Traders help each other out by communally and continuously commenting on their beliefs of where prices are headed.

On one occasion a trader screamed "*It's going up, it's going up!*," which was followed by another trader pointing out signals that confirmed this conclusion. In the end, the individual is the only one responsible for pushing the buy or sell button, but when experienced daytraders indicate their belief that stock prices are headed in a certain direction, the others will also respond. Thus, while the buy–sell responsibilities are inherently individual, the group at TradeCo collectively helps each other by having their personal analyses "confirmed" by other traders.

Had they worked as individuals instead of as a group, their view on the market would be more constrained, and they would not be able to take breaks. Knowledge is purposefully shared, and traders are socially stimulated to add their opinion to the community's view of what the stock market is doing, with the hope that it will lead to higher individual profits. This also allows them to collectively manipulate stock prices in certain directions. It thus appears that knowledge sharing in this community leads to communal benefits, made possible by instant communication, which may only be possible in a face-to-face situation.

Traders at TradeCo also have tacit agreements not to disrupt and negatively affect another trader's position. Each trader has enough buying power to manipulate stock prices, with each having several million dollars at their disposal. Assume that one trader, for instance, is 10,000 shares "long" (owns that amount of shares) in such a stock. Another trader could manipulate the price by aggressively selling that stock, so that the price would go down.

When asked to comment on why traders share their insights, despite a potential for internal competition, a trader said,

> *We have little effect on the world stage of the stock exchange, so we should not bother competing against each other when we can help each other instead. There's plenty of stock market out there for all of us. Also, if we don't help one another, the market is the laughing third party, scooping up what could otherwise have been our communal profits.*

Furthermore, individual traders can only "shake" and "move" market prices to some extent on their own. When cooperating in small task force groups, however, they are able to manipulate stock prices further to their advantage. This is especially useful in situations where they feel a price will continue to rise or fall after they have collectively given it a push. When the dollars these traders control are combined, the impact on the market can be enlarged. This is especially useful when they feel a stock price is at a price barrier, and pushing it over or under that barrier would lead to a strong shift in price. Ordinarily, they would then push the price in the direction they would like it to go. On occasions where the stock price gains more momentum beyond the daytraders' combined forces, the continued movement of the stock price leads to a profit when the trade is cashed in after this continued price movement.

It appears, therefore, that traders' knowledge of each others' whereabouts on the market is required and supported within this face-to-face group of traders. The stock market is a continuously moving target, and the ease, intensity, and richness of direct communication allows for such intensive and continuous dialogue. This would be severely hindered by even the best electronic systems.

5.4 Long-Term Communal Learning Benefits

In addition to the short-term benefits of sharing knowledge, there are also long-term reasons for working as a group in a colocated setting. There is a continuous need for learning and instant activities, as stock markets and stock price behaviors continuously evolve in unpredictable ways. It can be argued that complex market knowledge is transferred by continuous dialogue across the traders with profitable knowledge.

Being a daytrader is also a continuous learning process, where one must learn the generic adaptation and management of their portfolio with the market, and improve trading skills. As it is a continuous learning process, learning from each others' mistakes and insights is beneficial to all traders.

Knowledge in this case cannot be easily (if at all) captured and distributed through electronic systems. According to one very experienced trader,

> *I'm genuinely not concerned about making my strategies public, as it really is the trader and his experience that allow him to perceive the proper strategy for the moment and execute it properly. In order to be able to make money and see the opportunities the way I do, and even if someone memorized my reasons, the person first needs to bite the dust on the market and then learn from there with my help.*

Thus, as decisions are often split-second, tacit understanding, which can be directly and immediately applied in action, is essential.

Instead, knowledge sharing consists of hints provided at the right moment in time, when individual and collective activities can be combined in a particular market price situation. In an environment requiring instantaneous communication, the role of information technology is often limited to delivering numerous and instant market signals. Skype and IRC are only used for background reflections and long-term predictions between traders of various daytrader organizations; short-term communication and action is supported by verbal communication only. The high volatility of market knowledge with a very short expiry date makes a face-to-face community-of-practice essential.

One trader reported that he was a trainee for a very long time, until the traders who were sitting next to him quit their jobs because they were no longer able to adapt quickly to the community. The replacements located near him increased his capability.

> *When they left the company, other traders were relocated nearer to me, and watching them work helped me a lot. By continuously attempting to copy their strategy I managed to earn almost as much money as they did. They showed me why they acted upon which signals and helped me understand what I was doing wrong, as much as what to do and when.*

In this instance, observation of currently successful behavior appears to have stimulated the direction and speed of learning over a longer period of time. This tells us that socializing with the currently successful traders is key to individual success.

To summarize, learning is not merely of necessity for the inexperienced new entrants to the organization but is instead a continuous requirement. As the necessary knowledge for trading is complex, a more successful trader "teaching" another trader or a trainee requires quick and intense communication. This appears to be possible only where traders are colocated. In addition, it is argued that socialization between traders provides an impetus for knowledge sharing.

5.5 Detachment Leads to Diminished Profits

To further support this claim, there is evidence that those who are under-socialized or excluded from the group do poorly. For example, three traders seated in a more distant area of the office, in a quieter zone, believed they could perform better by just concentrating on the market. Instead, their profits diminished considerably with the passage of time. While they acknowledge that they were not earning as much as they did before, they attributed this to changed market behavior to which they had difficulties adapting, and not to a diminished ability to learn from fellow traders. According to the other traders, however, their increased distance has led to poorer profits than in the past because of their disassociation.

Next, we turn to our discussion of the results and implications for theory and practice.

6 DISCUSSION AND IMPLICATIONS

A problem with single case studies is that there is a limitation in generalizability and a risk of observer bias; a multiple site study helps guard against such a bias and adds confidence to findings by validating results across sites (Leonard-Barton 1990). Also, a cross-case analysis could specifically seek out a contrasting case to highlight differences between sites (Miles and Huberman 1994). Future research in the form of a multiple site study and cross-case analysis can help address these limitations.

Given this, the case offers a number of insights about the possibilities for IT mediation in this critical industry, where information technology appears able to individuate financial work. It illustrates the continued need for a colocated community of day stock traders. The results of socialization, short- and long-term learning, and decreased performance by isolated individuals provide some evidence for this conclusion.

Confirming our results, Millo et al. (2005) argue that the introduction of electronic markets has not replaced but transformed where the social takes place. This view contrasts with the possibility that the computerization of financial markets could lead to the detached and isolated individuation of financial work.

The current case contributes to the discussion on the role of technology in work transformation by delineating the specific areas where a community of practice is still required—in our case, to be an effective daytrader. In this case, technological intermediation through ECNs allowed quick delivery of information to the remote office of these daytraders, but face-to-face contact among traders was still required. This may suggest that the resistance for virtualized work is not only a nostalgic hope but a work-related need (Barrett 1999; Barrett and Walsham 1999).

The survey of the literature on community-of-practice theory and dispersed team work suggests that the intensity and richness of face-to-face communication allows for the transfer of complex knowledge. In contrast, electronic communication systems can impede such complex knowledge exchange. It suggests that knowledge transfers are best accommodated by working in a group where trust and collective goals are shared, so that insights are exchanged between novices and experts. This does not imply that there should be little or no IT, but that the final layer of strategic value—in this case, through communal learning and knowledge exchange to beat the markets—is accommodated through colocated teams using the capabilities of IT.

In particular, the literature on community of practice and dispersed team work emphasizes the importance of the social in "sticky" knowledge exchange. This not only arises as a stimulant for knowledge sharing, but also through the performing of individual and shared activities within a group. TradeCo's case provides empirical evidence where and why direct social contact during knowledge exchange are required to produce a community of practice that can "beat the markets."

The COP in this case provides a strangely supportive context and culture for knowledge transfer, with a common incentive or purpose, but where creative work can be individually recognized (Barrett et al. 2004). We build upon and extend this argument by providing an example where contextual circumstances lead to individual success through its dependence on communal coordinated (inter)activity, so that the incentive to share still leads to individual recognition through instant profits. As Barrett et al. (2004) state, "electronic-based contact will often require supplementing with direct face-to-face meetings for complex, delicate trust-based interactions" (p. 9). In our example, electronic-based systems did and probably would not provide for the social and learning benefits that are of crucial importance in successfully working with a fast-moving stock market.

Based on TradeCo's case, we argue that colocation remains of great importance in facilitating learning processes, which impedes the individualization, virtualization, and detachment of trading work that appears possible through information technology. The richness and social nature of the community allows knowledge and expertise to spread more completely and rapidly, thereby directing and accelerating the individuals' activities and capabilities within the community. This also provides for short-term benefits in coordination such as internal cooperation and external manipulation, and long-term benefits in the distribution of complex knowledge by facilitating the transfer about renewed market circumstances and general trading practices.

An implication of this research is that strategic knowledge can and does appear in the colocated practices where the new IT capabilities are embedded. This does not eliminate the importance and need for information technology, but illustrates its role in shaping the new and emergent face-to-face communities-of-practice around the introduction of IT. Our paper illustrates how electronic trading alters the sources of electronic and face-to-face communication. In many ways, computerized systems are contributing to a rearrangement of work that began with the earlier application of the telegraph to send stock prices outside of the stock exchange. Thus, more recent developments in information technology have only transformed and broadened the scope of where and how face-to-face work will be done.

References

Barrett, M. I. 1999. "Challenges of EDI Adoption for Electronic Trading in the London Insurance Market," *European Journal of Information Systems* (8:1), pp. 1-15.

Barrett, M. I., Cappleman, S., Shoib, G., Walsham, G. 2004. "Learning in Knowledge Communities: Managing Technology and Context," *European Management Journal* (22:1), pp. 1-11.

Barrett, M. I., and Scott, S. V. 2000. "The Emergence of Electronic Trading in Global Financial Markets: Envisioning the Role of Futures Exchanges in the next Millennium," paper presented at the 8th European Conference on Information Systems, Vienna, Austria, July 3-5.

Barrett, M. I., and Scott, S. V. 2004. "Electronic Trading and the Process of Globalization in Traditional Futures Exchanges: A Temporal Perspective," *European Journal of Information Systems* (13:1), pp. 65-79.

Barrett, M. I., and Walsham, G. 1999. "Electronic Trading and Work Transformation in the London Insurance Market," *Information Systems Research* (10:1), pp. 1-22.

Blackler, F. 1995. "Knowledge, Knowledge Work and Organizations: An Overview and Interpretation," *Organization Studies* (16:6), pp. 1021-1046.

Brown, J. S., and Duguid, P. 1991. "Organizational Learning and Communities-of-Practice: Toward a Unified View of Working, Learning, and Innovation," *Organization Science* (2:1), pp. 40-57.

Carlson, J. R., and Zmud, R. W. 1999. "Channel Expansion Theory and the Experiential Nature of Media Richness Perceptions," *Academy of Management Journal* (42:2), pp. 153-170.

Daft, R. L., and Lengel, R. H. 1986. "Organizational Information Requirements, Media Richness and Structural Design," *Management Science* (32:5), pp. 554-571.

Drucker, P. 1993. *Post-Capitalist Society*, Oxford, UK: Butterworth-Heinemann.

Finholt, T. A., Sproull, L., and Kiesler, S. 2002. "Outsiders on the Inside: Sharing Know-How Across Space and Time," in *Distributed Work*, P. Hinds and S. Kiesler (eds.), Cambridge, MA: MIT Press.

Fox, S. 1997a. "From Management Education and Development to the Study of Management Learning," in *Management Learning: Integrating Perspectives in Theory and Practice*., J. B. Burgoyne and M. Reynolds (eds.), London: Sage Publications.

Fox, S. 1997b. "Situated Learning Theory Versus Traditional Cognitive Learning Theory: Why Management Education Should Not Ignore Management Learning," *Systems Practice* (10:6), pp. 727-747.

Fox, S. 2000. "Communities of Practice, Foucault and Actor-Network Theory," *Journal of Management Studies* (37:6), pp. 853-867.

Ives, B., and Jarvenpaa, S. L. 1991. "Applications of Global Information Technology: Key Issues for Management," *MIS Quarterly* (15:1), pp. 33-49.

Kiesler, S., and Cummings, J. N. 2002. "What Do We Know about Proximity and Distance in Work Groups," in *Distributed Work*, P. Hinds and S. Kiesler (eds.), Cambridge, MA: MIT Press.

Knorr Cetina, K. 2002a. "Inhabiting Technology: The Global Lifeform of Financial Markets," *Current Sociology* (50:3), pp. 389-405.

Knorr Cetina, K. 2002b. "Traders' Engagement with Markets: A Postsocial Relationship," *Theory, Culture and Society* (19:5/6), pp. 161-185.

Knorr Cetina, K., and Bruegger, U. 2000. "The Market as an Object of Attachment: Exploring Postsocial Relations in Financial Markets," *Canadian Journal of Sociology* (25:2), pp. 141-168.

Knorr Cetina, K., and Bruegger, U. 2001. "Transparency Regimes and Management by Content in Global Organizations. The Case of Institutional Currency Trading," *Journal of Knowledge Management* (5:2), pp. 180-194.

Knorr Cetina, K., and Bruegger, U. 2002. "Global Microstructures: The Virtual Societies of Financial Markets," *American Journal of Sociology* (107:4), pp. 905-950.

Lave, J., and Wenger, E. 1991. *Situated Learning: Legitimate Peripheral Participation*, Cambridge, UK: Cambridge University Press.

Leonard-Barton, D. 1990. "A Dual Methodology for Case Studies: Synergistic Use of a Longitudinal Single Site with Replicated Multiple Sites," *Organization Science* (1:3), pp. 248-266.

McAndrews, J., and Stefanidis, C. 2000. "The Emergence of Electronic Communication Networks in the U.S. Equity Markets," *Current Issues in Economics and Finance* (6:12), pp. 1-6.

Miles, M. B., and Huberman, A. M. 1994. *Qualitative Data Analysis: An Expanded Sourcebook*, London: Sage Publications.

Millo, Y., Muniesa, F., Panourgias, N. S., and Scott, S. V. 2005. "Organised Detachment: Clearinghouse Mechanisms in Financial Markets," *Information and Organization* (15:3), pp. 229-246.

Mowshowitz, A. 2002. *Virtual Organization: Towards a Theory of Societal Transformation Stimulated by Information Technology*, Westport, CT: Greenwood Press.

North, D. C. 1991. "Institutions," *The Journal of Economic Perspectives* (5:1), pp. 97-112.

Orlikowski, W. J., and Baroudi, J. J. 1991. "Studying Information Technology in Organizations: Research Approaches and Assumptions," *Information Systems Research* (2:1), pp. 1-28.

Orr, J. 1987. *Talking about Machines: Social Aspects of Expertise*, Palo Alto, CA: Xerox Palo Alto Research Center.

Orr, J. 1990. "Sharing Knowledge, Celebrating Identity: War Stories and Community Memory in a Service Culture," in *Collective Remembering: Memory in Society*, D. S. Middleton and D. Edwards (eds.), Beverley Hills, CA: Sage Publications.

Royal, C., and Althauser, R. P. 2003. "The Labor Markets of Knowledge Workers: Investment Bankers' Careers in the Wake of Corporate Restructuring," *Work and Occupations* (30:2), pp. 214-233.

Sapsed, J., and Salter, A. 2004. "Postcards from the Edge: Local Communities, Global Programs and Boundary Objects," *Organization Studies* (25:9), pp. 1515-1534.

Scott, S. V., and Barrett, M. I. 2005. "Strategic Risk Positioning as Sensemaking in Crisis: The Adoption of Electronic Trading at the London International Financial Futures and Options Exchange," *Journal of Strategic Information Systems* (14:1), pp. 45-68.

Starbuck, W. H. 1992. "Learning by Knowledge-Intensive Firms," *Journal of Management Studies* (29:6), pp. 713-740.

Walsh, J. P., and Maloney, N. G. 2002. "Computer Network Use, Collaboration Structures and Productivity," in *Distributed Work*, P. Hinds and S. Kiesler (eds.), Cambridge, MA: MIT Press.

Walsham, G. 1993. *Interpreting Information Systems in Organizations*, New York: John Wiley and Sons.

Walsham, G. 1995. "Interpreting Case Studies in IS Research: Nature and Method," *European Journal of Information Systems* (4:2), pp. 74-81.

Walsham, G. 2006. "Doing Interpretive Research," *European Journal of Information Systems* (15:3), pp. 320-330.

Wenger, E. 1998. *Communities of Practice: Learning, Meaning and Identity*, Cambridge, UK: Cambridge University Press.

Wenger, E., McDermott, R. A., and Snyder, W. 2002. *Cultivating Communities of Practice: A Guide to Managing Knowledge*, Boston: Harvard Business School Press.

Wyckoff, R. D. 1919. *The Day Trader's Bible. Or. . . My Secrets of Day Trading in Stocks*, New York: Ticker Publishing.

Yin, R. 2003. *Case Study Research* (3rd ed.), Thousand Oaks, CA: Sage Publications.

About the Authors

Roger van Daalen Fuente is a Ph.D. student at Lancaster University Management School, Department of Management Science. His research interests include virtuality and social cognition as applied to the study of information systems in the finance industry. He holds a BScBA from RSM Erasmus University in Business Administration and a BSc in Industrial/ Organizational Psychology from the same university. He has worked as an entrepreneur in the IT services sector. Roger can be reached at r.vandaalen@lancaster.ac.uk.

Mike Chiasson is currently an AIM (Advanced Institute of Management) Innovation Fellow and a Senior Lecturer at Lancaster University's Management School, in the Department of Management Science. Before joining Lancaster University, he was an associate professor in the Haskayne School of Business, University of Calgary, and a postdoctoral fellow at the Institute for Health Promotion Research at the University of British Columbia. His research examines how social context affects IS development and implementation, using a range of social theories (actor network theory, structuration theory, critical social theory, ethnomethodology, communicative action, power-knowledge, deconstruction, and institutional theory). In studying these questions, he has examined various development and implementation issues (privacy, user involvement, diffusion, outsourcing, cyber-crime, and system development conflict) within medical, legal, engineering, entrepreneurial, and governmental settings. Most of his work has been qualitative in nature, with a strong emphasis on participant observation. Mike can be reached at m.chiasson@lancaster.ac.uk.

Paul Devadoss is a lecturer at the Department of Management Science, Lancaster University Management School, UK. He completed his Ph.D. in Information Systems at the School of Computing in the National University of Singapore. His research interests include enterprise systems and e-governments. In particular, he is interested in the social impacts of IT use in organizational settings and the managerial implications of technology use. He has previously published in journals such as *Decision Support Systems, MIS Quarterly Executive, Communications of the AIS, Information and Management*, and *IEEE Transactions on IT in Biomedicine*. Paul can be reached by e-mail at paul@devadoss.org.

12 BAZAAR BY DESIGN: Managing Interfirm Exchanges in an Open Source Service Network

Joseph Feller
University College Cork
Cork, Ireland

Patrick Finnegan
University College Cork
Cork, Ireland

Brian Fitzgerald
University of Limerick
Limerick, Ireland

Jeremy Hayes
University College Cork
Cork, Ireland

Abstract *As in many other sectors, competitive necessities are driving open source software companies to participate in cooperative business networks in order to offer the complete product and service offerings demanded by customers. This paper examines one such emerging business network archetype: an open source service network (OSSN). This type of business network is of particular interest as it not only addresses key challenges vis-à-vis OSS commercialization, but operates in a manner that overcomes exchange problems among participants by relying primarily on social mechanisms. The paper reveals the manifestation of social mechanisms in OSSNs and how these are used for coordinating and safeguarding exchanges between firms. Specifically, we illustrate the importance of (1) restricted access, (2) assessing the reputation of others, (3) a shared macroculture (goals and norms), and (4) collective sanctions for punishing firms who violate these goals and norms.*

Please use the following format when citing this chapter:

Feller, J., Finnegan, P., Fitzgerald, B., and Hayes, J., 2008, in IFIP International Federation for Information Processing, Volume 267, Information Technology in the Service Economy: Challenges and Possibilities for the 21st Century, eds. Barrett, M., Davidson, E., Middleton, C., and DeGross, J. (Boston: Springer), pp. 173-188.

Keywords Open source service network, social mechanisms, network governance, inter-organizational systems

1 INTRODUCTION

Advancements in the provision of open source software (OSS) have come to closely resemble complex product and service offerings in many other sectors. Davidow and Malone (1992) predicted that the challenges of the 21st century would require organizations to quickly and globally deliver a high variety of customized products and services, while Stafford (2002) documented that the market forces surrounding such offerings require organizations to deliver products and services to market so rapidly that they would have to align themselves in IT-mediated partner networks in order to meet customer requirements.

OSS is best defined by its central characteristic: the development, distribution, acquisition, and use of software based on collaboration and sharing rather than proprietary restrictions (Feller and Fitzgerald 2002). Commercial strategies, thus, focus on the provision of value-adding services around the software product (Fitzgerald 2006). This allows third-party service providers to engage in level-field competition with the software creators by leveraging the open nature of the source code. Such low barriers to entry in OSS service provision (see Woods and Guliani 2005) have led to the prolifera-tion of technology-driven OSS microfirms offering specialized products and services. However, such firms cannot always meet customer requirements for end-to-end system implementation and support.

Commercialization strategies for OSS point to a need to (1) build business models capable of a whole product (Moore 1999) approach to software and service delivery, and (2) marry community-driven development capabilities with high-levels of sector knowl-edge in vertical business domains. In other words, companies seek to *productize* open source software to meet the needs of both enterprises and end-users by offering support, implementation, modification, and related services (Woods and Guliani 2005). Such challenges suggest the need for new organizational forms, such as cooperative business networks of OSS companies (Fitzgerald 2006). Nevertheless, research on the commer-cialization of OSS has focused narrowly on revenue generation models (e.g., Krishna-murthy 2005; Markus et al. 2000), and has been dominated by studies of single firms, whether OSS start-ups such as RedHat and JBoss (e.g., Krishnamurthy 2005) or very large multinationals like Apple, IBM, and Sun (e.g., West 2003), neglecting network-based business models, although they are observable in practice.

We have identified one type of OSS business network, which represents what Clemons and Row (1992) term a "move-to-the middle," networks of organizations that interact in order to deliver value to the end consumer. We label such networks *open source service networks* (OSSNs). OSSNs typically seek to meet what Woods and Guliani (2005) describe as the challenge of productizing OSS by offering support, implementation, modification, and related services. They are characterized by their reliance on information and communication technologies to mediate collaboration and communication and have grown in an organic fashion strongly informed by the practice of community-based OSS projects. While being legal entities, they have avoided formal or legal contracts to govern interfirm coordination. This type of interfirm coordination has been termed *network governance* (see Jones et al. 1997).

This paper examines Zea Partners, a business network of firms developing software solutions and selling services around the Zope application server. The network enables participants to, collectively, undertake larger commercial contracts than they would be able to do individually. However, they primarily use social, rather than formal or legal mechanisms, to enable interfirm exchanges. Although this appears counterintuitive, the study demonstrates the role of social mechanisms in enabling network effectiveness. We use the work of Jones et al. (1997) on network governance and De Wever et al. (2005) on network effectiveness to investigate how social mechanisms are used to overcome exchange problems and enable the access to, and transfer of, strategic resources among network participants. The paper concludes by modeling how social mechanisms are used to overcome exchange problems in an open source service network.

2 BACKGROUND

2.1 Open Source Service Networks

We define an open source service network as

*A **network** of firms that collaborate in order to **service** customer software needs based on **open source** solutions.*

To clarify the definition, *network* is understood to mean a collaborative network with interdependencies between member firms and a shared identity. This differentiates OSSNs from non-interdependent groupings such as third-party directories and portals listing OSS firms (e.g., SourceForge.net Marketplace). Furthermore, the primary purpose of the network is to commercially *service* customer software needs based on open source solutions. This differentiates OSSNs from noncommercial groupings such as advocacy networks, nonprofit foundations development hosts (e.g., Apache Software Foundation), and research communities, as well as business networks offering solely proprietary solutions (e.g., Microsoft Partner Program).

OSSNs are interorganizational networks that facilitate the flow of resources among nonhierarchical and legally separate entities in order to meet challenges that are often beyond the capabilities of individual firms. Specifically, OSSNs exist to support member firms in their efforts to deliver whole products and productize open source software by offering support, implementation, modification, and related services to consumers, thus addressing the productizing challenge facing open source software outlined above. Therefore, OSSNs are relevant to the commercialization of OSS and consequently significant for practitioners.

As will be discussed later, research has shown that the effective operation of any network is dependent on overcoming exchange problems among participants. While many business networks rely primarily on formal or legal mechanisms, most OSS development communities rely on informal (social) mechanisms. Our observations of OSSNs reveal that they appear to be shaped by the informal ethos and operational style of the OSS communities from which they originated, and with which they continue to interact. Additionally, OSSNs have inherited much of the IT infrastructure characteristic of OSS communities. Such IT facilitates interaction and makes the communication and collaboration activities of a community persistently visible to all participants, thus enabling the use of social mechanisms to overcome exchange problems. However, OSSNs represent an intersection between the corporate-centered culture of traditional business networks and the community-centered culture of open source software. We, therefore, cannot assume that the operational effectiveness of OSSNs can be achieved using purely formal mechanisms as employed by other business networks; neither can we assume that our understanding of the operation of social mechanisms in OSS communities provides a suitable alternative.

2.2 Managing Exchanges in Business Networks and OSS Communities

The benefits of cooperative relationships between independent firms have been advocated for decades (Cash and Konsynski 1985; Finnegan et al. 2003; Henderson 1990; Kaufman 1966; Van de Ven 1976). The reasons for interorganizational cooperation include resource procurement and allocation (Alter and Hage 1993; Clemons and Row 1992; Galaskiewicz 1985), political advantages (Galaskiewicz 1985), risk sharing and acquiring expertise (Alter and Hage 1993), stability (Oliver 1990), legitimacy (Galaskiewicz 1985; Oliver 1990), and efficiency (Clemons and Row 1992; Oliver 1990). Frequently, participants in interorganizational networks believe that collaboration will result in adaptive efficiency, "the ability to change rapidly and at the same time provide customized services or products, and at low cost" (Alter and Hage 1993, p. 274).

Cooperative business relationships are social action systems as they exhibit the fundamental principles of any organized form of collective behavior; members aim to achieve collective and self-interest goals; the division of tasks and functions among members creates interdependent processes; the cooperative entity can act as a unit and has a separate identity from its members (Van de Ven 1976). Nevertheless, such relationships are typified by formal, structured coordination mechanisms and agreements in order for individual organizations to meet their own goals as well as the goals of the cooperative entity (Van de Ven and Walker 1984). While there is some evidence of the use of informal mechanisms to coordinate interorganizational relationships (Jones et al. 1997, 1998), the effectiveness of this approach has not been verified where the business network forms the basis of competitive strategy. Furthermore, the dominance of formal governance is influenced by the fact that the IT infrastructures implemented in the majority of business networks are designed to facilitate transactions (e.g., purchase orders, invoices, payment, etc.) between participants (Timmers 1999), and do not provide the visibility of social action for the effective operation of informal mechanisms.

Interfirm coordination characterized by organic or informal social systems, rather than bureaucratic structures within firms and formal contractual relationships between firms, is labeled *network governance* (Jones et al. 1997). In contrast to business networks, the use of informal social mechanisms has dominated investigations of exchange conditions and developer relationships in OSS communities (labeled *the bazaar* by Raymond 2001). In particular, the importance of (1) shared beliefs and values (i.e., macroculture) and (2) reputation has been repeatedly described. The OSS governance literature demonstrates that, in contrast to the cooperative business networks described previously, OSS development communities have been characterized by (1) a greater emphasis on collaboration and (2) the use of a wide range of technologies to enable social interaction between community members (see Bergquist and Ljungberg 2004; Feller and Fitzgerald 2002; Gallivan 2001; Markus et al. 2000; Sagers 2004; Szczepanska et al. 2005). As an exemplar of peer production, the OSS phenomenon characterizes an alternative model for organizing, a model that operates without reliance on markets, managerial hierarchies, property, and contracts. Peer production is characterized by the decentralized accumulation and exchange of information, and is seen as potentially superior to traditional hierarchy and market-based models (Benkler 2002, 2006). We consequently adopt a conceptual model for our study that illuminates the unique features of OSS as peer production rather than a traditional model of organizational behavior.

2.3 Conceptual Model

Jones et al. (1997) identify four preconditions for network governance: demand uncertainty, customized (asset-specific) exchanges, complex tasks executed under time-pressure, and frequent exchanges between partners. They argue that under such conditions, networks develop structural embeddedness, which they define, citing Granovetter (1992, p. 35), as the extent to which a "dyad's mutual contacts are connected to one another," creating both direct and indirect ties between parties. Jones et al. assert that when high levels of structural embeddedness are present, it enables the use of various social mechanisms to resolve exchange problems by coordinating and safeguarding exchanges within networks. These social mechanisms are

1. Restricted access: the "strategic reduction in the number of exchange partners in the network" (Jones et al. 1997, p. 927)
2. Macroculture: a "system of widely shared assumptions and values...that guide actions and create typical behavior patterns among independent entities" (ibid, p. 929)
3. Collective sanctions: "ways in which groups punish members who violate shared norms, values and goals" (ibid, p. 931)
4. Reputation: "estimations of one's character, skills, reliability and other attributes important to exchanges" (ibid, p. 932)

In resolving exchange problems in network governance, Jones et al. propose that coordination is supported by restricted access and macroculture, while safeguarding is supported by restricted access, collective sanctions, and reputation (Figure 1). Nevertheless, while the Jones et al. framework has been widely cited in a number of fields, neither the original framework nor subsequent applications of it provide measures of network effectiveness. However, research on interorganizational systems has provided a more considered conceptualization of network effectiveness. In particular, De Wever et al. (2005) posit that network effectiveness can be operationalized as the ability of networks to provide member firms with sustainable access to strategic or value-generating resources, specifically, the access to strategic resources and transfer or exchange of strategic resources in order to acquire them. By combining the work of Jones et al. and De Wever et al., we arrive at the conceptual model for out study (see Figure 1).

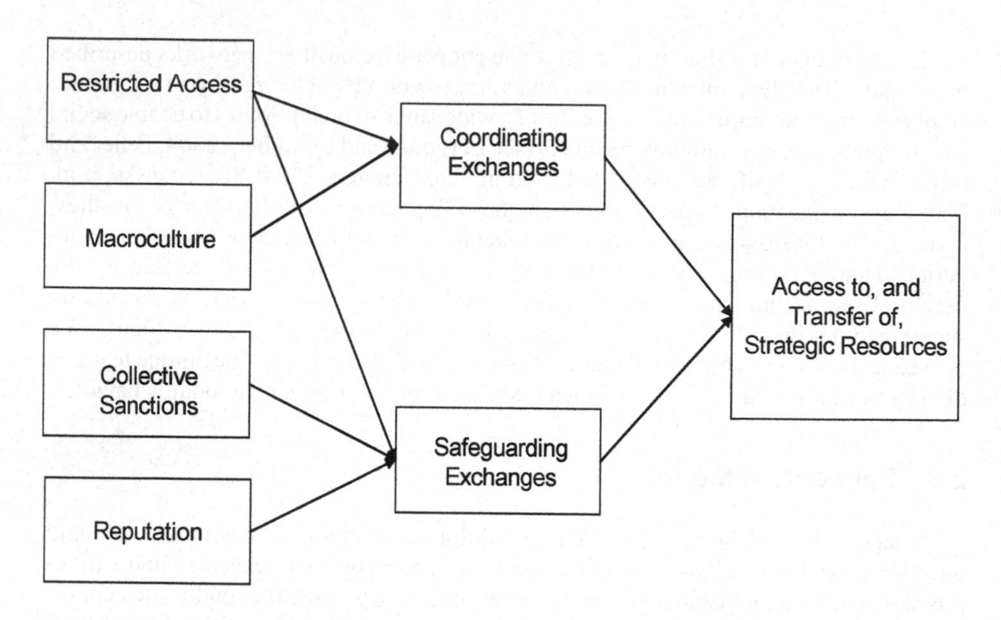

Figure 1. Conceptual Model

3 STUDY DESIGN

The objective of our study is to explore the OSSN phenomenon with the specific intention of examining how social mechanisms are used to overcome exchange problems, and thus facilitate network effectiveness in OSSNs.

We adopt a case study method consistent with that of Benbasat et al. (1987) and Yin (1994). We study the OSSN phenomenon in its natural setting, employing multiple data collection methods to gather information from a few entities, without employing experimental control or manipulation. Thus, we follow in the tradition of Eisenhardt (1989) and Madill et al. (2000) by seeking to reveal preexisting, relatively stable and objectively extant phenomena and the relationships among them.

Our OSSN study site, Zea Partners, was founded in 2003 as the Zope Europe Association (ZEA), and changed its name to Zea Partners in 2006. Headquartered in Belgium, Zea Partners operates as an international network of businesses that build software and deliver services around the application server technology called Zope; a well-known piece of open source software widely used for developing content management systems, intranets, portals, and related applications. Zea Partners consists of 25 firms in Europe, America, and Africa. The management team seeks project contracts on behalf of network members and performs network management activities such as marketing and project management. They also develop the network's business strategy in conjunction with the managing partners.

Data for the case study was gathered over a 17-month period and focused on the Zea management team and senior management in eight member companies. Data gathering techniques included 16 interviews and 4 intensive workshops. Interviews were generally of 1 to 2 hours in duration, with follow-up telephone conversations and e-mail used to clarify and refine issues that emerged during transcription. Interviews were complemented by comprehensive reviews of documents and presentations and discussions at the workshops.

Interview data was transcribed, generating 133 pages of field notes. Content analysis was undertaken using the grounded theory coding techniques proposed by Strauss and Corbin (1990). The first step (open coding) involved the data being examined line by line to ascertain the main ideas. These data were then grouped by meaningful headings (informed by the conceptual model illustrated in Figure 1) to reveal categories, subcategories, and properties. The next step (axial coding) involved determining hypotheses about the relationships between a category and its subcategories (e.g., conditions, context, action and interaction strategies, and consequences). The focus then turned to the data to assess the validity of these hypothesized relationships. Relational and variational sampling (see Strauss and Corbin 1990) were used to select data for this analysis. This process continued in an iterative manner and resulted in the modification of categories and relationships. Finally, selective coding was undertaken to identify the relationships between categories (constructs) using hypothesized conditions, context, strategies, and consequences. Discriminate sampling (Strauss and Corbin 1990) was used to select data to examine strong and weak connections between categories.

4 FINDINGS

The Zea Partners network illustrates the preconditions for network governance as predicted by extant research. The availability of contracts is subject to what Zea's founder calls "valleys and peaks" and the OSS domain experiences constant changes in knowledge and technology, leading to a requirement for information dissemination among firms. These factors produce what Jones et al. (1997) term *demand uncertainty*. The small size (typically fewer than 10 people), geographic and linguistic limitations, and specialized knowledge of member firms limit the size, location, and complexity of the contracts for which they can bid. The network helps overcome demand uncertainty by allowing companies to pool resources in order to compete for larger and/or global contracts. In competing on the basis of a whole product, the network allows partners to offer a full range of value-chain activities rather than concentrating exclusively on their own specialities. *Task complexity* is evident as producing the whole product for a wide range of markets and customers requires the inputs of specialists across a range of business functions. *Customized exchanges high in human asset specificity* are recognizable as member firms collaborate to produce a customized product or service that is not easily transferred. Interfirm routines are learned emergently through the collaborative process rather than by prior agreement. Due to geographical distance, Zea members *frequently exchange* knowledge in a digital environment. Ongoing interactions between member firms and the mutual sharing of partners, clients, and contacts with the wider OSS community all provide evidence of *structural embeddedness*.

The network's founder explained that for interfirm coordination, social mechanisms "are the only thing. This is an opt-in system. A large component of the perceived benefit is the reputation improvement from being in the network. Social mechanisms really underpin that." The manner in which social mechanisms were manifest in the network is shown in Table 1.

Table 1. Manifestation of Social Mechanisms in Zea Partners

Restricted Access	1. Limiting the number of firms in the network for strategic reasons. 2. Information on the competencies and activities of others is available to all members. 3. Firms wishing to join are assessed on the benefits of their skills/expertise to the network. 4. Firms must be known to, or have a prior relationship with, existing members in order to join the network. 5. Firms must demonstrate commitment to the ideals of open source software in order to join the network.
Macro-culture	1. Sharing a sense of belonging. 2. Having a common software development philosophy. 3. Agreeing on accepted ways of doing business. 4. Maintaining a common set of goals for the network. 5. Having a sense of mutual interest. 6. Sharing a sense of common destiny.
Collective Sanctions	1. Perception that the reputation of a member firm would be damaged if they behaved unacceptably. 2. Belief that firms may be excluded from projects if they behave unacceptably. 3. Threat that firms would be expelled from the network if they behave unacceptably.
Reputation	1. Assessing the competence and skills of member firms before doing business with them. 2. Evaluating the character and reliability of member firms before doing business with them. 3. Expecting others to fulfil their obligations because they are members of the network. 4. Considering it important to be regarded (by other members) as being professionally competent. 5. Considering it important to be regarded (by other members) as being reliable and trustworthy. 6. Fulfilling obligations to other members to maintain a firm's reputation within the network.

Access to the Zea Partners network is restricted based on a firm's *reputation* and performance in the OSS community as well as previous interactions with member firms. It is not just about keeping numbers low; it is a strategic restriction. Firms have to fit in with the *macroculture* and needs of the network, and exhibit a commitment to Zea Partners' success. The network's macroculture is typical of values associated with OSS communities: the involvement of users (customers) as equal partners, the visibility of actions, etc. However, this macroculture is evident not just as a shared set of values (including beliefs, language, etc.) as proposed by Jones et al. (1997) but also includes the mechanisms by which these ideas are expressed and shared among members. Consistent with extant research on social mechanisms, restricted access, reputation, macroculture, and collective sanctions were cited by study participants as enabling the coordination and safeguarding of exchanges between network members. However, our analysis revealed that the functioning of these mechanisms, and how they relate to coordination, adaptation, and safeguarding exchanges, followed a different pattern from the one proposed by Jones et al.

4.1 How Social Mechanisms Facilitate the Coordination of Exchanges

Restricted access facilitates coordinating exchanges by establishing routines for members to work together. The network's founder stated that the network is "still at an early enough stage that we are able to work together on a person-to-person basis" but it assists members by having "standing contracts, customer references etc on file…so the friction in assembling a team gets lowered." However, as the network grows he, acknowledges the need to develop "a common methodology, a common way of thinking about a problem, assigning work, tracking results, reporting bugs." Coordination costs are reduced by increasing visibility in the network, thereby facilitating coordination. The network's founder acknowledges the importance of transparency in the community when he says that

> the best barometer is the way people act in the community. If they are competent developers, then you can see them send mails to the mailing lists, do check-ins, and file bug reports, write papers for conferences, etc. There are a multitude of avenues available to show their competence. If someone is invisible from all those avenues, that says something.

The network has adopted deliberate membership requirements that facilitate a component-based approach to work allocation. The owner of Plone Solutions recognizes the importance of this sort of network design when he says that

> there is no single point of failure; you can swap out components or companies. If one company is not talented enough or does not have the domain knowledge to do this particular thing we normally have another company that has…it's very agile and flexible.

Restricted access also facilitates coordinating exchanges by giving network members a voice in decision making. The owner of Reflab says that members "also have the ability to steer…the whole product idea was something that we could approve or reject." The owner of Bubblenet reaffirms this point when he says that "[members] talk a lot…which gives me the chance to give my opinion and maybe shape part of the process." Finally, restricted access facilitates mutual adjustment by reducing the variances that parties bring to the exchanges and establishing a routine for working together. Zea's founder notes that "about 50 percent of the problem is solved by working the same way we worked [in OSS communities]. The companies have a set of tools, a way of working together, a common culture and ways of communication that we use on Zea projects."

Macroculture was found to facilitate coordinating exchanges in a number of ways. First, a culture of network agility is fostered among member firms in order to enable coordination. Zea's founder says that they "want to define this OSS business model idea.

Instead of having the cathedral model of Accenture, we want to have multiple players in multiple countries. We can move things around as new trends or specialities emerge." Coming from the OSS community, there is a preexisting macroculture that is important for ensuring new members fit into the network's culture. Zea's founder says that Zea Partners "are going to focus on people who've already decided they're interested in OSS, not people who don't understand value and values. The prime consideration is to make sure that the fabric that holds this experiment together doesn't get torn." Creating a sense of mutual interest or "good karma" is also seen as an important aspect of macroculture that facilitates coordination. The chief architect of Plone Solutions says that "we have not been taking Zea projects because there are other companies that need the work more. It's partly that we are busy and partly that we are trying to be nice to other people; they're good people and they should have more work than they currently do." Preventing the fear of lock-in among participants is an important factor in creating a sense of macroculture in the network. The owner of Bubblenet makes the point that "the part that seduces me is that someone who joins can leave easily. This is very important because it helps in reducing people's fears; this means they work better together." Part of the macroculture in the network is concerned with reducing information and knowledge asymmetries inside and outside the network. The director of Blue Fountain acknowledges the need for member firms to learn from each other when he says "it's very important that we coordinate, share best practice, learn from each other, leverage our collective experience." Zea's founder recognizes the importance of involving the client in the OSS community to reduce information asymmetry when he says that "the customer is a participant not a recipient. We want to engage with their in-house staff and teach them how to become OSS developers, how to participate in the community, ask questions, etc., even show them how to contribute back."

Contrary to the model proposed by Jones et al. (1997), our study revealed that coordinating exchanges is facilitated by the desire of members to demonstrate that they adhere to the values of the network by acting in a manner that is considered to be competent and professional. The consequences of not acting in this manner include damage to their reputation as well as exclusion from projects and/or the network (i.e. , collective sanctions). Thus, they strive to be flexible and transparent in their dealings with other members, and to abide by the norms that govern participation in open source development communities. We, therefore, conclude that collective sanctions, as well as restricted access and macroculture, facilitate coordinating exchanges.

4.2 How Social Mechanisms Facilitate Safeguarding Exchanges

The study illustrates that reputation and collective sanctions facilitate safeguarding exchanges as predicted by the Jones et al. (1997) model. Indeed, collective sanctions are seen to facilitate safeguarding exchanges in the same manner as they facilitate coordinating exchanges (see the previous section). However, the study revealed the social mechanisms that facilitate safeguarding exchanges include macroculture rather than restricted access as previously predicted.

Reputation facilitates the safeguarding of exchanges by acting as a prerequisite for participation. Zea's founder says that "getting in the network is only the first part. If you want to do transactions with others in the network, you need to show yourself to be competent and professional." Reputation also has an effect by rewarding good behavior. The owner of Bubblenet says that "getting a reputation in the community is very important; other companies will send me projects that they cannot handle." Reputation also has economic consequences. The owner of Bubblenet remarks that "my first contracts were subcontracts from other Zope companies. That gave me enough presence in the community to switch to my own contracts, but not ones as small as before. Being part of Zope Europe has helped a lot."

Macroculture facilitates the safeguarding of exchanges by balancing commercial interests with network objectives. The owner of Bubblenet says that "most of us are in this for the lifestyle rather than purely to make money; that's difficult to formalize." The owner of Zest Software says that "you get respect by helping other people; you earn the right to ask more questions and favors." Another aspect of the macroculture in the network is taking an ethical approach to commercial issues. The director of Blue Fountain says that "we tell our customers we do three things: transparency, integrity, and honesty. If there's a problem, we'll always put our hand up." The chief architect of Plone Solutions says that "we wouldn't leave a client hanging just because it didn't suit us to go back to that deployment. It's a value thing." Zea's founder acknowledges the importance of creating an environment of trust when he says that "there's a trust vector when it comes to organizing and assigning work that people can make money on. That person has to be perceived as being neutral and competent."

4.3 Enabling Network Effectiveness

Finally, the study also examined network effectiveness using the work of De Wever et al. (2005). Members consider the network effective as it allows them to *access and share strategic resources*. They cited the ability to access and leverage the skill base of other firms, share customer contacts, compete for larger contracts, share business experiences and expertise (e.g., project management), and enjoy an enhanced reputation as a result of membership. According to Zea's founder, such resource sharing allows Zea to "take the whole product and make it offerable by anyone in the network. It has so many benefits on profitability it's just amazing." Our analysis revealed coordination and adaptation to be essential for firms to offer the whole product; thus affecting access to and exchange of strategic resources. In addition, the exchange of strategic resources was dependant on such exchanges being safeguarded.

5 CONCLUSION

Our study has shown that a business network model can be particularly effective in delivering the whole product needed to commercially exploit peer produced software. OSSNs, thus, directly address key challenges in the commercialization of open source software, and are of interest to both research and practice. The IT infrastructure utilized by OSSNs builds upon the collaborative development and communication environments found in OSS communities, and is focused on enabling rich, transparent interactions between firms. This is in contrast to the transaction-oriented focus of other interorganizational systems. As a result of this type of IT infrastructure, the ways in which OSSNs overcome exchange problems contrast with other business networks in that social mechanisms play a primary role in coordinating and safeguarding exchanges among participants (see Table 2) and thus enable the access to and transfer of strategic resources (Figure 2).

Table 2. How Social Mechanisms Facilitate Coordinating and Safeguarding Exchanges

Social Mechanism	Coordination	Safeguarding
Restricted Access	Facilitating the establishment of routines for working together. Reducing coordination cost through increased visibility. Adopting deliberate membership. requirements facilitating a component-based approach to work allocation. Giving members a voice in decision making. Reducing variances that parties bring to the exchanges (facilitates mutual adjustment). Establishing routines for working together.	
Macroculture	Fostering a culture of network agility. Ensuring new members fit in to network's culture. Creating a sense of mutual interest (good Karma). Preventing fear of lock-in. Reducing information/knowledge asymmetries .	Balancing commercial interests with network objectives. Taking an ethical approach to commercial issues. Creating an environment of trust.
Collective Sanctions	Encourages firms to be transparent in their dealings with other members. Encourages firms to act in a manner that is regarded as competent and professional. Encourages firms to behave in line with network expectations.	Encourages firms to be transparent in their dealings with other members. Encourages firms to act in a manner that is regarded as competent and professional. Encourages firms to behave in line with network expectations.
Reputation		Acting as a prerequisite for participation. Rewarding good behavior. Ensuring that there are economic consequences of reputation.

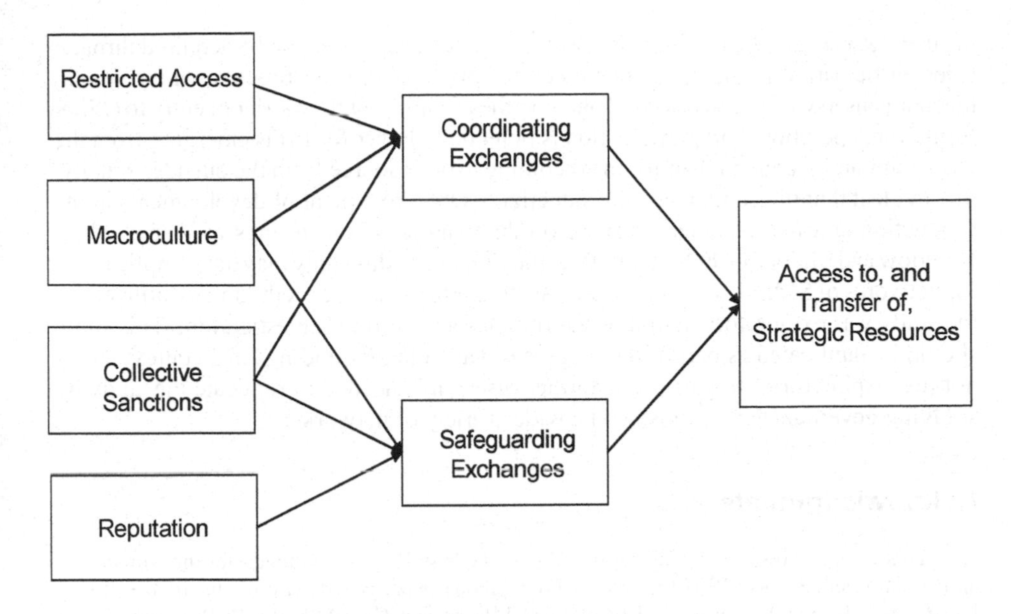

Figure 2. Overcoming Exchange Problems in OSSNs

A comparison of Figure 2 with extant research (as represented in Figure 1) reveals that the role of individual social mechanisms in facilitating coordinating and safeguarding exchanges is different from that envisaged by previous work (e.g., Jones et al. 1997, 1998). In particular, restricted access is not seen as facilitating safeguarding, while both macroculture and collective sanctions have greater influence than previously envisaged.

Our study found evidence that collective sanctions played a more significant role in coordinating and safeguarding exchanges than envisaged by previous work. Collective sanctions were evident as (1) damage to a firm's reputation, (2) exclusion from ongoing projects, or (3) expulsion from the network. Consequently collective sanctions were particularly potent given the importance of both reputation and the bidding leverage of the network to commercial success. Macroculture is seen to play a central role in the governance of OSSNs and, contrary to previous beliefs, has a role to play in both facilitating coordinating and safeguarding exchanges. Macroculture is central to effectiveness because (1) it allows firms to work easily without the imposition of formal agreements, and (2) the specific macroculture observed in OSSNs reflects that of OSS communities, there is an emphasis on collaboration, sharing, and the preservation of the *commons*. In addition, macroculture is evident not just as a shared set of values (including beliefs, language, etc.), but also includes the mechanisms by which these ideas are expressed and shared among members. This represents a richer view of culture, more typical of the thinking of Hannertz (1992) on cultural complexity in information societies than of Jones et al. (1997).

To conclude, our study suggests that as social mechanisms can be used to effectively overcome exchange problems in OSSNs, such networks have less overhead vis-à-vis coordination and safeguarding than if they relied exclusively on formal or legal mechanisms. Additionally, our study suggests that the competencies required to engage in such networks and participate in socially enabled exchanges can be acquired through prior participation in the open source community development process, where similar mechanisms have been observed. Together, these imply that the barrier of entry to OSSN formation, operation, and participation is potentially lower for firms emerging from the OSS community context than for firms coming to open source from the outside. Finally, it is evident that the challenges of productizing OSS are typical of developments in the production and use of other complex product and service offerings as discussed by Davidow and Malone (1992). Therefore, the findings of this study may have implications for networks of SMEs operating in other sectors. However, the methodology utilized for the study was exploratory, and thus the findings need further investigation. This study should be duplicated as part of the process of validating its findings in a context that is not just exploratory. In particular, further research is needed to replicate the study by studying governance mechanisms in a wider variety of networks.

Acknowledgments

This work has been financially supported by the Irish Research Council for the Humanities and Social Sciences (IRCHSS), the Science Foundation Ireland award to Lero – the Irish Software Engineering Research Centre, and by the EU FP6 project OPAALS (Open Philosophies for Associative Autopoietic Digital Ecosystems).

References

Alter, C., and Hage, J. 1993. *Organizations Working Together*, London: Sage Publications.

Benbasat, I, Goldstein, D. K., and Mead, M. 1987. "The Case Study Research Strategy in Studies of Information Systems," *MIS Quarterly* (11:3), pp. 369-386.

Benkler, Y. 2002. "Coase's Penguin, or Linux and the Nature of the Firm," *The Yale Law Journal* (112:3), pp. 369-446.

Benkler, Y. 2006. *The Wealth of Networks: How Social Production Transforms Markets and Freedom*, New Haven, CT: Yale University Press.

Bergquist, M., and Ljungberg, J. 2004. "The Power of Gifts: Organizing Social Relationships in Open Source Communities," *Information Systems Journal* (11:4), pp. 305-320.

Cash, J. I., and Konsynski, B. R. 1985. "IS Redraws Competitive Boundaries," *Harvard Business Review* (63:2), pp. 131-142.

Clemons, E. K., and Row, M. C. 1992. "Information Technology and Industrial Cooperation: The Role of Changing Transaction Costs," *Journal of Management Information Systems* (9:2), pp. 9-28.

Davidow, W. H., and Malone, M. S. 1992. *The Virtual Corporation*, New York: HarperCollins.

De Wever, S., Martens, R., and Vandenbempt, K. 2005. "The Impact of Trust on Strategic Resource Acquisition through Interorganizational Networks: Towards a Conceptual Model," *Human Relations* (58:12), pp. 1523-1543.

Eisendhardt, K.M. 1989. "Building Theories from Case Study Research," *Academy of Management Review* (14:4), pp. 532-550.

Feller, J., and Fitzgerald, B. 2002. *Understanding Open Source Software Development*, London; Addison-Wesley.

Finnegan, P., Galliers, R. D., and Powell, P. 2003. "Applying Triple Loop Learning to Planning Electronic Trading Systems," *Information Technology and People* (16:4), pp. 461-483.

Fitzgerald, B. 2006. "The Transformation of Open Source Software," *MIS Quarterly* (30:3), pp. 587-598.

Galaskiewicz, J. 1985. "Interorganizational relations," *Annual Review of Sociology* (11), pp. 281-304.

Gallivan, M. J. 2001. "Striking a Balance Between Trust and Control in a Virtual Organization: A Content Analysis of Open Source Software Case Studies," *Information Systems Journal* (11:4), pp. 277-304.

Granovetter, M. 1992. "Problems of Explanation in Economic Sociology," in *Networks and Organizations: Structure, Form, and Action*, N. Nohria and R. C. Eccles (eds.), Cambridge, MA: Harvard Business School Press, pp. 25-56.

Hannertz, U. 1992. *Cultural Complexity: Studies in the Social Organization of Meaning*, New York: Columbia University Press.

Henderson, J. C. 1990. "Plugging into Strategic Partnerships: The Critical IS Connection," *Sloan Management Review* (31:3), pp. 7-18.

Jones, C, Hesterly, W, and Borgatti, S. 1997. "A General Theory of Network Governance: Exchange Conditions and Social Mechanisms," *Academy of Management Review* (22:4), pp. 911-944.

Jones, C., Hesterly, W., Fladmoe-Lindquist, K., and Borgatti, S. 1998. "Professional Service Constellations: How Strategies and Capabilities Influence Collaborative Stability and Change," *Organization Science* (9:3), pp. 396-410.

Kaufman, F. 1966. "Data Systems that Cross Company Boundaries," *Harvard Business Review* (44:1), pp. 141-155.

Krishnamurthy, S. 2005. "An Analysis of Open Source Business Models," in *Perspectives on Free and Open Source Software*, J. Feller, B. Fitzgerald, S. A. Hissam, and K. R. Lakhani (eds.), Cambridge, MA: MIT Press, pp. 279-296.

Madill, A., Jordan, A., and Shirley, C. 2000. "Objectivity and Reliability in Qualitative Analysis: Realist, Contextualist and Radical Constructionist Epistemologies," *British Journal of Psychology* (91:1), pp. 1-20.

Markus, M. L., Manville, B., Agres, C. E. 2000. "What Makes a Virtual Organization Work?," *Sloan Management Review* (42:1), pp. 13-26.

Moore, G. 1999. *Crossing the Chasm,* Harper-Perennial, New York,.

Oliver, C. 1990. "Determinants of Interorganizational Relationships: Integration and Future Directions," *Academy of Management Review* (15:2), pp. 241-265.

Raymond, E. S. 2001. *The Cathedral & the Bazaar*, Sebastopol, CA: O'Reilly.

Sagers, G. 2004. "The Influence of Network Governance Factors on Success in Open Source Software Development Projects," in *Proceedings of the 25th International Conference on Information Systems*, R. Agarwal, L. Kirsch, and J. I. DeGross (eds.), Washington, DC, December 12-15, pp. 427-438.

Stafford, T. 2002. "Trust, Transactions, and Relational Exchange: Virtual Integration and Agile Supply Chain Management," in *Proceedings of the 8th Americas Conference on Information Systems*, R. Ramsower and J. Windsor (eds.), Dallas, TX, August 9-11, pp. 2365-2371.

Strauss, A., and Corbin, J. 1990. *Basics of Qualitative Research: Grounded Theory Procedures and Techniques*, Newbury Park, CA: Sage Publications.

Szczepanska, A. M., Bergquist, M., and Ljunberg, J. 2005. "High Noon at OS Corral: Duals and Shootouts in Open Source" in *Perspectives on Free and Open Source Software,* J. Feller, B. Fitzgerald, S. A. Hissam, and K. R. Lakhani (eds.), Cambridge, MA: MIT Press, pp. 431-446.

Timmers, P. 1999. *Electronic Commerce: Strategies and Models for Business-to-Business Trading*, New York: Wiley.

West, J. 2003. "How Open Is Open Enough? Melding Proprietary and Open Source Platform Strategies," *Research Policy* (32), pp. 1259-1285.

Woods, D., and Guliani, G. 2005. *Open Source for the Enterprise*, Sebastapol, CA: O'Reilly Media.

Van de Ven, A. H. 1976. "On the Nature, Formation and Maintenance of Relations Among Organizations," *Academy of Management Review* (1:4), pp. 24-36.

Van de Ven, A. H., and Walker, G. 1984. "The Dynamics of Interorganizational Coordination," *Administrative Science Quarterly* (29:4), pp. 598-621.

Yin, R. K. 1994. *Case Study Research, Design and Methods*, Newbury Park, CA: Sage Publications.

About the Authors

Joseph Feller received his Ph.D. from University College Cork, where he is currently a senior lecturer in Business Information Systems. He has published four books and his work has been published in a number of international journals and conferences, including *The Information Systems Journal, Journal of Strategic Information Systems, Journal of Information Technology Theory and Application, Software Process–Improvement and Practice, IEE Proceedings—Software*, and *Annals of Cases on Information Technology*. He has also published over 60 technical and practitioner articles, and is a frequent contributor to the Cutter Consortium. He was the program chair (with Alberto Sillitti) for the Third International Conference on Open Source Systems and was the founder and chair of the IEEE/ACM Workshop Series on Open Source Software Engineering (hosted by the International Conference on Software Engineering). Joseph can be reached at jfeller@afic.ucc.ie.

Patrick Finnegan received his Ph.D. from the University of Warwick, England, and is currently a senior lecturer in Information Systems at University College Cork, Ireland. His research on interorganizational systems and electronic business has been published in a number of international journals and conferences, including *Information Technology and People, DATABASE, The Information Systems Journal, The International Journal of Electronic Commerce, Electronic Markets,* and *Information Resources Management Journal.* He is currently an associate editor of *Information Systems Journal* and president of the Irish Association for Information Systems. Patrick can be reached at p.finnegan@ucc.ie.

Brian Fitzgerald received his Ph.D. from the University of London and currently holds the Frederick A Krehbiel II Chair in Innovation in Global Business and Technology at the University of Limerick, where he is also director of the Lero Graduate School in Software Engineering (www.lgsse.ie). His research interests include open source software, agile methods, and distributed software development. He has published 10 books and his work has appeared in most leading IS journals and conferences. He has served as a guest senior editor for several prominent journals, including *Information Systems Research, Communications of the ACM, European Journal of Information Systems*, and *Information Systems Journal.* Brian can be reached at brian.fitzgerald@ul.ie.

Jeremy Hayes is a lecturer in Business Information Systems at University College Cork. His research interests are in the area of electronic business models, interorganizational systems, and business agility. He has published his research findings at international conferences and journals, including the European Conference on Information Systems, the Collaborative Electronic Commerce Technology and Research Conference (CollECTeR Europe), IFIP 8.3, *Journal of Database Management*, and *European Journal of Operational Research.* Jeremy can be reached at j.hayes@ucc.ie.

13 EMERGING TECHNOLOGIES IN THE SERVICE SECTOR: An Early Exploration of Item-Level RFID on the Fashion Sales Floor

Claudia Loebbecke
Claudio Huyskens
Department of Media and Technology Management
University of Cologne
Cologne, Germany

Janis Gogan
Department of Information and Process Management
Bentley College
Waltham, MA U.S.A.

Abstract *This paper describes the early weeks of a live pilot of item-level RFID by METRO Group's German department store, Kaufhof. The RFID-Enabled Sales Floor utilizes UHF Gen2 RFID tags on fashion items, combined with RFID-enabled dressing rooms, intelligent displays, and smart mirrors. The pilot represents a pioneering attempt to conduct end-to-end UHF item-level tracking of items through the point-of-sale. Based on an exploratory case study, we reflect on the implications of ubiquitous item-level RFID technology and offer suggestions for further research on the socio-technical implications of this important product and process innovation.*

Keywords RFID, item level, supply chain, case study, pilot, retailing

We are moving RFID out of the innovation labs, beyond the supply chain, and into working retail stores.

Dr. Gerd Wolfram
Managing Director, METRO Group Information Technology, 2007

Please use the following format when citing this chapter:

Loebbecke, C., Huyskens, C., and Gogan, J., 2008, in IFIP International Federation for Information Processing, Volume 267, Information Technology in the Service Economy: Challenges and Possibilities for the 21st Century, eds. Barrett, M., Davidson, E., Middleton, C., and DeGross, J. (Boston: Springer), pp. 189-198.

1 INTRODUCTION

In recent years many companies have launched pilots to investigate uses of radio frequency identification (RFID) for improved operational control and managerial decision-making (Curtin et al. 2007). In the service sector in general, and more specifically in retailing, a number of companies have demonstrated the value of applying RFID tags to palettes, cases, or cartons to track goods across the supply chain (e.g., Angeles 2005; Loebbecke and Huyskens 2006; Sellitto et al. 2007). However, use of item-level RFID is in a nascent stage; thus far, only small-scale tests have been conducted, in lab-like settings by companies such as Prada (Raafat et al. 2007), Kaufhof and Gerry Weber (Loebbecke and Palmer 2006), Falabella (O'Connor 2007), Mi-Tu (Swedberg 2007a), and Throttleman (Swedberg 2007b). These early RFID trials revealed a variety of challenges, including network problems, poor integration, and lack of standardization with existing enterprise networks (Loebbecke 2006).

To capitalize on new standards set by EPCglobal and ISO, in September 2007, METRO Group and its sales branch Kaufhof, both pioneers in the use of emerging technologies in retailing (e.g., Loebbecke 2004, 2007), moved item-level technology from the lab to real-life on one entire floor in Kaufhof's department store in Essen, Germany (e.g., Wessel 2007). Kaufhof's RFID-Enabled Sales Floor is touted as the first real-life pilot of an end-to-end ultra-high frequency (UHF) generation 2 (Gen2) RFID application using all relevant EPCglobal standards. The purpose of the pilot is to evaluate the use of an emerging, commercially available, and standardized RFID technology on fashion items from their arrival at the store to check-out by shoppers. The system traces the flow of goods throughout the store and uses RFID-enabled smart dressing rooms, smart displays, and a smart mirror to improve the shopping experience.

With this paper, we answer calls for research on emerging information technologies (Curtin et al. 2007; Kendall 1997) with an exploratory study of the first weeks of the Kaufhof pilot. This paper describes the pilot preparations including the RFID item-level infrastructure and applications. Since the pilot is scheduled to run through 2008, it is premature to report on detailed outcomes. Instead, we will initiate an investigation of UHF item-level RFID. This will focus on the issues surrounding RFID as a product or process innovation in the light of ubiquitous computing and its socio-technical implications.

2 METHOD

An emerging technology such as item-level RFID cannot be separated from its context of use (Orlikowski 1992). In order to uncover contextual issues, we opted for an exploratory case study (Yin 2003) of a real-world item-level RFID pilot at Kaufhof. Data were gathered by means of public sources and in-depth interviews between December 2006 and November 2007, with managers from METRO Group, Kaufhof, and RFID technology provider Reva Systems. Interviews explored technical and organizational issues in the pilot preparations and the first few weeks of operation, with a focus on the interplay of technical aspects of item-level RFID and the sales floor social system (Scarborough 1995).

3 KAUFHOF'S RFID-ENABLED SALES FLOOR

3.1 Overview

In fall 2007, after 10 months of preparation, international retailer METRO Group and its department store chain Kaufhof, together with technology partners Checkpoint Systems, Impinj, and Reva Systems, moved item-level RFID for the second time from a laboratory environment to real-life when they launched the RFID-Enabled Sales Floor in the men's fashion department of a store in Essen, Germany.[1] Said Ashley Stephenson, Chairman and Co-Founder, Reva Systems, "The RFID network infrastructure controls the whole store experience. For a while, people have been tracking items from the shipper to the distribution center. This is the first time that the spotlight is on the retail sales floor."

In the pilot, all items are individually tagged with UHF Gen2 RFID and tracked from arrival at the store to check-out by shoppers. Sixty readers are installed in twenty locations. METRO Group and Kaufhof expect to "fill the data void that exists from the time that items are received at the store location until they are sold" (Dr. Gerd Wolfram, METRO Group Information Technology, 2007). Specifically, management aims to find answers to questions such as

- How long does it take until items arrive in the store?
- How long do items stay in the store before turnover?
- How does in-store shopping behavior relate to sales?
- What are the implications for store layout and item displays?

3.2 Pilot Participants

With 2006 sales of about €60 billion, METRO Group is the world's third largest retailing company. Its 270,000 employees operate in 2,400 outlets in 30 countries (www.metrogroup.de). Kaufhof Department Stores (Kaufhof), a sales division of METRO Group, operates 146 stores in Germany and Belgium. In 2006, it reported €3.6 billion sales, of which roughly half resulted from fashion (www.metrogroup.com). In 2003, Kaufhof became an RFID pioneer in the fashion industry when it initiated the Kaufhof-Gerry Weber RFID pilot, testing RFID at unit- and item-level using the high frequency spectrum across the fashion supply chain and in the store before wide-spread RFID standardization. Kaufhof learned that RFID was technically feasible and offered a positive business case as it enabled faster and more precise inventory management, enhanced supply chain processes, and improved customer service (Loebbecke and Palmer 2006). Kaufhof subsequently initiated full RFID roll-out on logistic units such as pallets and boxes in late 2004 (Loebbecke and Huyskens 2006) and envisioned the potential of RFID on item-level.

---

[1]The pilot is partially conducted under the European Commission funded BRIDGE project (www.bridge-project.eu).

3.3 Pilot Preparations

For the pilot, Kaufhof designed and prepared an entire floor holding men's fashion.

Readers and Antennas. Sixty RFID readers, each with a unique identification, and more than 100 antennas, were installed in the store, along with readers in the back room apparel and shoe storage areas, at the door connecting the back room and sales floor, in service elevators, and at doors covering goods shipments and receipts. On the sales floor, reader locations include dressing rooms, areas where shoppers try on shoes, the "return to floor" rack, shelves and racks for shoes, accessories and apparel, floor displays, promotion displays, checkout counter, checkout line, and exits (main door, customer elevators, and escalators). Together the reading zones cover about 30,000 individual fashion goods. All antennas are tuned for near-field operation in a clearly defined reading zone (hoping to avoid the "unwanted reads" problem). An early technical challenge was tuning antennas for this dense reader environment, in which each tag would be read by more than one reader. Display furniture and portions of checkout counters were shipped to technology vendor Impinj in Seattle, where engineers customized readers and antennas to achieve a read-rate better than the 95 percent METRO Group requirement.

Smart Displays. In the Gardeur shop-in-shop,[2] Kaufhof installed five smart shelves and tables, four smart dressing rooms, and a smart mirror. RFID-enabled smart shelves and tables help consumers quickly find the proper sizes. When shoppers approach a shelf or table containing dozens of folded shirts, they can find out if a shirt with their neck size and arm length is available without digging through all of the shirts. A reader on the display interrogates the shirts' RFID tags and updates a list of sizes for shoppers.

A reader in the smart dressing room recognizes the exact item a shopper brings in, and a touch screen displays additional product information, such as material, price, care instructions, and other available sizes and colors. In the future, smart dressing room software will offer suggestions regarding apparel coordination by, for instance, displaying a written description, photograph and/or video of an appropriate shirt or tie, ideally one that is available in the store.

The smart mirror outside the dressing room provides similar information. At shoppers' requests, the mirror will display the item in a different size or color and note its location (on the floor, in the stock room, or on order). Future plans call for software that will display descriptions, photographs, or videos of appropriate accessories that can be purchased.

In order for the RFID-enabled sales floor to be fully functional, all 60 readers in the store need to be connected to Kaufhof's enterprise applications. Tag acquisition processors (TAPs) and a standardized tag acquisition network (TAN) are provided by Reva Systems to connect the reader layer to Kaufhof's enterprise network. The TAPs centrally manage multi-vendor RFID readers as network elements. They combine real-time adaptive reader control, location-aware tag data processing, and standard-based data services, enabling them to deliver accurate, actionable RFID data to the application layer. The TAN manages readers in real-time as parts of an overall enterprise network architecture, relates data to a specific source and its location, and provides applications and back-

[2]Manufacturers design and staff specifically branded "shops-in-shops" within Kaufhof stores.

office systems with standardized data services. A key feature of the TAN is that it conducts singulation, which helps avoid problems when tags are covered by multiple readers (Krishna and Husak 2007).

Other Sales Floor Preparation. At Kaufhof's regional distribution center, UHF Gen2 RFID tags are attached to each item to be sold in the pilot department store.[3] With RFID readers on conveyors and packing-tables, Kaufhof scans the RFID tags as items are packed or moved. It relies on UHF dock-door portals to read the tags when items are shipped from the distribution center to the pilot department store.

As items arrive at the store, RFID portals in the back room are utilized for confirming their receipt. On the sales floor, employees hang clothes on racks and assign RFID tags on items to the RFID reader of the rack. Inventory software can then check whether items are located in their assigned racks.

To reassure shoppers regarding perceived privacy intrusion, signs and flyers posted throughout the store explain the RFID technology and its applications. At the cash register, employees inform shoppers that RFID tags can be discarded or retained for use as an electronic receipt for potentially returning an item.

3.4 In-Store Processes

Kaufhof can capture real-time data on various employee- and shopper-based activities.

RFID data —gathered when employees unload goods, store items in the back room, or move them to other floors—reveal which items are in the sales room and which ones need to be put on display. Inventory data inform employees of items to be taken off sales floor shelves (e.g., as seasons change). Dressing room data identify items that shoppers leave behind and direct employees to return them to the sales floor. Data provided via handheld readers help employees find specific items for customers.

RFID data help identify which items shoppers pick, purchase, or reject. Readers at escalators and in elevators show when shoppers move one or more items across floors. Readers on shelves identify when shoppers pick an item and whether shoppers take it to a dressing room or proceed directly to the checkout line. As shoppers approach checkout, an RFID reader installed under the counter reads each item's tag to update the sales system. If an item is placed on the counter but not purchased, Kaufhof requires an employee to read the tag using a handheld reader and thus informs the system.

3.5 Use Cases: Data Analysis Possibilities

Data captured at each reading instance of a tag as outlined above could be fed into a central data repository. Filling this data void is only the first step. Subsequently, Kaufhof will determine which of the data analysis possibilities and respective queries to the data repository given below will be technically sustainable and economically meaningful.

---

[3]In addition, Kaufhof outfits each item with a usual hangtag showing the price and bar-coded stock-keeping unit (SKU) number.

Concerning store deliveries and returns, Kaufhof could query which items have actually arrived on the store premises and which items have been checked into the store "returns" location. Subsequent analysis could reveal how long it takes before returns are picked up again and when they are shipped back to the distribution center. Weekly or even daily inventory counts could become possible. The system could report on daily, weekly, seasonal, or yearly turnover of individual items and could easily and quickly identify the fastest sell-through items and slow sellers, as well as those newly delivered items which are sold ahead of existing stock.

Kaufhof could also analyze whether an item was tried on prior to purchase and whether a purchased item was the first one a shopper picked and carried through the store (which might indicate that he entered the store intending to purchase that particular item). The system could reveal how much time shoppers spend in the check-out line before actually making a purchase, and whether an item sold was selected from a promotional display.

Further, the system could reveal which items shoppers try on and how many they buy, how long shoppers wait for a dressing room, what percentage of items are brought into dressing rooms, and how long rejected items stay off the sales floor. From such queries, Kaufhof could determine the actual locations for particular items. Kaufhof could derive insights about promising item positioning (shelves, racks, promotional displays) and location (near escalators or dressing rooms) and whether shoppers' behavior changes as a result of promotions, store layout, or employee actions. Also, Kaufhof could analyze how much time shoppers spend in line at checkout.

In terms of stocking decisions, Kaufhof could investigate how long items stay in the back room before moving to the sales floor, how many items are returned from the sales floor to the back room, and which items shoppers return to the store. With regard to theft prevention, the system may shed light on when and where missing items were last seen. Kaufhof could match clips of a security video, indexed via location and time, with RFID data.

4 CONCEPTUAL TRAJECTORIES

Occasionally, a new technology like UHF Gen 2 item-level RFID emerges with the potential to enhance an organization's ability to capture, analyze, store, distribute and retain valuable information for improved operations and analysis.

Early-stage pilots such as Kaufhof's RFID-Enabled Sales Floor offer clues as to how organizations could utilize RFID on the fashion floor. Organizations can gain reliable and useful data to inform management's decision-making. For instance, RFID-enabled traceability of items throughout the store may lead to more complete assortments, shorter waiting times, and enhanced customer experience, including the ability to find items that shoppers seek faster (e.g., Hardgrave et al. 2007). The marriage of item traceability with smart displays may facilitate interactions with shoppers and thus enable innovative services, competitive differentiation and improved profits. However, for these benefits to be achieved there are numerous challenges, which can be examined in future studies through several theoretical lenses.

Item-level RFID as product or process innovation. Research on product and process innovations (e.g., Benamati and Lederer 2001; Day and Schoemaker 2000; Jarvenpaa and Ives 1996; Van de Ven 2005) provides relevant insights for studying item-level RFID as an emerging technology. As a product innovation, there is potential for item-level RFID to rewrite the rules of competition and potentially alter entire industries (Gebauer and Segev 2000). However in IS research, emerging infrastructure technologies have been distinguished from product innovations (Robey 1986) and considered as process innovations (Zmud 1982). As a process innovation, item-level RFID could enable (or conversely, constrain) new models for managing and utilizing finer-grained data.

Item-level RFID raising ubiquitous computing to a new level. As this emerging technology moves to multi-application, enterprise-wide deployments, RFID readers could become the most numerous and densely deployed network devices, with both positive and negative ramifications. Item-level traceability can reveal new information about shoppers' interests and behaviors and increase organizations' monitoring capabilities (Price et al. 2005), extending the literature on ubiquitous computing and networked organizations. At the same time, ubiquitous RFID gives rise to privacy concerns (Shapiro and Baker 2001; Spiekermann and Ziekow 2006) and suggests investigating the counter-intuitive, adverse behavioral effects (Scheepers et al. 2006) more fully.

Socio-technical impacts of item-level RFID. As computing devices become ubiquitous, they give rise to new categories of applications and new relationships among users and between users and technologies. As item-level RFID becomes ubiquitous, new context-aware and/or location-based services will be developed. As managers, employees, and customers gain experience with these applications, there will be significant ramifications for organizations and industries (Lyytinen and Yoo 2001; Weiser 1999). To understand the ramifications of ubiquitous item-level RFID it is helpful to explore the socio-technical processes that lead to acceptance and seamless use in some contexts and not in others (Lyytinen and Yoo 2002). The mutual accommodation of technical features of item-level RFID and human processes requires further study, since prior research (DeSanctis and Poole 1994; Orlikowski 1992) finds that technology behaves as an influential agent in the organizational context. Thus, future studies may investigate how specific characteristics of item-level RFID constrain or enable users and explore new structures which emerge from its use. For example, it would be helpful to closely examine how shopper and employee behavior and attitudes influence the organizational change initiated by the implementation of item-level RFID technology. It would be of similar interest to investigate how the ubiquity of item-level RFID leads to challenges associated with customers' privacy and trust (Shapiro and Baker 2001; Weiser 1991). Finally, as item-level RFID also increases the amount of data available to managers, research may want to explore the resulting implications for companies' monitoring capabilities and control over assets (Curtin et al. 2007).

References

Angeles, R. 2005. "RFID Technologies: Supply-Chain Applications and Implementation Issues," Information Systems Management (22:1), pp. 51-65.

Benamati, J., and Lederer, A. 2001. "Rapid Information Technology Change, Coping Mechanisms, and the Emerging Technologies Group," *Journal of Management Information Systems* (17:4), pp. 183-202.

Curtin, J., Kauffmann, R., and Riggins, F. 2007. "Making the 'Most' out of RFID Technology: A Research Agenda for the Study of the Adoption, Usage and Impact of RFID," *Information Technology Management* (8:2), pp. 87-110.

Day, G., and Schoemaker, P. 2000. "Avoiding Pitfalls of Emerging Technologies," *California Management Review* (42:2), pp. 8-33.

DeSanctis, G., and Poole, M. 1994. "Capturing the Complexity in Advanced Technology Use: Adaptive Structuration Theory," *Organization Science* (5:2), pp. 121-147.

Gebauer, J., and Segev, A. 2000. "Emerging Technologies to Support Indirect Procurement: Two Case Studies from the Petroleum Industry," *Information Technology and Management* (1:1/2), pp. 107-128.

Hardgrave, B., Armstrong, C., Riemenschneider, C. 2007. "RFID Assimilation Hierarchy," in *Proceedings of the 40th Annual Hawaii International Conference on System Sciences*, Waikoloa, Big Island, HI, January 3-6, p. 224b..

Jarvenpaa, S., and Ives, B. 1996. "Introducing Transformational Information Technologies: The Case of the World Wide Web Technology," *International Journal of Electronic Commerce* (1:1), pp. 95-126.

Kendall, K. 1997. "The Significance of Information Systems Research on Emerging Technologies: Seven Information Technologies that Promise to Improve Managerial Effectiveness," *Decision Sciences* (28:4), pp. 775-792.

Krishna, P., and Husak, D. 2007. "RFID Infrastructure," *Communications Magazine, IEEE* (45:9), pp. 4-10.

Loebbecke, C. 2004. "Modernizing Retailing Worldwide at the Point of Sale," *MIS Quarterly Executive* (3:4), pp. 177-187

Loebbecke, C. 2006. "RFID in the Retail Supply Chain," in *Encyclopedia of E-Commerce, E-Government and Mobile Commerce*, M. Khosrow-Pour (ed.), Hershey, PA: Idea Group Publishing, pp. 948-953.

Loebbecke, C. 2007. "Piloting RFID Along the Supply Chain: A Case Analysis," *Electronic Markets* (17:1), pp. 29-37.

Loebbecke, C., and Huyskens, C. 2006. "Weaving the RFID Yarn in the Fashion Industry: The Kaufhof Case," *MIS Quarterly Executive* (5:4), pp. 169-179.

Loebbecke, C., and Palmer, J. 2006. "RFID in the Fashion Industry: Kaufhof Department Stores AG and Gerry Weber International AG, Fashion Manufacturer," *MIS Quarterly Executive* (5:2), pp. 15-25.

Lyytinen, K., and Yoo, Y. 2001. "The Next Wave of Nomadic Computing: A Research Agenda for Information Systems Research," *Sprouts: Working Papers on Information Environments, Systems and Organizations* (1:1), pp. 1-12.

Lyytinen, K., and Yoo, Y. 2002. "Issues and Challenges in Ubiquitous Computing," *Communications of the ACM* (45:12), pp. 63-65.

O'Connor, M. 2007. "Falabella Plans Second Item-Level RFID Pilot," *RFID Journal*, August 29 (www.rfidjournal.com/article/articleview/3585/1/1/definitions_on).

Orlikowski, W. 1992. "The Duality of Technology: Rethinking the Concept of Technology in Organizations," *Organization Science* (3:3), pp. 398-427.

Price, B., Adam, K., and Nuseibeh, B. 2005. "Keeping Ubiquitous Computing to Your-Self: A Practical Model for User Control of Privacy," *International Journal of Human-Computer Studies* (63:1/2), pp. 228-253.

Raafat, F., Sherrard, W., Meslis, L., and Windt, J. 2007. "Case Study: Applications of RFID in Retail Business," in *Proceedings of the 2007 Meeting of the South West Region of the Decision Sciences Institute*, San Diego, CA, pp. 695-701.

Robey, D. 1986. *Designing Organizations* (2nd ed.), Homewood, IL: Irwin.

Scarbrough, H. 1995. "Review Article on The Social Engagement of Social Science," *Human Relations* (48:1), pp. 1-11.

Scheepers, R., Scheepers, H., Ngwenyama, O. 2006. "Contextual Influences on User Satisfaction with Mobile Computing: Findings From Two Healthcare Organizations," *European Journal of Information Systems* (15:3), pp. 261-268.
Sellitto, C., Burgess, S., and Hawking, P. 2007. "Information Quality Attributes Associated with RFID-Derived Benefits in the Retail Supply Chain," *International Journal of Retail & Distribution Management* (35:1), pp. 69-87.
Shapiro, B., and Baker, C. 2001. "Information Technology and the Social Construction of Information Privacy," *Journal of Accounting and Public Policy* (20:4/5), pp. 295-322.
Spiekermann, S., and Ziekow, H. 2006. "RFID: A Systematic Analysis of Privacy Threats and a 7-Point Plan to Adress Them," *Journal of Information System Security* (1:3), pp. 2-17.
Swedberg, C. 2007a. "Hong Kong Shoppers Use RFID-Enabled Mirror to see what they Want," *RFID Journal*, September 4 (www.rfidjournal.com/article/articleview/3595/).
Swedberg, C. 2007b. "Throttleman Adopts Item-Level Tagging," *RFID Journal*, August 24 (www.rfidjournal.com/article/articleview/3580/).
Van de Ven, A. 2005. "Running in Packs to Develop Knowledge-Intensive Technologies," *MIS Quarterly* (29:2), pp. 365-378.
Weiser, M. 1991. "The Computer of the 21st Century," *Scientific American* (265:3), pp. 66-75.
Weiser, M. 1999. "Some Computer Science Issues in Ubiquitous Computing," *Mobile Computing and Communication Review* (3:3), pp. 12-21.
Wessel, R. 2007. "Metro Group's Galeria Kaufhof Launches UHF-Item-Level Pilot," *RFID Journal*, September 20 (www.rfidjournal.com/article/articleview/3624).
Yin, R. 2003. *Case Study Research: Design and Methods* (5th ed.), Thousand Oaks, CA: Sage Publications.
Zmud, R. 1982. "Diffusion of Modern Software Practices: Influence of Centralization and Formalization," *Management Science* (28:12), pp. 1421-1431.

About the Authors

Claudia Loebbecke holds the Chair of Media and Technology Management at the University of Cologne, Germany. In 2005-2006, she was elected President of the Association for Information Systems (AIS). She also worked and researched at the London School of Economics, Bentley College, CISR (at Massachusetts Institute of Technology), INSEAD, Copenhagen Business School, Erasmus University, Hong Kong University of Science and Technology, University of New South Wales, and McKinsey & Co. She is a senior editor for *Journal of Strategic Information Systems*, an associate editor for *The Information Society*, and is on the editorial boards of several IS and media journals. Claudia received a Master's and a Ph.D. in Business Administration, both from the University of Cologne, and an MBA from Indiana University, Bloomington, Indiana. She has published over 150 internationally peer-reviewed journal articles and conference papers (see www.mtm.uni-koeln.de/team-loebbecke-home-engl.htm). Claudia can be reached at claudia.loebbecke@uni-koeln.de.

Claudio Huyskens is a Ph.D. candidate in the Department of Business Administration, Media and Technology Management at the University of Cologne, Germany. He holds a Master's degree in Business Administration from the University of Cologne. His research covers RFID technology in supply chains, IT outsourcing via the Internet, and new media entrepreneurship. His publications appear in several journals (e.g., *MIS Quarterly Executive*, *European Journal of Information Systems*, *Electronic Markets*, *European Journal of Operational Research*) and conference proceedings (see www.mtm.uni-koeln.de/team-huyskens-home-engl.htm). Claudio can be reached at claudio.huyskens@uni-koeln.de.

Janis L. Gogan holds Ed.M., MBA, and DBA degrees from Harvard University. A member of the Information & Process Management faculty of Bentley College, Janis teaches IT management and strategy courses and has published about 80 refereed papers in conference proceedings and journals (such as *Communications of the Association for Information Systems, Electronic Markets, International Journal of Electronic Commerce, Journal of Information Systems, Journal of Information Technology*, and *Journal of Management Information Systems).* Her publications also include numerous book chapters, *Information Week* columns, and teaching cases which have been taught in the U.S., European, Australian, and Asian schools. Her current research examines two topics: the management of emerging information technologies and interorganizational information sharing under time pressure. Janis can be reached at jgogan@bentley.edu

14 THE COMPUTERIZATION OF SERVICE: Evidence of Information and Communication Technologies in Real Estate

Steve Sawyer
College of Information Sciences and Technology
Pennsylvania State University
University Park, PA U.S.A.

Fuyu Yi
College of the Liberal Arts
Pennsylvania State University
University Park, PA U.S.A.

Abstract *We explore the overlap between service and computerization using macro-level industrial data on the U.S. real estate market and five comparison industries (hospitals, financial services, legal services, machinery manufacturing, and fabricated metals). The macro-level data comes from the U.S. Bureau of Economic Analysis and the U.S. Census Bureau and we use it to develop insights on computerization and service relative to contributions to the U.S. gross domestic product. This analysis shows that while information and communication technology investments in real estate lagged comparison industries from 1969 to 1997, since then ICT investments in real estate have increased rapidly. At the same time, there has been a growth in the number workers even as the industry's contribution to GDP has grown. We identify two implications of these findings. First, ICTs are not being used are not as a substitute for labor. Second, the rapid growth in ICT investments has been absorbed into real estate quickly and well. Still, computerization in real estate continues, suggesting that process studies and more micro-analyses are critical next steps.*

Keywords Service, computerization, real estate, secondary data

Please use the following format when citing this chapter:

Sawyer, S., and Yi, F., 2008, in IFIP International Federation for Information Processing, Volume 267, Information Technology in the Service Economy: Challenges and Possibilities for the 21st Century, eds. Barrett, M., Davidson, E., Middleton, C., and DeGross, J. (Boston: Springer), pp. 198-209.

1 INTRODUCTION

In this paper, we draw on data regarding the take up and uses of information and communication technologies (ICTs) in the U.S. real estate and five other industries to speculate on implications for services sciences (Chesbrough and Spohrer 2006). We see the take up and uses of ICT as core to both computerization (e.g., Burris 1998) and to services (e.g., Rust and Miu 2006; Sheehan 2006). We define a service as the change in one's state—or the state of goods that one owns—brought about from activity of others, and jointly consummated. Our definition reflects elements common to the many definitions of service: the joint or mutual engagement by service provider and service consumer, the importance of knowledge in consummating the service action, a temporal connection between production and consumption of the service, the potential for the service to involve tangible and intangible goods, and reliance on ICTs.

This definition of service also makes explicit that shaping and sharing information among providers and services consumers is a core element of the coconstructed nature of a service. For that reason we draw on the work of real estate agents as an exemplar of informational service. We focus attention on the U.S. real estate industry as the rapid rise in the uses of ICT has been both visible and hotly debated. In 1997, very few real estate agents posted their listings online and very few consumers used online resources to pursue buying and selling of houses. In 2007, online access to house and housing information was extensive, with reports that more than 70 percent of all consumers began their house-hunting process by going online.

The contribution of this paper is to theorizing on the computerization of service. We use the evidence from the real estate market to raise issues relative to both the take up and uses of ICT in supporting services. To do this, we use macro-level data to develop a comparison of the take up and uses of ICT in real estate and five other industries: two manufacturing activities and three other professional services industries. Our comparison focuses on ICT investment, employment, and value-added to the U.S. gross domestic product over time.

This paper continues in three sections. In the next section, we report the selection of industries and data, and we explain the approach used to do the comparative analysis in these industries. In the third section, we report findings from this analysis. In the fourth section, we discuss these findings, speculate on their implications, and suggest future work.

2 RESEARCH APPROACH

In order to have a broader and at the same time representative set of industries for investigation, we selected real estate and five other industries using the 2002 U.S. North American Industry Classification System (NAICS) codes.[1] We chose three information-intensive and service-oriented industries (similar to real estate) and two more traditional manufacturing industries (to provide for comparison). For the two manufacturing industry sectors, we use fabricated metal products and machinery/manufacturing. These are mature industries, ones that have experienced tremendous changes in base technologies

[1]For information on the NAICS codes, see http://www.census.gov/epcd/naics02/.

over the past 100 years, but little change in the past 30 years. Including these industries provides a basis of comparison between real estate (as an information-intensive service industry) and classic industrial activity regarding computerization.

For services-oriented industries, we selected three that are also information-intensive: legal services, financial services (securities, commodity contracts, and investments), and hospitals. The service industries selected for comparison with real estate differ on the level of professionalization (e.g., the need for college or graduate degrees, extensive professional certification) relative to trades (that may or may not demand certification but likely require some level of training. We selected hospitals because this industrial sector also has also rapidly computerized.[2]

2.1 Data

We use secondary data because it is available and primary data is not. Secondary data at the industry level is often available from specific industrial trade associations (such as the National Association of Realtors and the American Manufacturing Association). However, these data are often proprietary, use different (and possibly incompatible) measures, and follow different procedures to gather; some industrial sectors do not have comprehensive associations or data collection. Another source of secondary data are nongovernmental agencies such as the Organisation of Economic Cooperation and Development (OECD),[3] which provides yearly reports by industry regarding a range of ICT expenditures.

A third source of secondary data is the U.S. federal government. For example, the Census Bureau provides data about ICT expenditures for the previous two years by industry and this is summarized in the bi-annual "Information and Communication Technology" report, with the most recent being released in April, 2007.[4] The U.S. Bureau of Economic Analysis (BEA) provides historical data on ICT expenditures by industry and so we chose to draw on their data, as detailed below.[5]

The first source of data used in this analysis is the BEA's "Historical-Cost Investment in Private Nonresidential Fixed Assets." The estimates are presented for detailed industries by asset type. The second source of data is the BEA's "Gross-Domestic-Product-(GDP)-by-Industry" data set. The BEA defines the value-added of an industry as "the gross output of an industry or a sector less its intermediate inputs; the contribution of an industry or sector to gross domestic product (GDP)." We use industry value-added to compare industries, acknowledging as we do that this is a granular measure. For this analysis, the focus is on change in value-add over time and change in value-add relative to investments in ICT. Third, we use the BEA's tables of industry employment based on full-time equivalents (FTE). Full-time and part-time employees

---

[2]We chose not to include in this comparative analysis either insurance carriers or travel agents, as these industries focus more on commodity sales (an airline ticket or insurance policy is more a commodity than is a house).

[3]Information on the OECD is available at http://www.oecd.org/.

[4]The Census Bureau's report, "Information ad Communication Technology: 2005," issued in April 2007, is available at http://www.census.gov/prod/2007pubs/ict-05.pdf.

[5]The information on the BEA in the following paragraph was obtained from its website, http://www.bea.gov/national/.

by industry are the number of employees on full-time schedules plus the number of employees on part-time schedules. FTE employees by industry are the number of employees on full-time schedules plus the number of employees on part-time schedules converted to a full-time basis. The number of FTE employees in each industry is the product of the total number of employees and the ratio of average weekly hours per employee for all employees to average weekly hours per employee on full-time schedules.

The time-range used for analysis spans 1969 to the latest available data (either 2005 or 2006). We chose 1969 because in this year IBM was forced to unbundle hardware from software (Carmel 1997; Wolff 2002). This unbundling triggered a rapid growth in IT investments across many industries and sectors of the economy. We provide separate tables for the subset of this period spanning 1997–2005/2006 because 1997 was the year when much of the multiple listing services data on homes was first made available via the Internet (on www.realtor.com), making this a watershed point in the computerization efforts in this industry. For the industries we are comparing, BEA value-add data are only available for the period 1977 to 2005. Relative to the occupational employment and wage estimate data, in order to achieve consistency in industrial comparison over time, we use NAICS FTE data for the period 1998–2005.

2.2 Analysis

We present our analysis in four figures. In Figures 1 through 3, we present data plotted against time[6] for each of the six industries. In Figure 4, we present the value-added by industry using chained value data to provide evidence of value-added by industry while accounting for inflation.

3 FINDINGS

Trends identified in the figures are summarized in Tables 1 and 2 and discussed below. We conclude this section by discussing three findings which we summarize in Table 3 and discuss below.

The summary provided in Table 1 makes clear that the real estate industry has been investing more in ICT since 1997 relative to the comparison group of industries. This is a marked change from the previous 30 years. The summary provided in Table 2 makes clear that the value-added (to GDP) by the real estate industry has increased relative to the comparison industries since 1997, and compares favorably with other service and information-intensive industries over the entire period for which we have data. There is a large, significant, and positive correlation between investments in ICT and increased value-added for all six industries.[7]

---

[6]The BEA also produces chain-type quantity indexes for investment in order to account for inflation over time. However, the price for computer and related products is decreasing dramatically over time. As a result, inflation is not a central concern.

[7]Pearson correlations (two-tailed, all significant at $p \leq 0.01$) are real estate = 0.872; legal services = 0.968; securities, commodity contracts, and investments = 0.544; hospitals - 0.990; fabricated metal products = 0.977; and machinery = 0.940. Correlations cannot imply causations.

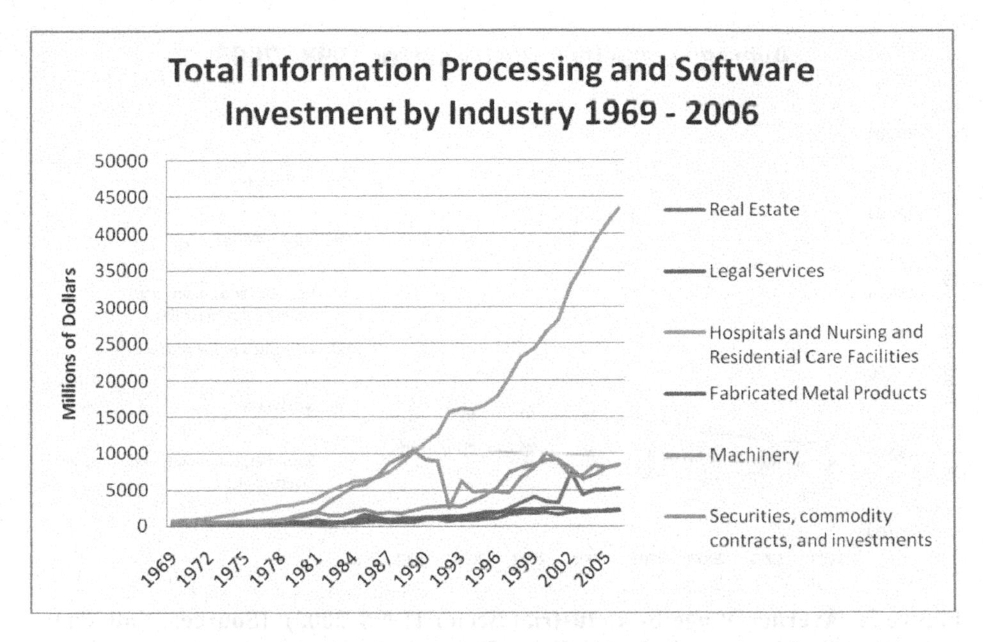

Figure 1. Total ICT (Equipment and Software) Investments by Industrial Sector (1969–2006). (Source: Historical-Cost Investment in Private Nonresidential Fixed Assets, BEA, http://www.bea.gov/national/FA2004/Details/Index.html)

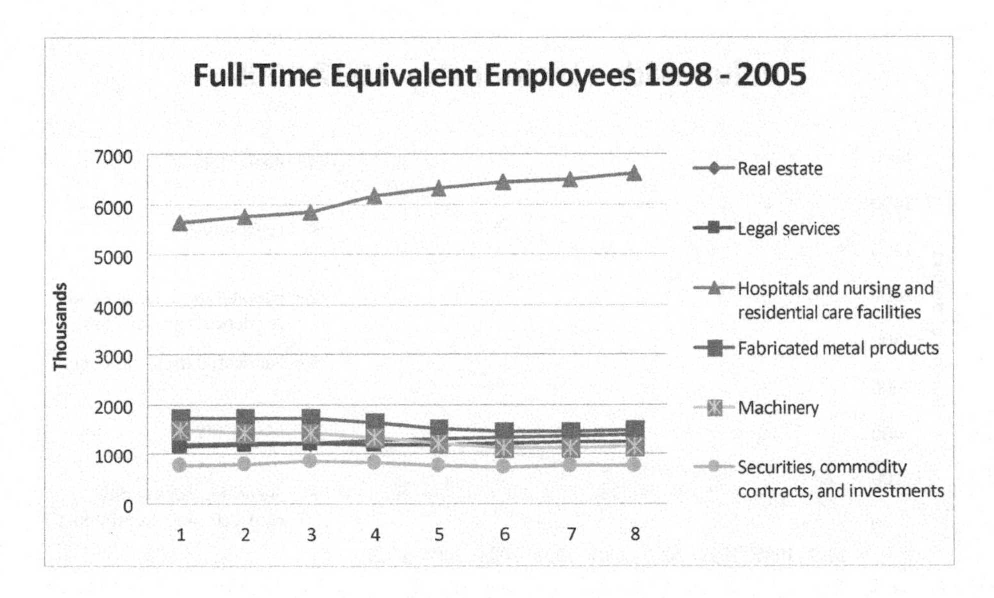

Figure 2. Full-Time Equivalent (FTE) Employment by Industrial Sector (1998–2006) (Source: Full-Time Equivalent Employees by Industry, Gross-Domestic-Product-by-Industry Accounts, 1947-2006, BEA)

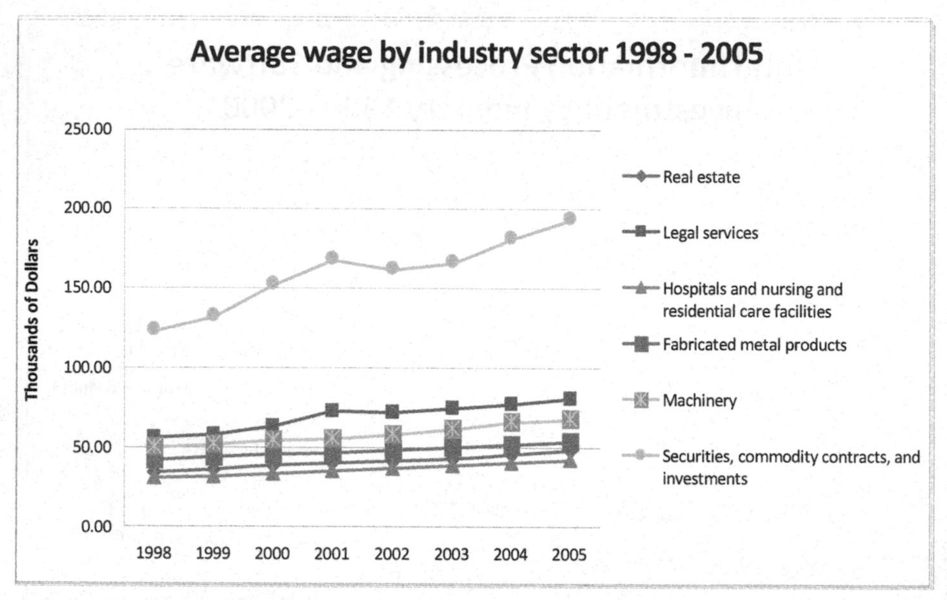

Figure 3. Average Wage by Industrial Sector (1998–2005) (Source: Full-Time Equivalent Employees by Industry, Gross-Domestic-Product-by-Industry Accounts, 1947-2006, and Components of Value Added by Industry Group, Gross- Domestic-Product-By-Industry Accounts, 1947-2006, BEA)

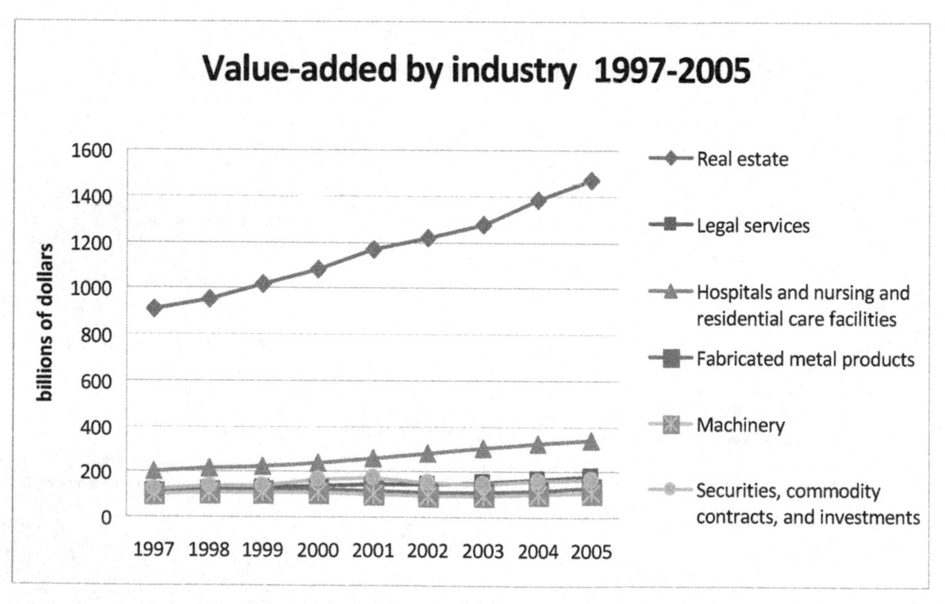

Figure 4. Value-Added (Chained Value) by Industrial Sector (with 2000 as Base), 1997–2005 (Source: Real Value Added by Industry, Gross-Domestic-Product-by-Industry Accounts, 1947-2006, BEA)

Table 1. Summary of ICT Investment Over Time for Six Comparative Industries

Industry	1969 – 1997		1997 - 2005		1969 - 2005	
	Ranking	**Average Rate**	**Ranking**	**Average rate**	**Ranking**	**Average rate**
Real estate	**6**	**7.86%**	**1**	**13.37%**	**6**	**15.62%**
Legal services	1	146.98%	4	3.77%	2	149.64%
Hospitals and nursing and residential care facilities	2	99.77%	2	12.92%	1	160.65%
Fabricated metal products	5	56.15%	6	0.42%	5	45.24%
Machinery	3	76.47%	5	1.06%	4	64.75%
Securities, commodity contracts, and investments	4	61.30%	3	9.92%	3	87.73%

Table 2. Value-Added Change Over Time for Six Comparison Industries

Industry	1977-1997		1997-2006		1977-2006	
	Ranking	**Average Rate**	**Ranking**	**Average Rate**	**Ranking**	**Average Rate**
Real estate	**4**	**18.17%**	**2**	**7.77%**	**4**	**23.27%**
Legal services	2	28.21%	3	7.25%	2	33.91%
Hospitals and nursing and residential care facilities	3	24.60%	1	8.51%	3	31.97%
Fabricated metal products	5	9.70%	5	2.59%	5	9.11%
Machinery	6	6.01%	6	1.66%	6	5.34%
Securities, commodity contracts, and investments	1	90.03%	4	5.01%	1	91.53%

Finding #1: Total ICT investments in real estate lagged most comparison industries until the late 1990s; since then it has been a leader. As depicted in Figure 1, total ICT investments in real estate lagged behind the comparison industries over the period 1969 to 2005. Over this time, total ICT investment in real estate grew at an annual rate of 6 percent per year. This stands in contrast to IT investment in hospitals, which grew at an average annual rate of 12 percent per year over the same period. However, as detailed in Table 1, over the last 10 years, the investments in ICT in real estate have grown substantially, moving real estate into the top of the comparison group of industries. There have been large increases in investments in both ICT equipment and software, and a growth in the number of ICT workers as a percentage of total FTE in real estate. Data in Figure 1 shows that total ICT investments in real estate rose rapidly in the 1990s relative to comparison industries.

Relative to investments in software, real estate is the industry with the most dramatic changes over the period 1969–2005. The first period of rapid change (from 5 percent in 1977 to 24 percent in 1980) is followed by a sharp decrease. A period of steady growth in the 1980s (from 5 to 10 percent per year of ICT investment in software) is followed by explosive growth (63 percent in 1992), dropping to between 33 and 40 percent of total ICT investment spending since.

Data presented in Figure 4 and Table 1 provide evidence that the value added to GDP by real estate is both substantial and increasing, even as ICT investments are growing. This is true for all other comparison industries, although the growth is greatest in the service industries. Real estate's value-adding growth is larger than inflation and, since 1997, greater than all comparison industries save for hospitals. This suggests that the real estate industry has been able to turn ICT investments into additional value-added at a rate greater than the other comparison industries. This stands in contrast to industries in this comparison group such as hospitals, which have had much larger investments in ICT but not the seven-fold increase in value added (since 1977).

Finding #2: Real estate FTE growth is the least of all service industries in the comparison group. Data in Figure 2 show that FTE employment in real estate has grown steadily at about 1.9 percent per year since 1969. The securities industry has had the largest growth (about 4.5 percent per year) and legal services industry has been the second largest (at 3.3 percent per year). Hospitals have also been growing (at about 2.8 percent per year over the period 1969–2005). Both fabricated metal products and machinery have a similar and general slightly downward trend. Moreover, compensation per FTE has been rising in all six comparison industries, with real estate similar to traditional manufacturing levels, and well below securities. The real estate average per FTE in 2005 was slightly greater than $50,000, or one-quarter of the average compensation per FTE in the securities industry.

4 DISCUSSION, IMPLICATIONS, AND FUTURE WORK

Building on the findings (summarized in Table 3), we discuss two implications. First, the modest growth of FTE in real estate, combined with the rapid computerization, suggests that computing is not being engaged as a labor-capital substitution. The second implication we raise is that while the take up and uses of ICT in real estate have been quickly absorbed, the effects of this take up and uses of ICT in real estate remain indeterminate.

We further note that the granularity of the secondary data and issues with measuring ICT.[8] Despite these concerns, such comparisons are common. Dumagan et al. (2002) and Wolff (2003), in fact, come to different findings on the roles that ICTs play in industry-level performance. Wolff finds little direct impact of investments in ICTs, save for the restructuring of work and production. Dumagan et al. find that spending on ICT does lead to industry-level differences in productivity, with services that invest more in ICT outperforming those that invest less. Still, these multiple interpretations suggest caution when relying solely on analysis and interpretation of secondary data.

Implication #1: Modest growth of FTE in real estate relative to rapid computerization. Data show there is growth in both ICT investment and FTE in all service industries. This pattern seems counter to the labor/capital substitution relationship that was (and remains) very clear in manufacturing and is often assumed as a basis for ICT investments. In contrast to this, we offer three possible reasons for the growth in both ICT investment and FTE in real estate, and more broadly for information services.

---

[8]Micro-level analyses such as those developed by Crowston et al. (2001) and Sawyer et al. (2003) help to interpret this macro-level data.

Table 3. Summary of Findings

1.	Total ICT investments in real estate lagged most comparison industries until the late 1990s; since then, it has been a leader.
2.	Real estate growth is the least of all service industries in the comparison industries.

The first reason for this dual growth may be that this is a transient artifact of the change-over from human-centered to more computing-centered work processes. The transient argument contends that FTE will drop away as ICTs begin to replace human knowledge work (like programmed trading, automated ticketing, or artificially intelligent loan originating decisions). However, using ICTs to replace knowledge work (as opposed to physical work) seems to be more complicated, particularly when the work flow is difficult to specify and very contingent, as it is in real estate (e.g., Crowston et al. 2001).

A second reason may be that there is a surge of FTE and ICT investment related to a particular technological innovation, and, when it is no longer new, ICT investment and FTE will drop. However if there is constant innovation and ongoing introduction of new ICTs, there may be a consistent need for new skills and more people to deploy and maintain the new ICTs. This means that FTE and ICT growth is not a temporary activity.

A third reason may be that take up and uses of ICT for known goals (so-called first-level effects) give way to uses of ICT for things not first imagined (the "second-level" effects; Sproul and Kiesler 1991). This suggests that ICT for informational services may be best seen as platform investments (e.g, Ciborra 1996).

Implication #2: ICTs have been absorbed into real estate quickly and well. Despite rapid increases in ICT investments since 1997, the real estate industry seems to have been able to absorb changes and innovations that arise from their take up and uses (when evaluated against the increasing value-added contribution to GDP over the same time). There are at least three reasons. First, it may be that the ICTs being taken up and used in real estate are so simple that there is a very small learning curve for any one individual. This translates into a short lag time between taking-up the new ICT to value creation, particularly when compared to other industries. A derivative of this reasoning may be that the tasks for which the ICTs are being used are also simple. However, the simple-tasks/simple-uses reasoning fails to account for the rise in FTEs and has no supporting empirical evidence.

A second reason to explain why ICTs have been absorbed into real estate so quickly may be due in part to the industry's communal investments in a shared information infrastructure embodied in the sharing of house listing data available in the 900 plus local multiple listing services (MLS) through common public portals. Data in Figures 1 and 2 suggest that there were extensive ICT investments in the late1990s, consistent with the time during which most local MLS invested in larger-scale digital and computing infrastructures to bring their agent's house-listing data online. This shared information infrastructure is a unique attribute of residential real estate that builds on the extensive history of the local and regional MLS, and the series of changes in real estate practices that have been engendered in part by the availability of online MLS access (see Sawyer et al. 2005). The implication of this reasoning for other information-centric services is that the specific and unique nature of real estate's information sharing infrastructure is critical to understanding

A third reason for this successful take up and use of ICTs into real estate may be a combination of both changing social experiences and the locus of innovation and change. By changing social experiences, we note that over the same time period (1997-2005) in which real estate was investing in ICTs, so were millions of consumers. By 2005, more than 70 percent of Americans were actively using the Internet, had computers in their homes, used them at work or school, and—perhaps more importantly—were aware that the Internet, computing, and digital information/communication had important implications for how people worked, lived and organized. A second aspect of this success in taking up and getting value from ICT arises from how real estate agents do their work. That is, much of the work done by real estate agents and professionals is localized and personal. As Crowston et al. (2001) detail, real estate processes are fluid and localized, putting the focus of innovation at the individual, not institutional level.

In general, the data presented and discussed here make clear that information-based services such as real estate rely on the concepts of coinvention (see Bresnehan and Greenstein 2001). This suggests that more micro-studies, focused on particular markets, technologies, and activities, are needed. These studies will amplify and help to explain variations among industries (due perhaps to differences in policy regimes, level of regulation, etc.). Melding macro-level analyses and micro-studies together will provide more empirical clarity and reduce the theoretical and conceptual preaching that seems to happen in the absence of good data.

A second area for future work would be to focus on the kinds of ICT being used in supporting services. For example, there is a clear trend in hospitals toward large-scale, enterprise-wide packages that provide a semiautomated administrative infrastructure to support operations. The uses of these types of systems have been documented in the manufacturing and fabricated products industry for nearly 30 years. However, in the legal industry, there is a growing use of personal computing and common shared information infrastructures, particularly as courts become more digitally enhanced. In the securities industry, there is a growth in large-scale programmed trading and individualized decision-support systems. And, relative to the empirical domain we use in the paper, the modest level of micro-scale data on ICT uses in real estate suggests computerization differs from these approaches.

Acknowledgments

This paper is much improved by comments to prior draft by Paul Bishop, Kevin Crowston, Rolf Wigand, Elizabeth Davidson, and three anonymous reviewers. This work is supported in part by funding from the U.S. National Association of Realtors and the U.S. National Science Foundation.

References

Bresnehan, T., and Greenstein, S. 2001. "The Economic Contribution of Information Technology: Towards Comparative a User Studies," *Journal of Evolutionary Economics* (11), pp. 95-118.

Burris, B. 1998. "Computerization of the Workplace," *American Review of Sociology* (24), pp. 141-157.

Carmel, E. 1997. "American Hegemony in Packaged Software Trade and the 'Culture of Software,'" *The Information Society* (13:2), pp. 125-142.

Chesbrough, H., and Spohrer, J. 2006. "A Research Manifesto for Services Science," *Communications of the ACM* (49:7), pp. 35-49.

Ciborra, C. 1996. "The Platform Organization: Recombining Strategies, Structures, and Surprises," *Organization Science* (7:2), pp. 103-118.

Crowston, K., Sawyer, S., and Wigand, R. 2001. "The Interplay Between Structure and Technology: Investigating the Roles of Information Technologies in the Residential Real Estate Industry," *Information Technology & People* (14:2), pp. 163-183.

Dumagan, J., Gill, G., and Ingram, C. 2003. "Industry Level Effects of IT Use on Productivity and Inflation," in *Digital Economy*, D. Evans (ed.), Cheltenham, UK: Edgar Algar, pp. 41-60.

Metka, S., Jaklic, A., and Kotnik, P. 2006. "Exploiting ICT Potential in Service Firms in Transition Economies," *The Service Industries Journal* (26:3), pp. 287-299.

Rust, R., and Miu, C. 2006. "What Academic Research Tells Us About Service," *Communications of the ACM* (49:7), pp. 49-54.

Sawyer, S., Crowston, K., Wigand, R., and Allbritton, M. 2003. "The Social Embeddedness of Transactions: Evidence from the Residential Real Estate Industry," *The Information Society* (19:2), pp. 135-154.

Sawyer, S., Wigand, R., and Crowston, K. 2005. "Redefining Access: Uses and Roles of Information and Communications Technologies in the Residential Real Estate Industry from 1995-2005," *Journal of Information Technology* (20:4), pp. 3-14.

Sheehan, J. 2006. "Understanding Service Sector Innovation," *Communications of the ACM* (49:7), pp. 42-47.

Sproul, L., and Kiesler, S. 1991. *Connections: New Ways of Working in the Networked Organization*, Cambridge, MA, MIT Press.

Wolff, E. 2002. "Productivity, Computerization and Skill Change," *Federal Reserve Bank of Atlanta Economic Review*, pp. 63-78.

About the Authors

Steve Sawyer is a founding member and an associate professor at the Pennsylvania State University's College of Information Sciences and Technology. Steve holds affiliate appointments in the department of Management and Organization, the department of Labor Studies and Employer Relations, and the program in Science, Technology and Society. Steve does social and organizational informatics research with a particular focus on people working together using information and communication technologies. Steve can be reached at sawyer@ist.psu.edu.

Fuyu Yi is a master's student in the Pennsylvania State University's Department of Labor Studies and Employer Relations. Fuyu's research focuses on the implications and uses of information and communication technologies relative to work and organizing. Fuyu's email is fuy101@psu.edu.

Part 3:

IT-Enabled Change in Public Sector Services

15 E-GOVERNMENT AND CHANGES IN THE PUBLIC SECTOR: The Case of Greece

Dimitra Petrakaki
Lancaster University
Lancaster, U.K.
Wolverhampton Business School
Wolverhampton, U.K.

Abstract *During the past few decades, many governments around the globe have orchestrated e-government projects in order to improve the way they operate and provide public services to citizens. Apart from the opportunities they open up, e-government projects bring about changes in the well-established practices of the public sector. This paper illustrates some of these changes by exploring a Greek e-government initiative. As the case illustrates, e-government requires an output orientation and business-like behavior from officials, enables constant electronic control, and leads to the standardization of official's knowledge. Drawing upon these changes, we propose a framework of the transformations that e-government brings about in the work roles, nature of work, forms of knowledge, modes of control, and source of accountability of officials.*

Keywords E-government, public sector change, power, performance, standardization

1 INTRODUCTION

During the past few decades, many governments around the globe have reformed the ways in which they operate and provide services to citizens. These reforms have been influenced by ideas incorporated in the electronic government agenda. E-government emerged in the 1990s with the aim of harnessing the benefits of the deployment of information and communications technologies (ICTs) to the public sector. ICTs mediate at two levels: they are used internally, for the reorganization and streamlining of

Please use the following format when citing this chapter:

Petrakaki, D., 2008, in IFIP International Federation for Information Processing, Volume 267, Information Technology in the Service Economy: Challenges and Possibilities for the 21st Century, eds. Barrett, M., Davidson, E., Middleton, C., and DeGross, J. (Boston: Springer), pp. 213-227.

government procedures (Hazlett and Hill 2003; Moon 2002; Vintar et al. 2003), and they are deployed so as to improve public service provision, enable citizens' interactions with government, increase citizens' awareness about government's function and results, and to achieve accountability and transparency (Basu 2004; Gil-Garcia et al. 2007; Kumar and Best 2006; Martin and Byrne 2003; Moon 2002; Vintar et al. 2003; von Haldenwang 2004; Zhang 2002). Frequently, e-government projects are orchestrated by the best practices of the private sector, such as process reengineering, performance measurement, contractualization, and establishment of internal markets (Bloomfield and Hayes 2004; Ciborra 2003; Hoggett 1996; Hood 1991; Tan and Pan 2003; Vintar et al, 2003). The adoption of the best practices of the private sector is thought to eliminate many problems of the public sector such as red tape, time lags, irresponsiveness, and unaccountability (Clarke and Newman 1997; du Gay 2000; Farrell and Morris 2003).

E-government, however, is not reduced to the managerial and information technologies that typically drive its implementation (Hazlett and Hill 2003; Vintar et al. 2003). Rather, its initiation presupposes, to a major or lesser extent, the organization and reorganization of the policies, practices, and procedures that surround the function of the public sector (Cordella 2007). In that way, e-government opens up opportunities for improvements in the function of the public sector but at the same time challenges the regulatory, historical, normative and socio-cultural context within which it emerges.

The aim of this paper is to illustrate some of these changes and challenges by drawing upon a Greek e-government initiative, Citizens Service Centers (CSCs). CSCs were developed as one-stop shops that mediate between citizens and public organizations for the quick and continuous service of citizens. Drawing upon the Greek case, the study illustrates the type of changes that surround e-government by placing specific focus on the practices of officials. The paper is structured as follows: The next section tracks the changes that e-government conditions on the public sector. The methodology that underpinned the research process is presented, followed by presentation of the case. First, we provide an account of the historical construction of the Greek public sector. Next, we present the establishment of CSCs and the transformations that accompanied it. In section five, we discuss the changes that e-government brings about in officials' practices. The paper ends with some concluding remarks.

2 THEORETICAL FRAMEWORK: E-GOVERNMENT AND THE PUBLIC SECTOR

In this section, we first present the ideal type of organization for the public sector by drawing upon Weber's (1948) work. Next, we revise the literature on the function of one-stop shops in e-government initiatives. The section ends with a presentation of previous studies on the changes that e-government brings about to officials' work practices.

2.1 From the Bureaucratic Public Sector to E-Government

The Weberian bureaucracy (Weber 1948) constitutes an iconic example of the organization of the public sector. In its ideal form, the Weberian bureaucracy is conditioned upon certain principles. First, it is comprised of offices (the bureau) each of which has certain designated jurisdictions and specific legal rules and regulations, which officials are responsible for knowing and abiding (Weber 1948). The law constitutes, in this case, officials' form of expertise and government's source of power against which officials' practices are evaluated (Campbell 1993; du Gay 2000; Goodsell 2005; Minson 1998). Second, the hierarchical structure of bureaucracy indicates differences in status, expertise, and jurisdictions and illustrates authority relations of super and sub-ordination (Clark and Newman 1997). Third, the position in the hierarchy is the source of officials' responsibility. Specifically, officials are loyal to their superiors and liable for their subordinates' actions (Blau 1970; Campbell 1993; Goodsell 2005; Minson 1998; Weber 1948). Fourth, the Weberian bureaucracy presupposes impersonality. Officials are disconnected from the authority that their job position entails and obstructed from using the bureau in opportunistic ways (Weber 1948).

Despite its rigid and often dehumanizing principles, the Weberian bureaucracy constitutes a virtuous forms of organization (du Gay 2000; Weber 1948). This is because impersonality, adherence to rules, and abstinence from personal emotions eliminate discrimination and favoritism, obstruct opportunistic behaviors, and ensure equal and democratic public service provision (Blau 1970; Clarke and Newman 1997; du Gay 2000).

Yet, the principles of the Weberian bureaucracy are ideal and not essential. Bureaucracies vary according to the context within which they emerge. For that reason, their function is often perceived as being ineffective, inflexible, and surrounded by illegal phenomena such as corruption and bribery. It was within this context that e-government emerged, with the hype to engender accountability and effectiveness in the public sector.

2.2 One-Stop Shops in E-Government

E-government is perceived as being a trigger for change, yet it serves different purposes and is differently deployed in each country (Yildiz 2007). This study focuses on the e-government initiatives in which one-stop shops are used as governmental access channel (Chadwick and May 2003) or intermediaries (Heeks 2002) between citizens and public officials.

In the e-government literature, one-stop shops are presented as single points, either geographical or electronic (e.g., portals, central/local government websites), where government departments integrate for public service provision (Ho 2002; Illsley et al. 2000; von Handenwang 2004). Citizens typically turn to them in order to submit all of their, administrative in nature, requests quickly and at no cost (Cowell and Martin 2003; Ho 2002; Illsley 2000; Lenk 2002). One-stop shops are, therefore, intended to substitute, with a single contact point, the multiple transaction points that citizens have with the public sector. In that way, they also dissolve heterogeneity and differentiation in public service delivery (Huang and Bwoma 2003; Martin and Byrne 2003; Tan and Pan 2003).

One-stop shops take various forms. They can be shops that provide multiple services from a single point or access points that bring together multiple organizations (Illsley et al. 2000). In their ideal form, one-stop shops are organized in a front and back office. The front office is responsible for coming in contact with citizens and submitting any requests or enquires they might have. The back office is liable for processing citizens' information.

The establishment of one-stop shops presupposes elimination of departmental and organizational boundaries and reorganization of public administrative procedures in accordance with citizens' needs (Kunstelj and Vintar 2004). Technologies such as business process reengineering and customer relationship management are frequently deployed in order to achieve these purposes (Bloomfield and Hayes 2004).

The one-stop shop approach in public service delivery is supposed to lead to efficiencies because it enables fast, integrated, and customer-oriented public service provision (Illsley et al. 2000; Kunstelj and Vintar 2004; Roche 2004). Also, it engenders transparency in the function of the public sector by drawing clear lines of responsibility between the front and back office staff (Wilkins 2002). It is in that sense a general method of improving public service delivery rather than a technique that is solely adopted in e-government initiatives. The rest of the section focuses on the changes that one-stop shops in e-government initiatives and e-government initiatives in general condition in the work practices of public officials.

2.3 E-Government and Changes in the Public Sector

Many studies report the failure of e-government initiatives to engender transformations in officials' work practices. Some of these failures lie with officials' reluctance to perceive the beneficial impact of IT on their work practices (Gil-Garcia et al. 2007), heterogeneous and incompatible information systems (Ciborra 2005), and the "silo mentality" that conditions departmentalism (Bloomfield and Hayes 2004; Davidson et al. 2005; Turner and Higgs 2003).

Of the studies that have accounted for the changes that e-government triggers in the public sector, some focus on issues of power and politics. Specifically, Purnendra (2002) argued that e-government provides autonomy to officials by allowing them to access, manage, and process information. Also, according to Purendra, e-government imposes on officials the imperative for extensive use of ICTs and conditions fears and anxieties that come from the computerization of officials' work and the possibility for downsizing. Further, according to Zhang (2002), e-government provides the potential for exercising electronic forms of surveillance over officials' work. Specifically, by centralizing information in public networks, governments control at a distance both the type of information that is made accessible and the people who are eligible to access it. This allows governments to intervene, take corrective actions, and exercise sophisticated power. Moreover, Davison et al. (2005) illustrated that the transition from government to e-government requires public officials to develop new capacities. Officials, for instance, are asked to improve their performance by sharing their knowledge with colleagues and accelerating the processing of information and the provision of public services.

On the other hand, Bovens and Zouroudis (2002) claimed that e-government transforms officials from "street-level bureaucrats" to "system level bureaucrats." Specifically, e-government creates three types of civil servants: those who process data, those who communicate with citizens (interface staff), and those who manage the whole process. Further, they argued that e-government initiatives presuppose the codification of laws, reducing in that way officials' autonomy and freedom to exercise their discretion. Also, according to Cordella (2007), e-government draws upon the new public management agenda in order to impose managerial practices on officials' work.

Furthermore, Kumar and Best (2002), who studied the establishment of a one-stop shop in India, pinpointed three changes in officials' work practices. First, the one-stop shop reduced the direct contact that officials had with citizens. Second, officials were obliged to use new technology without getting appropriate training. Third, officials lost their freedom to decide how, when, and to whom public services are provided. This type of decision making was now displaced to the staff who worked in this one-stop shop Internet kiosk. The latter also diminished officials' opportunities for corruption and obstructed their rent-seeking behavior.

Despite these interesting conclusions, e-government studies have neglected to account for how e-government influences the historically shaped practices of officials. This methodological omission obstructs, in turn, our further conceptualization of the changes that surround e-government projects (Heeks and Bailur 2006). The paper will track these changes by examining a Greek e-government project within the historical context of the Greek public sector.

3 RESEARCH METHODOLOGY

The research drew upon the qualitative paradigm and particularly social constructionism (Berger and Luckman 1966). According to this paradigm, social phenomena are constructed by various discourses and material artifacts. These in turn are accompanied by cognitive and normative aspects that, when adopted, construct subjectivities, guide the conduct of individuals, and reproduce the institutions within which they are created (Berger and Luckman 1966, p. 111). The cognitive aspects define the roles and responsibilities we undertake, for instance, as parents, spouses, and employees. The normative aspects indicate the "right" behaviors we are expected to internalize and the "wrong" to avoid in order to be accepted as legitimate members of the reality of which we are a part. The adoption and reproduction of these cognitive and normative aspects render institutions legitimate and unquestioned (Suchman 1987). In accordance with that, this paper will consider e-government as an initiative that is socially constructed through the deployment and reproduction of various managerial and technological means and discourses.

The research was conducted with the use of interviews and observation along with document collection. Particularly, we reviewed laws concerning public administration from the 1950s until the present, governmental regulations about CSCs' function and their collaboration with public organizations, newspaper articles on public administration along with government documents that were produced by the Ministry of Interior, Public Administration and Decentralization (MIPAD) such as presentations and reports. We also reviewed the Greek sociological literature in order to account for how the Greek public sector has historically been constructed.

The research project was carried out in three periods. The first was between October and December 2005, the second between March and May 2006, and the last in August 2006. The first period was mainly but not exclusively devoted to interviewing and observing. We interviewed officials from the Greek MIPAD, two politicians, who were implicated due to their position in the regulation of CSCs, the vendor who implemented and maintains the technological platform for the CSCs, and supervisors and staff of the CSCs and made observation of the latter's daily practices. Interviews were semi-structured, recorded, and lasted, on average, for an hour. In some cases, second interviews were conducted. The second and third period were mainly devoted to the observation of CSCs' staff's collaboration with civil servants from various public sector organizations and interviews with civil servants. We visited five CSCs located in the two biggest cities in Greece. In order to understand how CSCs functioned, we observed their practices for at least four hours a day, each day, for a period of two months. In parallel with the observations, we held discussions with CSCs staff when they were dealing with citizens and also during their breaks. Notes of these discussions were recorded in a diary.

Interview transcripts, research diaries, and documents were gathered, carefully read, matched, compared, and organized into large themes. Each theme was then further analyzed and divided into subthemes. These were then discussed, compared with the literature, and developed. The current study draws mainly upon information that derived from documents and interviews with the MIPAD officials and civil servants.

4 THE CASE OF E-GOVERNMENT IN GREECE

This section describes the Greek e-government initiative. First, we present a brief historical account of the way in which the Greek public sector has been constructed. Then, we describe the establishment of Citizens Service Centers (CSCs).

4.1 Greek Public Administration

The Greek public sector consists of the central and local government, public and quasi-public organizations, and independent administrative authorities. The people who work in the public sector, civil servants, are permanently employed in public organizations. Additionally, there are a number of people who work on a contract basis in order to cover urgent and temporary needs. Independently of their employment status, all civil servants are legally accountable to the law (i.e., the constitution, civil servants' code).

The Greek public sector was founded in 1830s after the independence of the state from the four centuries of Ottoman subjection (Argyriadis 2000). It was modeled on the French public sector, which functioned in accordance with the principles of the Weberian bureaucracy (Argyriadis 2000; Weber 1948). Through time, the function of the Greek public sector has been pertained by hierarchical relations, affluence of rules and regulations, and an increased number of departments. Also, from its foundation until now, the public sector has displayed certain characteristics.

First, it has been the largest employer. Specifically, in the 19th century, the government employed an increasing number of citizens in the public sector. It did so in order to address socio-economic problems, such as unemployment and poverty, that followed the war of independence (Mouzelis 1978; Tsoukalas 1986). The government continued the same practice in the 1950s. The aim at that time was to address the problem of internal immigration that civil war conditioned (Argyriadis 2000; Tsoukalas 1987). The public sector continued to expand until it reached its peak at the end of the 1980s. During that decade, the socialist government promoted the rhetoric of participation and democratization and imposed social criteria for employment in the public sector (Avgerou 2002).

Second, the function of the public sector has been surrounded by the development and operation of client relations (Avgerou and McGarth 2007; Legg 1969; Mouzelis 1978). The latter are relationships of mutual exchanges that are developed between unequal, in terms of authorities and/or resources, parts in order to satisfy their interests (Brinkerhoff and Goldsmith 2004; Kaufman 1974; Lemarchand 1972). Client relations prevailed from the 1830s (state's independence) until now in all echelons of the public administration (i.e., between civil servants and citizens and politicians and civil servants). Many times, client relations entailed the exchange of political support for employment in the public sector. Also, client relations took various forms through time. For instance, in the 19th century, client relations were personal and direct exchanges between local governors and citizens, whereas in the 20th century, client relations were mediated relationships between political parties and citizens.

Third, the growth of the public sector rendered civil servants a powerful group or, as Tsoukalas (1987, p. 115) argued, "a state within the state." Every government wanted to gain civil servants' sympathy, ensuring in that way their political support. For that reason, governments were concerned that any changes they brought about wouldn't threaten the interests of civil servants. This attitude, however, triggered a chain of consequences. The favorable treatment of civil servants led to a lack of effective mechanisms to control their behavior. This in turn conditioned illegitimate phenomena such as development of client relations and corruption (Ballas and Tsoukas 2004). Also, officials' favorable treatment meant lack of effective performance measurement techniques. Officials' performance was estimated by their supervisors and as a result was highly subjective and partial (Ballas and Tsoukas 2004). Moreover, the affluence of laws in combination with their abstract and contradictory character allowed civil servants to exercise a high degree of discretion when interpreting the law. This is thought to condition arbitrariness and irresponsibility. Finally, the growth of the public sector has been accompanied by low salaries, lack of incentives, limited jurisdictions, and unclear responsibilities, which conditioned civil servants' dissatisfaction, alienation, and low morale (Argyriadis 2000; Avgerou 2002; Lipsky 1980).

It was in this context that the first steps toward the modernization of the Greek public sector were taken. At the time, various laws were passed about the introduction of management practices into the function of the public sector, privatization of public organizations, and depoliticization of employment procedure. The majority of these measures did not respond to the intended way, yet they diffused the imperative to change the public sector and paved the way to the establishment of CSCs.

4.2 The Establishment of Citizens Service Centers

Citizens Service Centers were established in 2002 as one-stop shops. They were developed by the Ministry of Interior, Public Administration and Decentralization (MIPAD) and they were managed by a group of officials from the MIPAD who constituted the CSCs' project team. CSCs are staffed with individuals who work on fixed term contracts, which are renewable subject to satisfactory performance. According to the law (L.3013/2002), the role of CSCs is to mediate between citizens and public organizations for the continuous and fast service of citizens. CSCs accept citizens' requests for public services, then request that they to be processed by the relevant public organizations, and when they have been processed, pass the outcome (such as certifications, licenses, copies of public documents, etc.) to citizens. CSCs collaborate with public organizations either through fax, e-mail, post, courier, or, mainly, the personal transfer by the CSC staff to the relevant public organization. Despite the establishment of CSCs, citizens can still turn directly to public organizations in order to submit their administrative requests.

CSCs were conceptualized by the MIPAD as an amalgamation of the best practices of the private sector. An official from the MIPAD said, "*bureaucracy came to a halt...it couldn't be further improved. On the other hand, the private sector had experience, so we had to look at the practices of the private sector.*" In particular, CSCs adopted the branch system of the banking sector, the logistic system of automotive industry, and the customer orientation of fast-food outlets. Their function was also supported by a centralized computer system that monitored the outputs of each CSC and each CSC staff. Table 1 is a synopsis of the best practices of the private sector that were transferred to CSCs. The table is an outcome of the accounts of MIPAD officials during interviews. These practices are discussed below.

To begin with, the system of branches, which the banking sector deploys, ensures that customers submit all of their banking requests in any branch, independently of its location. In the Greek public sector, however, public organizations have been a monopoly in public service provision. Citizens are anticipated to know the variety of public services that are provided by each public organization and turn to each specialized public department in order to submit their enquiries. Yet most citizens either lack of this knowledge or are indifferent to it. Further, officials from the MIPAD argued that citizens are used to qualitative services from the private sector and expect the same quality standards from the public sector. For that reason, the MIPAD created homogeneous governmental shops across the country "*that have the label of the state and within which citizens can receive any public service.*" These shops, CSCs, didn't substitute but supplemented public organizations. In other words, CSCs are an alternative public service provider and consequently an option to citizens.

Second, the MIPAD studied and imitated the logistic system of automotive industry. The latter partitions the production of cars, which takes place centrally, from their delivery, which is undertaken by local dealers. Similarly, the MIPAD intended through CSCs to divide the supply and provision of public services. Traditionally, civil servants came in

Table 1. The Best Practices of the Private Sector Transferred to CSCs

Best Practices of the Private Sector	Practices Transferred to CSCs
Banking sector: system of branches	Multiple public services from single, homogeneous points
Automotive industy: logistics system	Supply versus provision Mediated public service provision
Fast-food chains: customer service	Standardization of administrative procedures
Information and communication technologies (ICTs)	Electronic performance monitoring system

direct contact with citizens and were solely responsible for public service provision. With the establishment of CSCs, however, they are rendered "suppliers" of public services; their role is restricted to checking and processing documentation and issuing public documents or services (certifications, licenses, social benefits, etc). Also, their new role presupposes various responsibilities: civil servants are anticipated to be professionals when they process citizens' cases, experts in and responsible for their work object. They also have to be quick and effective, to be able to explain in a simple and understandable way anything necessary to CSCs' staff, and to care for the results of their actions.

Third, the customer service of fast food outlets was used as a prototype for the development of CSCs. As an official from the MIPAD argued, fast food outlets are "*homogeneous shops...have a recognizable label and a standard menu that people know*." This allows them to provide food of standard type, quantity, and quality. Drawing upon the fast-food outlets the officials from the MIPAD decided to "*create shops of McDonalds' type*" in which standardized services would be provided. Specifically, the MIPAD simplified and standardized 1,000 administrative procedures in order to eliminate any unnecessary information and documentation from the process of public service provision. Because of standardization, citizens would also receive public services or documents instantly, without queues and time lags. The standardization of administrative procedures was accompanied by a number of regulations that required civil servants to adopt the standardized procedures and work in collaboration with CSCs' staff for the quick processing of citizens' cases. Civil servants are also forbidden from requesting different documents from those stipulated by standardization and given a maximum time limit of 50 days to process the requests submitted by CSCs' staff. Non-compliance with the above obligations is thought to be a breach of civil servants' duties and is legally punishable.

Further, the establishment of CSCs was accompanied by the development of a central computer system, in which updated administrative information, administrative forms, and general details about citizens is kept. CSCs' staff use the system on a daily basis by uploading citizens' requests and downloading administrative forms and information. Also, the system incorporates an MIS application that keeps daily statistics about the outputs of each CSC and staff, such as their complete and pending requests and the time that it takes to deliver a submitted request back to citizens (response time). These statistics are solely available to CSCs' project team in the MIPAD in order to measure the productivity of CSCs' staff and civil servants. The productivity of CSCs' staff is directly estimated through the credentials that staff has to log into the system and is calculated based upon the number of requests that CSCs' staff submit per day. The productivity of civil servants refers to the time that it takes them to process citizens' requests. Yet, because few public organizations run information systems that are connected to CSCs' system, civil servants' productivity is indirectly estimated through the pending requests and response time of CSCs. These indicators show, respectively, which public organizations fail to process citizens' requests and how long it takes them to process and provide the public service or document back to CSCs' staff.

For the officials of the MIPAD, the MIS constitutes a great innovation in the public sector. This is because it provides, for the first time, objective information to estimate civil servants' performance based upon their outputs: "*technology gives the opportunity for the first time to control public administration....In the past, performance was measured by supervisors and, thus, all estimations were subjective and biased.*" Indeed, out of this information, the CSCs' project team constructed average performance indicators. The latter were then the criterion against which actual performances were compared and, in case of deviations, decisions for intervention and correction were made. For instance, the MIPAD suspended the function of some CSCs that had low productivity and intervened into the function of various public organizations in order to advise them on how to become more effective. Finally, CSCs' computer system was perceived as being an intermediary that would unite, in the long run, CSCs with the "traditional" public sector.

5 ANALYSIS AND DISCUSSION

This section discusses the changes that CSCs induced to civil servants' practices along with their implications. As we shall see, CSCs changed civil servants' work roles, nature of work, forms of knowledge, mode of control, and source of accountability. These types of changes are illustrated in Table 2 and constitute the analytical framework for the discussion that follows.

First, the establishment of this one-stop shop split the (previously united) process of public service provision into two, processing and delivery, and attributed the ***role*** of suppliers to civil servants. This partitioning would achieve two results: (1) the elimination of the contact point between civil servants and citizens would reduce any opportunities for corruption and client relations (Kumar and Best 2002); (2) the supply–provision split would bring work efficiencies. The reduction in the number of citizens that turn to public organizations would free up more time for civil servants to focus solely on their work object. Because of that, the supply–provision split could also be approached as an effort to increase the status of civil servants, which historically has been devaluated, as a professional group.

Second, the establishment of CSCs influenced the ***nature of civil servants' work.*** The establishment of CSCs constructed a quasi-market in which civil servants, who used to be a monopoly in public service provision, would compete with CSCs' staff for the provision of public services to citizens. Specifically, civil servants would need to promote their competitive advantage so as to create, maintain, and expand the number of citizens that still turn to them. Their clientele, rather than their monopoly, constitutes the source of their legitimacy. Moreover, the development of a quasi-market is under-

Table 2. The Types of Changes to Work Practices Brought by E-Government

Types of Changes	Chang
Work role	Officials ➔Suppliers
Nature of work	Monopoly ➔ Quasi-market
Forms of knowledge	Implicit knowledge ➔ Explicit knowledge
Mode of control	Periodic, legal, and hierarchical ➔Constant, individualistic, and output-oriented
Source of accountability	Law and hierarchy ➔ output orientation

pinned by the idea that civil servants are neither guards of the public interest nor public service providers, but mere entrepreneurs. This is manifested by the fact that civil servants have to process information within specific time limits and respond to performance averages, which are electronically monitored by the ministry. Further, the development of a quasi-market in the provision of public services aimed to limit the authorities of civil servants by diffusing part of them (e.g., the right to deliver public services) to CSCs' staff (Bovens and Zouroudis 2002). In that way, civil servants' opportunities to develop client relations with citizens would be considerably reduced (Kumar and Best 2002).

What is more, the aim of standardizing administrative procedures was to transform civil servants' abstract, idiosyncratic and tacit ***knowledge*** into standardized and explicit knowledge. This is mainly because it is anticipated that civil servants will draw upon the standardized procedures, rather than their knowledge and judgment, in order to make decisions (Bovens and Zouroudis 2002), eliminating the autonomy of civil servants to freely exercise discretion and leading, contrary to what was previously said, to the deprofessionalization of their occupation.

In addition, the establishment of CSCs introduced new, managerial in nature, forms of power. Specifically, the electronic performance monitoring system that was deployed in the Greek one-stop shop displaced the periodic, legal, and hierarchical ***mode of control*** and imposed one that is constant, individualistic, and output-oriented. The performance monitoring system constituted a *panopticon* (Foucault 1977) in public service provision (Zhang 2002). The continuous monitoring, along with the development of average performances, was intended to impose the imperative for continuous self-control on civil servants. This is because their awareness of being the object of monitoring would make civil servants internalize the performance averages and continuously strive to become better performers. Further, by rendering performance visible, inefficiencies would be corrected and arbitrariness and civil servants' autonomy would be eliminated.

Moreover, the standardization of administrative procedures was intended to ***normalize*** civil servants' conduct. Standardization would eliminate any fragmentation or differentiation in the processing and provision of public services by bringing each public service down to its necessary documentation/information. In that way, public service provision became a preprocessed and predictable activity that could be undertaken by anyone who lacked work experience and/or knowledge on the function of the public sector (Bovens and Zouroudis 2002). Also, by fostering homogeneity, the aim of standardization was to obstruct civil servants' autonomy, which, as the historical account indicated, was for many years beyond the control of the government. In that way, the well-established problem of client relations that civil servants typically developed with citizens or political parties would also be eliminated. So, although we agree with Purnendra (2002) that e-government shakes the established power relations in the public sector, we argue in line with Zhang (2002) that, rather than providing autonomy, e-government is a means for delicate and constant exertion of power over officials.

Finally, the establishment of CSCs influenced officials' ***source of accountability.*** Typically, officials' source of accountability was the law and the hierarchy. The establishment of CSCs added a further source of unaccountability that is output-oriented in nature. Specifically, officials are rendered responsible for quickly processing citizens' requests and effectively collaborating with the staff from the one-stop shop. This orientation toward outputs is justifiable within the Greek public sector that had long enabled officials to be politically partisian and exercise unlimited discretion. Yet, we need to anticipate that by paying attention solely to quantifiable activities (for instance, the number of citizens that are served or the time to process citizens' requests), important prerequisites for public service provision, such as human judgement and sensitivity, are excluded.

The above types of changes to officials' practices similarly influence the process of public service provision. Specifically, e-government renders the process of public service provision predictable, quantifiable, and standardized or, as Ritzer (1993) and Garson (1988) would say, intends for the McDonaldization of public service provision. First, e-government eliminates personal contact and renders public service provision a mediated procedure. It does so by standardizing public services and by treating citizens as cases that need to be fit into predefined categories (Ciborra 2005; Fountain 2001). Moreover, public services are provided in a fast, standardized but hardly customized way by multiple public service providers, which compete in (quasi) internal markets. Further, we can speculate that these effects influence citizens who, far from being considered as the locus of social and political rights, are recast as responsible individuals that are aware of their needs, anticipate fast and qualitative service, and are able to make effective choices among competing public service providers.

Overall, the type and scope of changes that e-government and one-stop shops bring about to officials' practices are contingent and situated rather than definite (Yildiz 2007). They depend, not least, upon the historical context within which e-government is initiated. History, as the case illustrated, doesn't only provide the rationale upon which e-government draws but also opens up avenues for comprehending not only what e-government is (and is not) but also what e-government does. Specifically, our findings indicate that e-government initiatives bring about five types of changes to the nature of work, roles, mode of control, knowledge, and source of accountability of officials, and similarly influence the provision of public services. These changes, as they are encapsulated in Table 2, constitute a methodological framework, which, together with the history of each public sector, can assist governments and researchers in the conceptualization of the (potential) consequences, risks, and challenges of e-government.

6 CONCLUSION

The paper discussed the changes that e-government brings about to civil servants' practices by drawing upon a Greek one-stop shop. As the study indicated, e-government reorganizes the public sector in accordance with the best practices of the private sector; for instance, through internal markets, service standardization, performance measurement, and electronic surveillance. At the same time, e-government challenges the power relations that are embedded in the public sector. It attributes new roles and authorities to public officials and in doing so directs their conduct toward achieving and performing. Further, it increases the sources of accountability and control by focusing not on the process but the outcomes of public service provision, limiting in that way the autonomy and discretion of public officials. Moreover, the initiation of e-government intends to address well-established problems of bureaucracies, in our case client relations and accountability. These effects on public service provision are not deterministic but dependent upon the history of each public sector and the way in which e-government is conceptualized and carried out by each government and why. This indicates that whether or not and how e-government influences officials' conduct and, as a result, public service provision is first and foremost a political question.

Acknowledgments

The author wishes to express her gratitude to Dr. Niall Hayes and Professor Lucas Introna for their support in the completion of this article.

References

Argyriadis, D. "Facets of Administrative Change in Greece," in *Reports of Experts on Public Administration 1950-1998*, A. Makrydimitris and N. Michalopoulos (eds.), Athens: Papazese Publications, pp.373-428 (in Greek; first published in 1970).

Avgerou, C. 2002. *Information Systems and Global Diversity*, Oxford: Oxford University Press.

Avgerou, C., and McGarth, K. 2007. "Power, Rationality and the Art of Living Through Socio-Technical Change," *MIS Quarterly* (31:2), pp. 295-315.

Ballas, A., and Tsoukas, H. 2004. "Measuring Nothing: The Case of the Greek National Health System," *Human Relations* (57:6), pp. 661-690.

Basu, S. 2004. "E-Government and Developing Countries: An Overview," *International Review of Law Computers* (18:1), pp. 109-132.

Berger, P., and Luckmann, T. 1966. *The Social Construction of Reality: A Treatise in the Sociology of Knowledge*, New York: Penguin Books.

Blau P. 1970. "Weber's Theory of Bureaucracy," in *Max Weber*, D. Wrong (ed.), Englewood Cliffs, NJ: Prentice Hall

Bloomfield, B., and Hayes, N. 2004. "Modernisation and the Joining-Up of Local Government Services in the UK: Boundaries, Knowledge and Technology," paper presented at Information, Knowledge and Management: Reassessing the Role of ICTs in Private and Public Organizations, Bologna, Italy, March 3-5.

Brinkerhoff, D., and Goldsmith, A. 2004. "Good Governance, Clientelism and Patrimonialism: New Perspectives on Old Problems," *International Public Management Journal* (7:2), pp. 163-185.

Bovens, M., and Zouroudis, S. 2002. "From Street-Level to System-Level Bureaucracies: How Information and Communication Technology is Transforming Administrative Discretion and Constitutional Control," *Public Administration Review* (62:2), pp. 174-184.

Campbell, C. 1993. "Public Service and Democratic Accountability," in *Ethics in Public Service*, R. Chapman (ed.), Edinburgh: University Press, pp.111-133.

Chadwick, A., and May, C. 2003. "Interaction between States and Citizens in the Age of the Internet: E-Government in the United States, Britain and the European Union," *Governance: An International Journal of Policy, Administration and Institutions* (16:2), pp. 271-300.

Ciborra, C. U. 2003. "Unveiling E-Government and Development: Governing at a Distance in the New War," working paper 126, Lancaster University (http://is2.lse.ac.uk/wp/pdf/WP126.PDF).

Ciborra, C. U. 2005. "Interpreting E-Government and Development: Efficiency, Transparency or Governance at a Distance?," *Information Technology and People* (18:3), pp. .260-279.

Clarke, J., and Newman, J. 1997. *The Managerial State: Power, Politics and Ideology in the Remaking of Social Welfare*, London: Sage Publications.

Cordella A. 2007. "E-Government: Towards the E-Bureaucratic Form?," *Journal of Information Technology* (22), pp. 265-274.

Cowell, R., and Martin, S. 2003. "The Joy of Joining-Up: Modes of Integrating the Local Government Modernization Agenda," *Environment and Planning C: Government and Policy* (21:2), pp. 159-179.

Davison, R., Wagner, C., and Ma, L. 2005. "From Government to E-Government: A Transition Model," *Information Technology and People* (18:3), pp. 280-299.

du Gay, P. 2000. *In Praise of Bureaucracy: Weber, Organization, Ethics*, London: Sage Publications.

Farrell, C., and Morris, J. 2003. "The Neo-Bureaucratic State: Professionals, Managers and Professional Managers in Schools, General Practices and Social Work," *Organization* (10:1), pp. 129-156.

Foucault, M. 1977. *Discipline and Punish: The Birth of the Prison*, London: Penguin Books.

Fountain, J. 2001. "Paradoxes of Public Sector Customer Service," *Governance: An International Journal of Policy and Administration* (14:1), pp. 55-73.

Garson, B. 1988. *The Electronic Sweatshop: How Computers Are Turning the Office of the Future into the Factory of the Past*, New York: Simon and Schuster.

Gil-Garcia, R., Chengalur-Smith, I., and Duchessi, P. 2007. "Collaborative E-Government: Impediments and Benefits of Information-Sharing Projects in the Public Sector," *European Journal of Information Systems* (16), pp. 121-133.

Goodsell, C. 2005. "The Bureau as Unit of Governance," in *The Values of Bureaucracy*, P. du Gay (ed.), Oxford, UK: Oxford University Press, pp. 17-40.

Hazlett, S., and Hill, F. 2003. "E-Government: The Realities of Using IT to Transform the Public Sector," *Managing Service Quality* (13:6), pp. 445-452.

Heeks, R. 2002. "E-Government in Africa: Promise and Practice," *Information Polity* (7:2/3), pp. 97-114.

Heeks, R., and Bailur S. 2007. "Analyzing eGovernment Research: Perspectives, Philosophies, Theories, Methods and Practice," Government Information Quarterly (24:2), pp. 243-265.

Ho, T. A. 2002. "Reinventing Local Governments and the E-Government Initiative," *Public Administration Review* (62:4), pp. 434-444.

Hoggett, P. 1996. "New Modes of Control in the Public Service," *Public Administration* (74:1), pp. 9-32.

Hood, C. 1991. "A Public Management for All Seasons?," *Public Administration* (69:1), pp. 3-19.

Huang, Z., and Bwoma, P. 2003. "An Overview of Critical Issues of E-Government," *Issues of Information Systems* (4:1), pp. 164-170.

Illsley, B. M., Lloyd, M. G., and Lynch, B. 2000. "From Pillar to Post? A One Stop Shop Approach to Planning Delivery," *Planning Theory and Practice* (1:1), pp. 111-122.

Kaufman, R. 1974. "The Patron–Client Concepts and Macro-Politics: Prospects and Problems," *Comparative Studies in Society and History* (16:3), pp. 284-303.

Kumar, R., and Best, M. 2006. "Impact and Sustainability of E-Government Services in Developing Countries: Lessons Learned from Tamil Nadu, India," *The Information Society* (22:1), p.p. 1-12.

Kunstelj, M., and Vintar, M. 2004. "Evaluating the Progress of E-Government Development: A Critical Analysis," *Information Polity* (9), pp. 131-148.

Legg, K. 1969. *Politics in Modern Greece*, Stanford, CA: Stanford University Press.

Lemarchand, R., and Legg, K. 1972. "Political Clientelism and Development: A Preliminary Analysis," *Comparative Politics* (4:2), pp. 149-178.

Lenk, K. 2002. "Electronic Service Delivery: A Driver of Public Sector Modernization," *Information Polity* (7), pp. 87-96.

Lipsky, M. 1980. *Street-Level Bureaucracy: Dilemmas of the Individual in Public Services*, New York: Russell Sage Foundation.

Martin, B., and Byrne, J. 2003. "Implementing E-Government: Widening Lens," *Electronic Journal of E-Government* (1:1), pp. 11-22.

Minson, J. 1998. "Ethics in the Service of the State," in *Governing Australia: Studies in Contemporary Rationalities of Government*, M. Dean and B. Hindess (eds.), Cambridge, UK: Cambridge University Press, pp. 47-69.

Moon, J. 2002. "The Evolution of E-Government among Municipalities: Rhetoric or Reality?," *Public Administration Review* (62:4), pp. 424-433.

Mouzelis, N. 1978. *Modern Greece: Facets of Underdevelopment*, Athens: Eksantas (in Greek).

Purnendra, J. 2002. "The Catch-Up State: E-Government in Japan," *Japanese Studies* (22:3), pp. 237-255.

Ritzer, G. 1993. *The McDonaldization of Society: An Investigation into the Changing Character of Contemporary Social Life*, Thousand Oaks, CA: Pine Forge Press.

Roche M. 2004. "Complicated Problems, Complicated Solutions? Homelessness and Joined-Up Policy Responses," *Social Policy and Administration* (38:7), pp. 758-774.

Suchman, L. 1987. *Plans and Situated Actions: The Problem of Human–Machine Communication*, Cambridge, UK: Cambridge University Press.

Tan, C., and Pan, S. 2003. "Managing E-Transformation in the Public Sector: An E-Government Study of the Inland Revenue Authority of Singapore," *European Journal of Information Systems* (12:4), pp. 269-281.

Tsoukalas, K. 1986. *Social Development and State: The Composition of Public Space in Greece* (2nd ed.), Athens: Themelio (in Greek).

Tsoukalas, K. 1987. *State, Society, Employment in Post-War Greece* (2nd ed.), Athens: Themelio (in Greek).

Turner, P., and Higgs, G. 2003. "The Use and Management of Geographic Information in Local E-Government in the UK," *Information Polity* (8:3/4), pp. 151-165.

Vintar, M., Kunstelj, M., Decman, M., and Bercic, B. 2003. "Development of E-Government in Slovenia," *Information Polity* (8:3/4), pp. 133-149.

von Haldenwang, C. 2004. "Electronic Government (E-Government) and Development," *European Journal of Development Research* (16:2), pp. 417-432.

Weber, M. 1948. "Bureaucracy," in *From Max Weber: Essays in Sociology*, H. H. Gerth and C. W. Mills (eds.), London: Routledge.

Wilkins, P. 2002. "Accountability and Joined-Up Government," *Australian Journal of Public Administration* (61:1), pp. 114-119.

Yildiz, M. 2007. "E-Government Research: Reviewing the Literature, Limitations, and Ways Forward," *Government Information Quarterly* (24), pp. 646-665.

Zhang, J. 2002. "Will the Government 'Serve the People'? The Development of Chinese e-Government," *New Media and Society* (4:2), pp. 163-184.

About the Author

Dimitra I. Petrakaki is a doctoral student at Lancaster University (UK) and a lecturer in Information Systems at the University of Wolverhampton (UK). Her research interests revolve around the relation of knowledge and power in the initiatives of public sector reformation and e-government. She can be reached at d.petrakaki@lancaster.ac.uk.

16 BANDWITHING TOGETHER: Municipalities as Service Providers in a Policy Environment

Andrea H. Tapia
Julio Angel Ortiz
College of Information Sciences & Technology
Pennsylvania State University
University Park, PA U.S.A.

Abstract *In this paper we have highlighted three things. First, that public organizations are engaged in a technologically driven servitization of their traditional service products. Second, that public organizations must approach this servitization differently because of the decision-making role of citizens in the process. Third, it is essential to study the influence citizens have on this process through studying the public policy process around decisions concerning the delivery of technology and technologically driven services. In this paper we provide an overview of four constituencies and their relationships involved in municipal wireless broadband policy. The idea that the growth of information technology-dependent services and activities (education, healthcare, and Web 2.0, for instance) are dependent on wide-scale availability of broadband access, and that local governments are jumping into this market to establish the necessary infrastructures for such services, makes this a very hotly contested space. While national and international political issues are debated on the Internet daily, new avenues for very local, political speech and action on the Internet seem to go hand-in-hand with municipal wireless broadband issues. The creation of public policy, while normally seen as a top-down process, has always drawn varying input from the outside through avenues such as lobbying, town meetings, referendums, and public action. We claim that in the case of municipal wireless broadband policy, policy efforts have been turned upside-down, with the majority of policy making now happens at the local level.*

Keywords Municipal wireless, servitization, public policy, broadband access

Please use the following format when citing this chapter:

Tapia, A. H., and Ortiz, J. A., 2008, in IFIP International Federation for Information Processing, Volume 267, Information Technology in the Service Economy: Challenges and Possibilities for the 21st Century, eds. Barrett, M., Davidson, E., Middleton, C., and DeGross, J. (Boston: Springer), pp. 229-248.

1 INTRODUCTION

In the past 20 years, the world's economy has moved toward the creation of a service economy in which the service sector has grown in importance, the percentage of service companies is growing in relation to manufacturers, and products have a higher service component. In the management literature, this is referred to as the *servitization of products* (Vandermerwe and Rada 1988).

Information and communication technologies (ICTs) have played a large role in this servitization of products and companies. ICTs have allowed for the converting of products into services and the enhancing of traditional products with information service components. ICTs have also fostered new forms of communication and information exchange between supply chains of intangible services as well as increased communication directly between customers and service providers.

While this change in the private sector has been well studied, similar changes in the public sector have remained virtually untouched. Perhaps this is because the public sector has often been construed as principally a service provider, not having to reinvent from a manufacturing base. The core competencies often associated with local governments are nearly all services, including the provision of arts, recreation, water, sewage, electricity, library, public housing, transportation, police, fire, environment, health, and economic development, to name a few. However, through further development and use of ICTs, the public sector has transformed its traditional offline services into online services, developed new online services, and entered new markets not previously associated with the public sector.

We focus our attention on this last form of public servitization transformation. We are interested in how local governments have come to see themselves as direct providers of telecommunication services. Approximately 400 municipalities in the United States are developing and deploying affordable wireless broadband networks. These cities are deploying their own wireless broadband networks with the intention of meeting the increasing information needs of city employees and citizens and simultaneously increasing the efficiency and effectiveness of the delivery of those services.

ICT-enabled services are human intensive services that are delivered over telecommunication networks, specifically the Internet. The idea that the growth of IT-dependent services and activities are dependent on wide-scale availability of broadband connectivity, and local government officials are affording wireless broadband to establish the infrastructure necessary for such services, makes this a complex and very local telecommunications space. Although the entrance of governments into the telecommunications market is not ahistorical (e.g., they already provide telephones, healthcare, education, and so on.), the strong push-back from the telecommunication incumbents and other for-profits about municipal service providers is something unique to this service area.

Again, perhaps unlike servitization in the private sector, public sector entrance into the telecommunications market has been hotly contested. In 2005, telecommunication incumbents and state legislatures proposed and enacted legislation to restrict, and in some cases prohibit, municipalities from entering into this space. In 2006 and 2007, less than half of the proposed legislation had passed, and none passed without negotiation and compromise between the parties. Simultaneously, five U.S. federal-level bills have been proposed in congress, addressing the municipal broadband issue, none as yet having made it out of committee. A new Telecommunications Act is expected in 2010, making up for the obsolescence of the Telecommunications Act of 1996, which will significantly alter the political landscape at the state and municipal levels. However, until then, policy making concerning this issue is happening primarily at the local level as municipalities, interest groups, local providers, and state representatives negotiate the best means to provide the most people with the best delivery of broadband and online government services.

Furthermore, what makes this issue so interesting is that this is a technological debate being held in a technologically enabled forum. This is compounded by the fact that while all parties involved seem to agree on the goal, very few agree on the means to that end. Additionally, while national and international political issues are debated on the Internet daily, new avenues for very local, political speech and action on the Internet seem to go hand-in-hand with municipal wireless broadband issues. The creation of public policy, while normally seen as a top-down process, has always drawn varying input from the outside through avenues such as lobbying, town meetings, referendums, and public action. We claim that in the case of municipal wireless broadband policy, policy efforts have been turned upside-down, with the majority of policy making now happening at the local level.

Consistent with research approaches in public policy research, in this paper we provide an overview of the different constituencies and their relations surrounding municipal broadband policy in the United States. It is acknowledged from the outset that the perspective adopted, out of necessity, takes an instrumental and narrow view of what is public policy. We are interested in the process of policy, rather than the policy itself. To meet our goals of understanding this complex policy arena, we have developed three research questions that have guided our work.

(1) Who are the constituents that generate broadband policy for municipalities?
(2) What is *unique about their relations* considering the lack of federal policy, the role of the Internet, and bottom-up political action?
(3) What are the mechanisms that facilitate input from these diverse constituencies into a more top-down, traditional policy making setting?

We will provide a general introduction to and highlight the potential role and usefulness of analyzing formal/informal and top-down/bottom-up policy approaches in municipal broadband network research. The paper will provide a brief literature review of public policy process, and will describe recent municipal wireless developments and how these developments are being used to inform policy in the United States. We will then review the evidence on the documentation of these networks, focusing on four archetypes of policy camps driven by different dynamics. The remaining sections of the paper will be devoted to a discussion of municipal constituencies and their relations with regard to public policy.

2 THE CONTEXT: MUNICIPAL WIRELESS BROADBAND NETWORKS

As full participation in civic, commercial, and social life is tied to Internet and computer literacy and access, broadband access is becoming a necessity rather than a luxury. During the 1990s, this trend was clearly recognized by the U.S. government, which championed the Internet and used the power of the federal government to encourage its growth. However, while the United States has made significant gains in broadband adoption, a first step in closing this gap, it still lags far behind other countries (Bleha 2005). The United States also trails other countries in terms of the average speeds available over their broadband connections. Recent commentary has characterized U.S. broadband among the "slowest, most expensive, and least reliable in the developed world" (Bleha 2005, p. 111).

Recently, cost-saving wireless technologies have unsurprisingly replaced wired technologies. As the demand increases and more users join the wireless community, wireless technologies become faster, more robust, and cheaper (Lehr et al. 2004). As a result, over 400 cities in the United States have announced plans to deploy wireless broadband networks, claiming that these networks would enhance economic development, provide for additional tourism, support city services and personnel, and perhaps decrease the digital divide.

Municipalities enjoy certain advantages in this space. While local governments do not have control over state and federal policies, they do have control over local policies. These local policy efforts can influence communications infrastructure deployment, business and residential demographics that shape demand, and the nature and quality of existing infrastructure, all of which can have a direct impact of the development and deployment of municipal wireless networks (Gillett and Lehr 1999). Given existing municipal assets such as buildings, rights of way, and structures that can house wireless antennas, yet another incentive is that municipalities may enjoy a lower cost of broadband infrastructure deployment.

As municipal wireless broadband deployments have become more high profile, private sector providers have expressed a number of concerns. Private providers express concern that cities providing wireless broadband service have an unlimited base from which to raise capital, act as a regulator for local rights of way and tower permitting, own public infrastructure necessary for network deployments including street lights, and are tax-exempt organizations. In addition, it has been argued that these broadband networks may cost more than the cities anticipate, resulting in money and attention being diverted away from other public interests. Another fear is that if these networks are allowed to flourish, municipalities will have unfair regulatory and economic advantages (Lenard 2004).

Many telecommunications companies have sought legislative relief at the state level to regulate or restrict a municipality's ability to provide wireless broadband services to the public. With no guidance from the Telecommunications Act of 1996, the Supreme Court sided with the Federal Communications Commission and various ILEC (incumbent local exchange carrier) lobbyists in its decision in *Nixon v Missouri Municipal League*, to allow states to bar their subdivisions from providing telecommunications services. The opinion gave states the authority to determine when and where municipalities can deploy communications services.

Currently most states have legislation proposed, pending or passed that prohibits municipalities from providing telecommunication services directly or indirectly. In some cases state legislatures have prevented municipalities from expanding existing networks. In other cases, state legislatures have not out-right prohibited the development and deployment of municipal broadband networks, but they have created organizational and bureaucratic barriers causing these networks to be curtailed, reconfigured, or resized.

Thus the current setting is that most municipalities are caught between citizens, local businesses, and their own employees who are demanding high quality, affordable, universal broadband Internet service and their state legislators and incumbent telecommunications companies who seek to keep the offering of telecommunications services out of public hands, yet cannot, or will not, comply with local citizen and business demands. In some cases, municipalities have entered into public–private partnerships in which they do not offer broadband service directly, but instead offer rights of way or government employee contracts, among other things, to either an outside nonprofit or local Internet service provider to offer the service on their behalf. These negotiations usually result in hybrid organizations offering service to consumers at reduced prices, covering more square miles, and reaching underserved populations, as well as complying with some of the more restrictive state policies.

3 DIMENSIONS OF THE PUBLIC POLICY PROCESS

The perspective of this paper is a public policy one. This is in contrast to other social research perspectives, which are often based on a narrow disciplinary focus. A public policy focus acknowledges the complexity of the policy decision-making process. For our purposes, we examine the process of creating broadband policy as a series of choices, rather than a study of costs and benefits (March and Olsen 1989). Our working definition of public policy is "an action which employs governmental authority to commit resources in support of a preferred value" (Considine 1994, p. 3). Federal, state, and local governments usually set the rules of the political arena and, consequently, determine the outcome of public policies. It is in this light that public policy is a reflection of a government's political agenda, encompassing both specific policies (i.e., universal access policy) or those wider in scope (i.e., economic policy).

During the past decade, the role of governments in public policy has been steadily changing. Increasing emphasis (often implicit) is being placed on setting overall direction through policy and planning, on engaging stakeholders and citizens, and on empowering stakeholders or partners to deliver programs and services (Dunn 1994; Fischer 1995; Patton and Sawicki 1993). At the same time, the environment for policy and planning has increased in complexity (Demsetz 1999). The ownership of issues is often unclear, especially when more than one department and often more than one level of government are involved. Local governments and grassroots communities also increasingly claim ownership of policy issues and processes. In this complex environment, the demand for good public policy development steadily increases, as must the capacity of managers, policy analysts, planners, and others involved in the design and delivery of policies and programs.

An array of definitions regarding public policy exist (Brooks 1989; 1978; Dye 1972; Frederich 1963) and there appears to be a lack of consensus about what constitutes public policy. Nonetheless, most definitions recognize policy as an action, which employs governmental authority to use resources in support of a preferred value. Briefly stated, public policy is a choice or decision made by federal, state, and local governments that guides subsequent actions. It is outcome-oriented and operates on the basis of general criteria, namely, mission statements, organizational values, political priorities, and so on.

Public policy issues can be separated into two main categories: those that are on the public policy agenda, and those that are not. If an issue is on the public policy agenda, it has a high profile, and a formal process is likely to be in place. If an issue is not on the public policy agenda, the job of the stakeholders/community is to provide information and education, and to take other steps to raise awareness and get it on the agenda. According to Gerston (1997), an issue will appear, materialize, and continue on the public policy agenda when it meets one or more of three criteria. Gerston argues that the issue must have sufficient intensity (the magnitude of the impact is high), scope (a significant number of people or communities are affected), and/or time (it has been an issue over a long period).

The literature divides public policy into two basic types: vertical policy and horizontal policy (Chung 2001; Guiraudon 2000; Lafferty and Hovden 2003; Williams and Griffin 1996). Vertical policy is developed within an organization that has authority and resources for implementation. Vertical policy is what we think of as the normal or traditional way in which policy decisions are made. Vertical policy is developed within a single organizational structure and generally starts with broad overarching policy. Horizontal policy is developed by two or more organizations, each of which has the authority or ability to deal with only a part of the situation. The distinction reflects how clearly a mandate rests with one department, unit, or agency, and its capacity to address the root cause of the issue with existing resources. Horizontal policy, often referred to as integrated policy, is developed between parts of an organization, or among organizational components that are in similar hierarchical positions. There is a great deal of discussion today about horizontal policy issues (sometimes referred to as *crosscutting issues*) and the challenges that organizations face in dealing effectively with them (Kubler et al. 2003).

The first dimension in our discussion of the process of policy decision making is rational versus political models (Grindle and Thomas 1990; Majone 1989; Stokey and Zeckhauser 1978). The rational model explicates policy making as a problem-solving process, which is balanced, objective, rational, and very analytical. In this model, decisions are made in sequential steps, starting with identifying the issue, and ending with a list of actions to address it. Because policy is categorized by objective analysis of choices and separation of the policy from the implementation, some argue that a linear model is inaccurate (Clay and Schaffer 1986). This split between decision making and implementation is generally "attributed to decision-makers sense that politics surrounds decision-making activities while implementation is an administrative activity" (Grindle and Thomas 1990, p. 1170). As a result, policy making and policy implementation are best appreciated as a chaos of purposes and accidents; a combination of concepts and tools from different disciplines can be used to put order into the chaos (Sutton 1999). This contrasting model is termed the *political model* and is the most prevalent view of the policy process (Majone 1989; Stone 1997). The main rationale behind the political model is the influence on political decisions through the passing of laws, regulations, and legislation. In short, this model is instrumental and tactful in achieving quickly the objectives of the government.

In yet another dimension, the policy-making process can be viewed as either top-down or bottom-up (Chrispeels 1997; Christman and Rhodes 2002). Top-down approaches to public policy are authoritarian and coercive in nature; bottom-up approaches are viewed as grassroots movements and in objection to existing (or pending) policy.

The *bureaucratic process model* starts with the policy message at the top and sees implementation as occurring in a chain (Dunsire 1978). The policy is seen as a paramount and resistance to it tends to be seen as irrational, and as a barrier to implementation. The bottom-up approach, in contrast, involves the citizenry in decision-making processes considered for adoption at the top at different stages of development. The depth and breath of the latter framework is to involve citizens in the governance process and, by so doing, regenerate, reinforce, or retract laws and regulations that affect practices impacting on their lives. It reflects the ways in which local people understand their situation, strategize, and formalize a plan to implement new laws. Nevertheless, the policy-making process remains essentially a top-down process involving political figures (e.g., elected officials, political appointees, advisors, etc.) (Majone 1989). This said, however, some scholars argue public policy is more accurately understood to be produced at the intersection of top-down and bottom-up forces (Baumgartner and Jones 1993; Goggin et al. 1990; Sundquist and Davis 1969).

Our third dimension of the policy process is formal versus informal. Formal policy is written, either as rules or ordinances and typically represents the policy goals of most governments. Formal policy is published as a distinct policy document that is publicly available. On the other hand, informal policy is unwritten and is conceived as a set of "understood" practices that individuals must follow. Informal policy is unpublished and there is little assurance that government officials will adhere to it, as well as tremendous potential for variations in understandings about how the policy should apply.

From this brief review of the dimensions of public policy, we can draw several conceptual tools to apply to the municipal broadband sphere. The policy process can be seen as having both vertical and horizontal action and action that is both rational and politically motivated and construed. For our purposes, we see these actions in terms of their origins, top-down versus bottom-up, and in terms of the form, formal versus informal. There is a great deal of overlap between each of these aspects of policy. For instance, vertical policy can be viewed as a political and top-down approach for implementing both formal and informal policy. As we will see in this paper, broadband policy for municipalities is no exception.

4 METHODOLOGICAL APPROACH

In June 2005, we created a dynamic and evolving database of all municipal/community wireless initiatives in the United States. As of June 2007, the database contains more than 400 entries. The data that we have collected spans multiple categories including information on the shape, form, uses, and technologies of the municipal/community network itself; the business plan and/or service delivery plan; the status of the development/deployment of the network; the social impacts of the network; and the marketing language used by the owners and users of the network.

This database has been populated through a variety of methodologies. In most cases, information was obtained through the use of the Internet, using crawling techniques via municipal/community sponsored websites, press releases, public documents, and online news and web logs. In addition, when information proved scarce or dubious, the municipality or community was called, and information was supplemented and verified via phone. Drawing from one subsection of fields from the database, we have compiled all texts from these cities. While the documents analyzed do not form a complete picture of the intentions of the city or its representatives, as they are specific in time and space in the experiences of that city, they were read literally in terms of discursive event. (For a complete presentation and analysis of these data, see Ortiz and Tapia 2008.)

Building on the these public policy documents, theoretical underpinnings, and our evolving database, we develop a conceptual framework. In this paper, we attempt to provide an overview of the different public policy constituencies and their relations that generate municipal broadband cities in the United States. We do not attempt to summarize academic, industry, or public policy literature. We strive to categorize (not summarize) the current types of public policies by way of a conceptual framework. This framework describes four constituencies involved in municipal wireless broadband policy and not necessarily the process of policy making.

Figure 1, with the various actors/intermediaries (websites, policy documents, etc.), informs our analysis. It presents a matrix of the sources of input to policy making and implementation. The cell entries are incomplete and only illustrative. The arrow represents the path traveled from the Type 4/Broadcast Interaction to the Type 1/Law Building quadrants. In other words, grassroots, bottom-up driven policy efforts aim to inform and influence formal, top-down policy. The lines between each quadrant are not solid as there is a great deal of overlap between each type. Each quadrant is explained below.

4.1 Type 1: Law Building (Top-Down + Formal Quadrant)

This approach is distinguished from other approaches in that it is embedded within a hierarchical structure initiated and executed by an organization's top leadership by way of laws, statutes, ordinances, and policy legislation. It is the ideal goal desired in policy implementation as it is needed at times for instituting major changes in direction. The Telecommunications Act of 1996 is an example of a top-down, formal approach. The Act was the first major overhaul of telecommunications law in over 60 years.

4.2 Type 2: Rhetorical Inscription (Top-Down + Informal Quadrant)

In contrast to the top-down, formal approach, this quadrant is composed of formal policy that comes in the form of the spoken and written word; it is rhetoric that appears in governmental addresses (president and mayor speeches, for instance), government press releases, and commentaries. For example, the State Corporation Commission, the Federal Trade Commission, the Federal Communications Commission, and the Federal Reserve Board follow standards of policy. Standards of policy are informal rules and practices that are part of their governing documents and thus a part of the rules of the rights and obligations governing their members. Members of these organizations envisage a gradual transition from informal standards of policy to a more organized, formal standard. The transition is a systemic process to ensure consistently high standards of policy development.

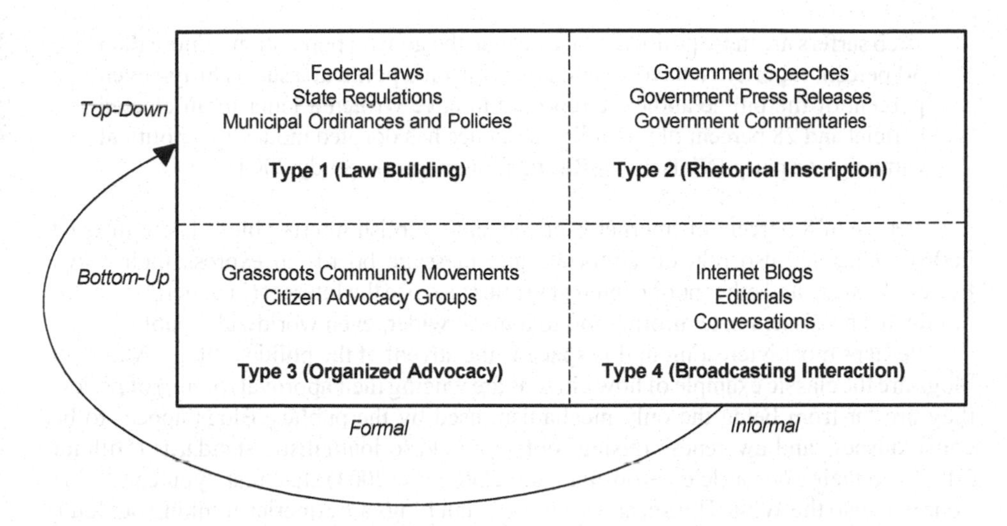

Figure 1. Public Policy Conceptual Framework

4.3 Type 3: Organized Advocacy (Bottom-Up + Formal Quadrant)

This quadrant represents organized and established public interest groups, referred to as advocacy groups here. They can be grassroots groups or professional lobbyists; the only requirement is that they are organized, established, and comprised of citizens not employed, elected, or serving the current government. Public interest groups bring together and speak for individuals, groups, and organizations who have common interests, views, and concerns (Pross 1986). Advocacy groups represent the general public by educating and influencing public policy decision makers. Advocacy groups' methodologies encompass lobbying efforts, writing, or voicing concerns to public officials. For example, in early spring 2006, the FCC opened an investigation that studies a significant number of U.S. television stations; it is believed they were broadcasting pro-Bush propaganda and airing it as bona fide, normal news (Buncombe 2006). The FCC began their probe after a report produced by an advocacy group, the Center for Media and Democracy. Without the existence of such a group, perhaps the FCC would have never opened the investigation.

4.4 Type 4: Broadcast Interaction (Bottom-Up + Informal Quadrant)

The participants in this quadrant are often solitary individuals or small, unorganized groups or newly forming efforts comprised of citizens unaffiliated with the government or other, more-formalized organizations. Often these participants serve a watchdog function. While these participants have always been present (e.g., Ben Franklin's Widow Silence Dogood) information and communication technologies like the Internet have produced or enabled an explosion of online, individualized, personal, political activity. Nielsen reports that

> Web surfers are more politically active than the general population…more than 50 percent reported to have signed a petition on a political issue. Thirty-seven percent of the online audience reported to have written a letter to an elected official and 28 percent of the online audience has donated money to a political campaign or party (Nielsen/NetRatings Enumeration Study 2004).

An average of 8 percent of Internet users actually publish Internet blogs (Pew Internet 2006). Citizens also rely on editorials and message boards to express their view. Political groups and other public interest groups also use the Internet (in general) to communicate and disseminate information to a much wider, even worldwide, public.

Perhaps most interesting in this space is the advent of the political blog. Although blogs are the classic example of how citizens are voicing their approval (or not) of policy, they are far from being the only mechanism used by the public. Blogs appear to be consciousness and awareness raising tools not held to journalistic standards (Gillmor 2004). In their *Time* article, Groosman and Hamilton (2004) state that "you can't blog your way into the White House, at least not yet, but blogs are America thinking out loud, talking to itself, and heaven help the candidate who isn't listening." A particular case in point is the recent popularity of news and opinion web logs or blogs from a liberal perspective that serve as an alternative to the discussions of talk radio, which tends to be primarily for a conservative audience. As a result, such blogs will set the agenda for their readership on certain topics since other media did not assume that role (Morris 2001).

5 ANALYSIS OF THE MUNICIPAL WIRELESS BROADBAND POLICY MATRIX

Using the types identified in the model described above, we present the following analysis of the current position of the municipal wireless broadband policy constituencies and their relations. The matrix presented in Figure 2 operates with four dimensions.

5.1 Type 1: Law Building

Congress passed the Telecommunications Act of 1996 with the intention of promoting competition by further deregulating the industry. Despite the effort by forward-thinking leaders in 1996, they were in actuality wholly unprepared for the exponential advancement of high-speed, affordable Internet technologies, and perhaps even more unprepared for the revolutionary changes its adoption into society would create. The 1996 act did not address broadband or municipal entry into providing such services, and, therefore, the act was toothless in regulating the issue on a federal level: in essence, it left regulation to the state governments. States, however, were also unprepared and began passing legislation in reference to municipal entry.

Since the existing federal laws were not applicable, state legislatures began considering how to respond to the objections of private sector providers. The legislative initiatives made use and continue to make use of a variety of tools that ostensibly aim to ensure that

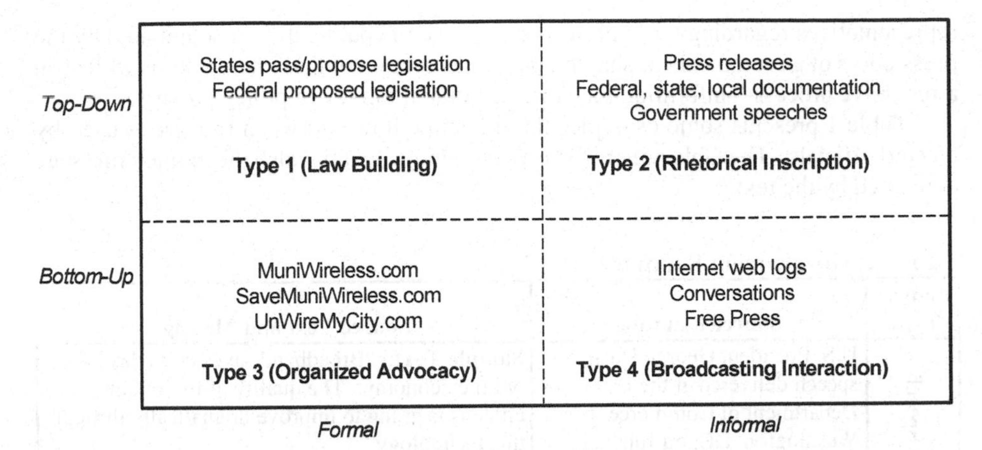

Top-Down
Bottom-Up
States pass/propose legislation
Federal proposed legislation
Type 1 (Law Building)
Press releases
Federal, state, local documentation
Government speeches
Type 2 (Rhetorical Inscription)
MuniWireless.com
SaveMuniWireless.com
UnWireMyCity.com
Type 3 (Organized Advocacy)
Internet web logs
Conversations
Free Press
Type 4 (Broadcasting Interaction)
Formal
Informal

Figure 2. Mobile and Wireless Network Policy Constituencies Matrix Mapped to Types

(1) a majority of local residents are behind the initiative
(2) the broadband project will not negatively affect a city's finances
(3) the broadband deployment does not compete or competes on a level playing field with private carriers

Several tools have been used to achieve these objectives. First, to ensure that a majority of the residents support the initiative, several states included a requirement that municipalities hold hearings and/or referenda about the broadband deployment. These activities also went some way to answering the second concern, that the project did not negatively affect finances. In addition to reporting to the public, some states have also required plans be submitted for approval to a state entity or agency. Tools used to achieve the third objective included a variety of stipulations ranging from providing the local exchange carrier (LEC) the right of first refusal to outright prohibition of competing with LECs (Tapia et al. 2005, 2006). Currently, there are approximately 35 state laws governing or affecting municipal broadband. As states have considered legislation, so has the U.S. Congress. Congressional leaders have agreed for some time that the Telecommunications Act of 1996 needs to be rewritten to reflect the many developments in telecommunications that have occurred during the last decade. As of June 2006, there were more than a dozen bills relating to Internet and broadband adoption.

5.2 Type 2: Rhetorical Inscription

A key principle of rhetoric is that words can have power or force over listeners. Some stipulate that rational manipulation of the economy through rhetoric requires that the tone of presidential remarks be optimistic, constructive, positive, and a practical assessment of the current and future state (Lewis 1997; Ragsdale 1984).

In the case of municipal wireless broadband policy, little research has been conducted which actually examines the role of public statements made by government representatives regarding these networks. We see the public discourse initiated by the press corps of governments as shaping public perception of these networks as well as in a recursive process—informing the more traditional top-down policy process.

Table 1 presents some examples that illustrate how and when rhetoric is used by elected officials. The table not only shows the sample text but also the implicit message conveyed by the text.

Table 1. Government Sample Texts

Source Type	Source Example	Source Text and Message
Presidential Speech	U.S. President George Bush, speech delivered at the U.S. Department of Commerce in Washington, DC, on June 24, 2004 Source: http://www.whitehouse.gov	**Sample Text**: "Broadband saves costs throughout the economy. The quality of life of our citizens is going to improve dramatically through this technology." **Implicit Message**: We should promote broadband access as a way to help U.S. workers become more productive and improve the economy.
Municipal Document	Request for proposals document issued by the City of Long Beach, CA, on January 27, 2006 Source: http://www.muniwireless.com/reports/docs/LongBeachRFP.pdf	**Sample Text**: "The City believes that the advancement of WiFi technologies has presented significant opportunities for local jurisdictions. The City believes that the deployment of a city-wide WiFi network will allow residents and businesses to experience significant economic and social benefits through increased options for broadband Internet connectivity." **Implicit Message**: Broadband is good for their citizenry and new technology will promote social inclusion.
Press releases	Mayor Sam Teresi's statement, press release issued by the City of Jamestown, NY, on January 12, 2005 Source: http://www.tropos.com/pdf/jamestown.pdf	**Sample Text**: "Today, we have effectively unwired our downtown area using Tropos Wi-Fi equipment. We plan to use the newly-formed Wi-Fi network to improve public safety, increase the productivity of our city workers out in the field. [Our city] is the first municipality in New York State to utilize this newest technology known as Wi-Fi for public safety communications." **Implicit Message:** Wi-Fi enhances public safety tools for all.
Commentary	Texas Agriculture Commissioner Susan Combs, commentary on http://www.statesman.com/search/content/auto/epaper/editions/today/business_245237edf10221dd0093.html (printed on March 2, 2005)	**Sample Text**: "For economic development, it is a death blow in the 21st century if you don't have broadband. If I wanted to encourage some company to move to small-town Texas, they will ask about education and housing. And then they will ask about broadband." **Implicit Message**: Without broadband, cities will not survive in the new digital global economy.

5.3 Type 3: Organized Advocacy

We can assess the potential for public interest groups to participate in the policy development process by looking at the case of Texas House Bill 789. In partnership with local incumbents, Texas introduced state legislation that considered prohibition of Municipal Wireless Broadband Networks but the legislation, as written, failed. The failure of this legislation is often attributed to the organized activities of several public interest groups. Citizen activists formed organized and established groups that interacted on their behalf. The following are simple descriptions intended to provide an overview of these organizations:

- SaveMuniWireless.com is a coalition of Texas organizations and citizens concerned about legislation that would outlaw any municipal involvement in networks or information services. They seek to ensure any legislation supports competition and innovation and helps Texans, not just the incumbent communication providers. Their action is intimately linked with a mailing list called TxMuni-Action, which was instrumental in the defeat of anti-municipal wireless provisions in Texas HB 789.

- Muniwireless.com reports on municipal wireless and broadband projects. The site is maintained by Esme Vos, a consultant from Amsterdam. The bipartisan repository is consulted by vendors (e.g., Tropos, Red Line Communications, NextPhase Wireless), think tanks, public officials, academics, and consultants. Aside from fee-based request for proposal alerts, the site also includes recent press releases, magazine articles, reports, and international documents.

- UnWireMyCity.com keeps a repertoire of the MWN debate; the site is operated by John Cooper, a technology consultant from Texas. The site is tailored to municipal chief information officers, city and borough managers, and mayors considering community Wi-Fi. With white papers and resources, the web forum aims to provide information for discouraging the use, design, and deployment of community Wi-Fi.

5.4 Type 4: Broadcasting Interaction

The Internet serves as both an interactive and a broadcast medium at the same time. This unique aspect of the medium has allowed those with political opinions to express them so that they can be widely read and responded to. Despite a lack of being formally trained, vetted, and published, multitudes of political voices are sounding out on topics. Municipal wireless broadband initiatives are no different.

- http://roisforyou.blogspot.com. Craig Settles, a technology consultant, created this website in order to help people understand how to use wireless technology, save money, make money and run a better business. The blogger is able to do this by writing reports and how-to books, and by providing workshops to end-users. The blog also supports the readers of his book, *Fighting the Good Fight for Municipal Wireless: Applying lessons from Philadelphia's Wi-Fi Story*. The blog allows his readers to share their opinions about municipal Wi-Fi.

- http://www.jhsnider.net/telecompolicy. As the current director of research in the Wireless Future Program at the New America Foundation, Dr. J. H. Snider offers a blog about telecom policy in general. Readers interested in municipal Wi-Fi and spectrum policy can post their comments by following the "Municipal Wi-Fi" link. This blog is important as it allows ordinary citizens to voice their opinion (good or bad) to a key public interest group.

- http://www.gigaom.com. GigaOM is a maintained by Om Malik, a senior writer for *Business 2.0* magazine in San Francisco, CA. The site seeks to engage the public in discussing broadband's impact on our society. Some of the this blog's contributors include Daniel Berninger (a senior analyst at a research institution), Jackson West (a writer for a technical magazine), and Robert Young (business entrepreneur).

- The Free Press is a nonpartisan, nonprofit organization that promotes a more equitable and democratic media policy in the United States. On June 13, 2006, Ben Scott, policy director for the Free Press, testified before the U.S. Senate Committee on Commerce, Science and Transportation regarding, in part, the Communications, Consumers' Choice and Broadband Act of 2006 (S. 2686). During his testimony, he supported the goal of this legislation to expand consumer choice and access to broadband services. The Free Press serves as an informal organization that urges activists to file informal requests to the government. Their goal is to expand diverse, local voices on the national front.

6 DISCUSSION

Public policy can significantly impact the formation and development of municipal broadband efforts. We assert that a study of the municipal broadband public policy process can help us better understand the extent and patterns of the municipal broadband policy space. These findings support the results of recent research that noted that studies of public policy process (Sabatier 1999; Schlager and Blomquist 1996; Sutton 1999) and municipal broadband (Gillett et al. 2004; Lehr and McKnight 2003; Lehr et al. 2005; U.S. Department of Commerce 2002) are essential for advancing knowledge within the field.

Figure 3 depicts the public policy process developed by the authors and applied to the development of municipal broadband in the United States. This matrix recognizes four key relationships that operate continuously and simultaneously in the interaction between each quadrant. Note that all arrows have some relation to quadrant 4.

Most important in this analysis is the growing political presence of quadrant 4 and the role the Internet has played in moving individuals to such a key, central location in the policy formation process. As noted above, around the municipal broadband issue there has been little policy direction from the federal level. This is unusual in telecommunication policy in the United States. This has left a political vacuum into which Internet activists, bloggers, and policy entrepreneurs have stepped in.

Arrow A: Collective Action. Each of these arrows represents a flow of information and people, and the creation of relationships to facilitate the movement of both between the categories. With arrow A, we see movement between individuals and organizations, between the informal and the formal. We see this arrow as depicting individuals seeking

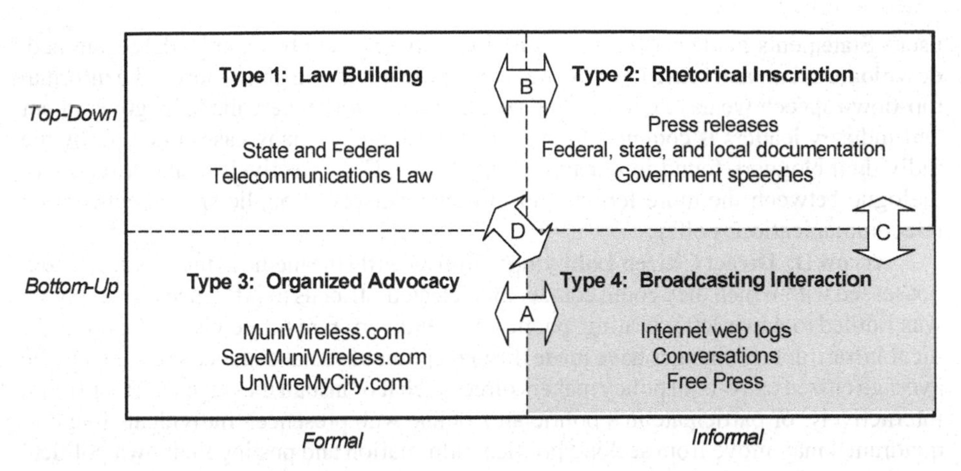

Figure 3. Flow of Information for the Four Quadrant Model

political information concerning municipal broadband networks individually at first, then moving toward more collective action. Technology plays a large role in this arrow as more and more individuals first seek political information via the Internet, may make their first political statements via the Internet and then once finding like-minded others, may act politically with others in a more formalized way. This arrow is double headed. Type 4 provides a conduit into type 3, and vice versa. Individuals hear about the issues and ideas from webloggers in type 4 and then move to join something more formalized in type 3. Type 3 is a means to organize type 4—channeling energy and money into more formalized lobbying attempts.

Arrow B: Policy Inscription. Policy is influenced by the language used by government and industry leaders. Inscription is a process by which various stakeholders who have political, social, or economic interest in a socio-technical artifact attempt to protect and ensure their own interests regarding the artifact. Often this is done through the process of defining the artifact through the use of language (Akrich 1992; Akrich and Latour 1992; Callon 1992; Latour 1992). It is obvious that these municipal networks are being defined differently to different relevant social groups. The municipal governments, Internet service providers, state and federal legislators, device designers, and potential users all represent relevant social groups who have political interest in shaping these networks. In this case, public discussions, press releases, and public commentaries on municipal networks feed directly into public opinion and the policy process itself. For example, rhetorical analysis points to the fact that municipal networks are being textually defined in the public sphere as a solution to the complex problem of the digital divide. Municipal officials have made textual efforts to "inscribe" these networks with concepts of social inclusion, utility status, social revitalization, and equality. It is possible that these municipal labeling efforts reflect a need to sway taxpayers toward a more favorable stance concerning the development of these networks, especially in the light of continued state and federal legislation that threatens continued municipal deployment.

Arrow C: Appropriating the Master's Tools. The rhetorical outputs from the type 2 quadrant are not the ethereal audio or video recordings of the past, due in part to technological changes, political text, audio, and video are now available to every Internet user. Statements made concerning municipal networks can be examined, reexamined, downloaded, cut, and reused by anyone with political will and know-how. The informal top-down speech we have come to call rhetorical inscription here made by government and industry leaders is commented on, repurposed, and in many cases mocked, by the individual bloggers found in our type 4 quadrant. This effectively crates a recursive dialogue between the more formal and informal aspects of public speech concerning telecommunications policy.

Arrow D: Direct Citizen Lobbying: Until recently the methods the average citizen possessed with which they could contact their elected officials to influence formal policy was limited to paper letter writing, phone calls, and very rare office visits. Again, technical infrastructure changes have made the perception of social distance short, in that the average citizen can e-mail policy makers directly, chat or instant message with them more interactively, or participate in a politician's online web presence. Individuals found in quadrant 4 may move from seeking political information and posting their own political opinions to actively lobbying policy makers concerning telecommunications policy. An example of this direct citizen action can be seen through the Electronic Frontier Foundation.

> When our freedoms in the networked world come under attack, the Electronic Frontier Foundation (EFF) is the first line of defense…defending free speech, privacy, innovation, and consumer rights today.…EFF fights for freedom primarily in the courts, bringing and defending lawsuits even when that means taking on the US government or large corporations (Electronic Frontier Foundation, http://www.eff.org/about).

However, the EFF also claims to provide individual citizens with the tools to be their own lobbyist through their "Action Center…Here you can contact your representatives on impending legislation that will have a direct effect on your civil liberties online."

These efforts have already met with limited success. This public momentum has driven many of the state legislative efforts to end in compromise, resulting in less stringent, negotiated legislation. Many of the state-level bills that have passed have done so through amendment and compromise. States with pending legislation have learned from others and have created bills that are far more flexible in what municipalities can and cannot do in their efforts to implement more affordable broadband. Similarly, policy decision makers have also learned more about incumbent providers and what roles they should and should not serve in the delivery of services.

There is an increased desire among citizens to participate in decisions that will affect them, and an increasing need for the policy development process to be informed by input from diverse sources, especially from those involved or affected. For many, the threat of poor and expensive broadband service was enough of a catalyst to encourage them to act politically, and the Internet provided that vehicle.

7 CONCLUSIONS

In this paper we have highlighted three things. First, that public organizations are engaged in a technologically driven servitization of their traditional service products.

Second, that public organizations must approach this servitization differently because of the decision-making role of the citizens in the process. Third, it is essential to study the influence citizens have on this process through studying the public policy process around decisions concerning the delivery of technology and technologically driven services.

Municipalities are complex organizations engaged in the service economy. In some ways, the public sphere is the quintessential service provider, addressing citizen needs for public safety, recreation, elections, and building permits. The public sector, like the private sector, has struggled to transform its service delivery to citizen customers using ICTs with the goals of lowering costs, becoming more effective and efficient, and becoming more responsive to citizen customer needs.

In addition, the public sector, like the private sector, has used these technological affordances to jump into new markets. Unlike the private sector, this municipal jump to provide telecommunications services has provoked public and private outcry and initiated legislative efforts to curtail the move. Essentially, while both public and private organizations have moved to offer more services online and in diverse service-oriented markets, in the public realm policy matters.

If municipalities are going to encourage citizens to go online to apply for permits, register to vote, seek employment, or pay utility bills, among other public services, then they also must ensure that citizens have the infrastructure to do so. Broadband services have come to be seen as a public utility, a necessity for citizens to fully participate in civic life, as well as a responsibility of local government. Unlike the private sector, municipalities operate in the collective in which they are governed by the will of the people and strategic and operational decisions are often made through public policy. While private organizations are affected by, and seek to influence, public opinion, they are not governed by it.

Public policy is an essential piece to understanding the development and deployment of ICT-based services by municipalities.

Little research has been conducted which actually examines different constituencies and their relations in the role of policy in planning and implementing municipal broadband initiatives in the United States. Recent research (Davis et al. 2002) suggests that IT-dependent services are development drivers for at-risk communities. While others have called for new ways of providing telecommunications services in the wake of government broadband interventions, few have examined the choices and trade-offs made by different constituencies and their relations. The ecology of political rhetoric among public officials is more interesting and more relevant to policy than the old stereotypes would suggest. The data clearly indicate that municipal broadband projects are very convoluted and complex, not a one-dimensional sphere, and those different groups have varied preferences as to where and how they influence and formulate policy on that sphere.

Although this analysis provides some answers, other questions remain. This paper provided a description of the interactions existing between the four types of municipal wireless broadband constituents (law building, rhetorical inscription, organized advocacy, and broadcasting interaction). Therefore, future research might consider addressing the following questions: What interactions have priorities over others? What sequence of interactions has been observed? Considering different municipalities have different policy making processes than others, which cities have been more successful than others? What contributed to these differences? Also, the missing, or unidentified arrows seem potentially more interesting that those identified in the matrix. For instance, future studies could examine the interaction between Type 1 (Law Building) and Type 3 (Organized Advocacy) in that the statements and actions taken by organized groups might influence legislators. Similarly, organized groups might influence Type 2 (Rhetorical Inscription).

In the situation of municipalities offering telecommunication services, the boundaries between the public sphere and the private sphere become permeable. Both public and private organizations experience servitization stemming from technological changes, market forces, and citizen-customer expectations. The most important difference between these two change processes is the role of the public in the decision-making process. It is essential to study the role of public constituent groups in the decisions about how to deliver electronic services and services electronically to citizens because of their key role in the process.

References

Akrich, M. 1992. "The De-Scription of Technical Objects," in *Shaping Technology/Building Society*, W. E. Bijker and J. Law (eds.), Cambridge, MA: MIT Press, pp. 205-224.

Akrich, M., and Latour, B. 1992. "A Summary of Convenient Vocabulary for the Semiotics of Human and Nonhuman Assemblies," in *Shaping Technology/Building Society*, W.E.Bijker and J.Law (eds.), Cambridge, MA: MIT Press, pp. 259-264.

Baumgartner, F. R., and Jones, B. D. 1993. *Agendas and Instability in American Politics*, Chicago: University of Chicago Press.

Bleha, T. 2005. "Down to the Wire," *Foreign Affairs*, May/June, pp. 111-125.

Brooks, S. 1989. *Public Policy in Canada: An Introduction*, Toronto: McClelland and Stewart.

Buncombe, A. 2006. "FCC Investigates Fake News," *The Independent UK – Online Edition*, May 31 (http://www.alternet.org/mediaculture/36878/).

Callon, M. 1992. "The Dynamics of Techno-Economic Networks," in *Technology Change and Company Strategies: Economic and Sociological Perspectives*, R. Coombs, P. Saviotti, and V. Walsh (eds.), London: Academic Press, pp. 72-102.

Chrispeels, J. H. 1997. "Educational Policy Implementation in a Shifting Political Climate: The California Experience," *American Educational Research Journal* (34:3), pp. 453-481.

Christman, J. B., and Rhodes, A. 2002. *Civic Engagement and Urban Improvement: Hard to Learn Lessons from Philadelphia*, Philadelphia: Research for Action.

Chung, J. 2001. "Vertical Support, Horizontal Linkages, and Regional Disparities in China Typology, Incentive," *Issues and Studies* (37), pp. 121-148.

Clay, E. J., and Schaffer, B. B. 1986. *Room for Manoeuvre: An Explanation of Public Policy in Agriculture and Rural Development*, London: Heinemann.

Considine, M. 1994. *Public Policy: A Critical Approach*, Melbourne, Australia: MacMillan.

Davis, C., McMaster, J., and Nowak, J. 2002. "IT-Enabled Services as Development Drivers in Low- Income Countries: The Case of Fiji," *The Electronic Journal on Information Systems in Developing Countries* (9:4), pp. 1-18.

Demsetz, J. 1999. "Industry Structure, Market Rivalry and Public Policy," *Journal of Law and Economics* (16), pp. 1-9.

Dunn, W. N. (ed.). 1994. *Public Policy Analysis: An Introduction*, Englewood Cliffs, NJ: Prentice-Hall, Inc.

Dunsire, A. 1978. *Implementation in a Bureaucracy*, Oxford, UK: Martin Robertson.

Dye, T. R. 1972. *Understanding Public Policy*, Englewood Cliffs, NJ: Prentice Hall.

Fischer, F. 1995. *Evaluating Public Policy*, Chicago: Nelson-Hall Publishers.

Frederich, C. J. 1963. *Man and His Government*, New York: McGraw-Hill.

Gerston, L. N. 1997. *Public Policy Making: Process and Principles*, New York: M. E. Sharpe.

Gillett, S. E., and Lehr, W. H. 1999. "Availability of Broadband Internet Access: Empirical Evidence," paper presented at the 27th Annual Telecommunications Policy Research Conference, Alexandria, VA, September 25-27, 1999.

Gillett, S. E., Lehr, W. H., and Osorio, C. 2004. "Local Government Broadband Initiatives," *Telecommunications Policy* (28), pp. 537-558.

Gillmor, D. 2004. *We the Media: Grassroots Journalism by the People, for the People*, New York: O'Reilly Media, Inc.

Goggin, M. L., Bowman, A. O. M., Lester, J., and O'Toole, L. 1990. *Implementation Theory and Practice*, New York: HarperCollins.

Grindle, M., and Thomas, J. 1990. "After the Decision: Implementing Policy Reforms in Developing Countries," *World Development* (18:8), pp. 1163-1181.

Groosman, L., and Hamilton, A. 2004. "Meet Joe Blog," *Time*, June 24 (http://www.time.com/time/magazine).

Guiraudon, V. 2000. "European Integration and Migration Policy: Vertical Making as Venue Shopping," *Journal of Common Market Studies* (38), pp. 249-269.

Kubler, D., Schenkel, W., and Leresche, J. P. 2003. "Bright Lights, Big Cities? Metropolisation, Intergovernmental Relations and the New Federal Urban Policy in Switzerland," *Schweizerische Zeitschrift für Politikwissenschaft* (9:1), pp. 261-282.

Lafferty, W. M., and Hovden, E. 2003. "Environmental Policy Integration: Towards an Analytical Framework," *Environmental Politics* (12:3), pp. 1-22.

Latour, B. (ed.). 1992. *Where are the Missing Masses? The Sociology of a Few Mundane Artifacts. Shaping Technology/Building Society*, Cambridge, MA: MIT Press.

Lehr, W., and McKnight, L. 2003. "Wireless Internet Access: 3G vs.WiFi?," *Telecommunications Policy* (27:5/6), pp. 351-370.

Lehr, W., Osorio, C., Gillet, S., and Sirbu, M. 2005. "Measuring Broadband's Economic Impact," paper presented at the 33rd Research Conference on Communication, Information and Internet Policy (TPRC), Arlington, VA, September 23-25; revised October 4 (http://itc.mit.edu/itel/docs/2005/MeasuringBB_EconImpact.pdf).

Lehr, W., Sirbu, M., and Gillet, S. 2004. "Municipal Wireless Broadband: Policy and Business Implications of Emerging Access Technologies," unpublished manuscript, Massachusetts Institute of Technology (http://itc.mit.edu/ itel/docs/2004/wlehr_munibb_doc.pdf).

Lenard, T. M. 2004. "Government Entry into the Telecom Business: Are the Benefits Commensurate With the Costs?," *Progress on Point*, February (http://www.pff.org/issues-pubs/pops/pop11.3govtownership.pdf).

Lewis, D. A. 1997. "The Two Rhetorical Presidencies: An Analysis of Televised Presidential Speeches 1947-1991," *American Politics Quarterly* (25:3), pp. 380-395.

Majone, G. 1989. *Evidence, Argument, and Persuasion in the Policy Process*, New Haven, CT: Yale University Press.

March, J. G., and Olsen, J. P. 1989. *Rediscovering Institutions: The Organizational Basis of Politics*, New York: The Free Press.

Morris, H. J. 2001. "Blogging Burgeons as a Form of Web Expression," *US News & World Report* (130), p. 52.

Nielsen/NetRatings. 2004. "Three Out of Four Americans Have Access to the Internet," March 18 (http://www.netratings.com/pr/pr_040318.pdf).

Ortiz, J.,and Tapia, A. 2008. "Deploying for Deliverance: The Digital Divide Discourse in Municipal Wireless Networks," *Sociological Focus* (accepted for publication).

Patton, C. V., and Sawicki, D. S. (eds.). 1993. *Basic Methods of Public Policy Analysis and Planning*, Englewood Cliffs, NJ: Prentice Hall.

Pew Internet. 2006. "Home Broadband Adoption 2006," May 28 (http://www.pewinternet.org/pdfs/PIP_Broadband_trends2006.pdf).

Pross, A. P. 1986. *Group Politics and Public Policy*, Toronto: Oxford University Press.
Ragsdale, L. 1984. "The Politics of Presidential Speechmaking." *American Political Science Review* (78), pp. 971-984.
Sabatier, P. A. (ed.). 1999. *Theories of the Policy Process*, Boulder, CO: Westview Press.
Schlager, E., and Blomquist, W. 1996. "A Comparison of Three Emerging Theories of the Policy Process," *Political Research Quarterly* (49), pp. 651-672.
Stokey, E., and Zeckhauser, R. 1978. "Thinking about Policy Choices and Putting Analysis to Work," Chapter 1 in *A Primer for Policy Analysis*, New York: W. W. Norton and Company, pp. 320-329.
Stone, D. A. 1997. *Policy Paradox: The Art of Political Decision Making*, New York: W. W. Norton.
Sundquist, J. L., and Davis, D. W. 1969. *Making Federalism Work*, Washington, DC: Brookings Institution.
Sutton, R. 1999. "The Policy Process: An Overview," ODI Working Paper No. 118, Overseas Development Institute, London.
Tapia, A., Stone, M., and Maitland, C. 2005. "Public-Private Partnerships and the Role of State and Federal Legislation in Wireless Municipal Networks," paper presented at the 33rd Research Conference on Communication, Information and Internet Policy, Washington, DC, September 23-25.
Tapia, A., Maitland, C., and Stone, M. 2006. "Making IT Work for Municipalities: Building Municipal Wireless Networks," *Government Information Quarterly* (23:3/4), pp. 359-380.
U.S. Department of Commerce. 2002. "Understanding Broadband Demand: A Review of Critical Issues," Office of Technology Policy, September 23 (http://www.technology.gov/reports/TechPolicy/Broadband_020921.pdf)..
Williams, J., and Griffin, T. 1996. "Evolution of Vertical Policy: US Steel's Century of Commitment to the Mesabi," *Journal of Industrial and Corporate Change* (5:1), pp. 147-173.
Vandermerwe, S., and Rada, J. 1988. "Servitization of Business: Adding Value by Adding Services," *European Management Journal* (6:4), pp. 315-325.

About the Authors

Andrea H. Tapia is an assistant professor at the College of Information Sciences and Technology at Pennsylvania State University. Andrea is a sociologist with expertise in social research methods and social theory, applying those to the study of information and communication technologies and their context of development, implementation, and use. Her guiding research question is, "What is the role that technology plays in institutional patterns of power, hierarchy, governance, domination and resistance?" The domains of interest in which she asks this question are government and public institutions, virtual worlds, and social policy. Andrea's work has been funded by the U.S. National Science Foundation, the U.S. Department of Defense, the United Nations, and Penn State's Schreyer's Honors College. Her work has appeared in *The Information Society, Government Information Quarterly, Database for Information Systems Research, Communications of the ACM, Science Technology and Human Values*, and *Information Technology and People*. Andrea can be reached at atapia@ist.psu.edu.

Julio Angel Ortiz recently completed his Ph.D. at the College of Information Sciences and Technology at Pennsylvania State University. He is a scholar of technological and telecommunications policy and its impact on social problems such as digital inequality. He applies sociotechnical theories and social science research methods to the study at the intersection between social policy, industry, and communities. He is an Alfred P. Sloan and KPMG Fellow. Julio's thesis was entitled *The Impact of Municipal Wireless Broadband Networks (Mu-Fi) on the Digital Divide: A Tale of Five U.S. Cities*. He can be reached at jortiz@ist.psu.edu.

17 ANALYZING PUBLIC OPEN SOURCE POLICY: The Case Study of Venezuela

Edgar Maldonado
Andrea H. Tapia
College of Information Sciences and Technology
Pennsylvania State University
University Park, PA U.S.A.

Abstract *This research examines public open source software (OSS) adoption policies using a framework built upon the analysis of information and communication technology (ICT) policies. The legislative and objective framework is used to picture the formal public OSS policies applied in Venezuela. Preliminary results indicate negligence for the inclusion of the private sector in the migration plan. Future research looks for an analysis in situ of the activities carried out in the country, and the validation of the framework.*

Keywords Open source software, open source development, national policy, Venezuela, Latin America

1 INTRODUCTION

The goal of this research is to demonstrate our thinking about the role of government in the service economy. Although this relationship is complex, we have chosen to highlight three aspects of this role: the government in the role of regulator, the government in the role of consumer, and the government in the role of thought leader.

It is often true that a national government will take an active role in the development and maintenance of an information-based service economy within its borders with the goal of furthering economic development. The shape this intervention takes differs from nation to nation. Governments, as primary decision makers, provide a set of measures

Please use the following format when citing this chapter:

Maldonado, E., and Tapia, A. H., 2008, in IFIP International Federation for Information Processing, Volume 267, Information Technology in the Service Economy: Challenges and Possibilities for the 21st Century, eds. Barrett, M., Davidson, E., Middleton, C., and DeGross, J. (Boston: Springer), pp. 249-257.

that, at least in their intent, will lead the country to continued economic and social development (Pieterse 2001, p. 3). Of particular interest is when a national government uses its powers to mandate the use of open source software and systems with the goal of influencing the course of development of the service economy within its borders.

The number of countries with policies that endorse or support the use of OSS in some way is significant and growing. The Center for Strategic and International Studies, in its report titled "Government Open Source Policies," identified 268 OSS public policies initiatives (CSIS 2007). However, very few governments have set their policy bar at the far end, making the move to OSS mandatory. Among all initiatives, six have been categorized as *mandatory* at a country level: Belgium, Denmark, Norway, Taiwan, Peru, and Venezuela. In addition to these six, the Minister of Economic Affairs of the Netherlands announced he Netherlands in Open Connection plan (Ministry of Economic Affairs 2007). Among other claims, the official document sets a deadline for the adoption of open source software in all ministries as April 2009. The Philippines is considering House Bill #1716, which not only will make the use of OSS mandatory in public settings, but it will change the Philippines intellectual property legislation, making software a non-patentable invention (Casi 2007).

The reasons for this strong interventionist approach to OSS often fall into two main categories: overall social benefits and pragmatic arguments. The social benefits of using OSS are economic independence, the potential of OSS to contribute to solving the universal access problem, to increase local control and local economic growth, and to improve transparency and democratic accountability. The pragmatic arguments point out the interoperability, security, and cost of OSS (Ghosh 2005). In addition, OSS is seen as having potential for developing countries to create specific solutions for their particular needs, eliminating or reducing foreign dependencies (Weber 2003; Weerawarana and Weeratunga 2004).

This research examines the public OSS policies of Venezuela, which, because of high revenues from oil exports, finds itself in the unique position of having the economic ability to support an OSS migration plan for the entire nation. Venezuela (ranked 83rd in the world in 2006-2007 by The Networked Readiness Index Ranking) is a developing country leapfrogging into the global information and service economy. The country has been rapidly crafting telecommunications policies and establishing ministries to enact and enforce them. Venezuela established its Ministry of Science and Technology in 1999 and its Ministry of Telecommunications and Informatics in 2007.

In the case of Venezuela, we see the national government influencing the growth of the technologically enabled service economy (1) in the role of regulator, as it make laws that mandate the use of OSS in public offices, (2) in the role of consumer, as an important sector of the working population are government employees and are becoming the consumers of this change, and (3) in the role of thought leader, as it attempts to persuade Venezuelan citizens to adopt OSS inside and outside the public sphere.

2 ANALYTICAL FRAMEWORK DEVELOPMENT

The circumstance of governments mandating OSS through legislation is relatively new and thus frameworks for analyzing these initiatives are few. One such framework is offered by the Centre for Strategic and International Studies, which characterized ublic

OSS policies using two dimensions: jurisdiction and level of coercion. Jurisdiction includes state and local policies, and national level policies. The level of coercion is divided into four levels, in diminishing order: mandatory, preference, advisory, and research and development (CSIS 2007). Although this framework is useful, it does not help clarify the purpose of the policies or the role the government intends to take in terms of the development of the economy.

With this research, we seek to illustrate a new analytical framework that combines the two dimensions of the CSIS categorization into one and introduces policy goals and the government's role into the analysis. Both jurisdiction and level of coercion obtain their power via legitimate authority in the form of legislation. Therefore, in order to merge these two aspects into a single dimension, we find we need a means of categorizing legislation.

To accomplish this, we draw from Kelsen (2002), who introduced a model to explain the logical structure of law where general rules are found at the top and specific rules are found at the bottom. The rules at the bottom inherit their authoritative power from the top. The logical structure of law can be likened to a pyramid with the most fundamental and authoritative rules (the *Grundnorm*) at the top and the most particular rules (those applying to particular concrete situations) at the base. Kelsen's pyramid is represented with the constitution as a fundamental law at the top, general laws in the middle, and other forms of legal instruments at the bottom (for a detailed Kelsen's pyramid of Venezuela's law, see Naranjo 1982).

Building on Kelsen, we have created a truncated legislative pyramid with three levels: law level, decree level, and norm level. The law level includes any specific law (organic, special, ordinary, etc.). The decree level includes all decrees passed by executive positions (president and governors). Finally, the norm level consists of resolutions, orders, or other administration procedures usually passed by ministers or local authorities. Figure 1 illustrates a representation of the pyramid.

While adopting Kelsen categorization allows us to place policies in a continuum of most to least authoritative-coercive power, it does not address the nature of the policies themselves, especially in respect to information technologies.

In order to situate OSS polices according to their relationship with information technologies and development, we draw on the work of Alabau (1997). Nations seeking the development of a sector in their region typically follow guidelines given by the public sector. Alabau divides these government initiatives, aimed at growing the ICT sector, into four categories: (1) the stimulation of ICT initiatives in the private sector, (2) the stimulation of ICT initiatives in the public sector, (3) ICT deployment as part of territorial planning, and (4) regulation of ICT-related activities. Initiatives that fit the first category are those aimed at fueling the private sector by creating more jobs or improving competitiveness, and to all citizens in general by improving the quality of life. In the second category the policies seek to improve public activities by the introduction of ICTs. In the category of territorial planning are those initiatives in charge of the planning of a city or town; the planning is usually done with long-term thinking and the ICT deployment is an essential part of the overall infrastructure. Finally, the regulation category includes those initiatives that seek to achieve ideal circumstances for the development of ICTs (full competition, universal access, interoperability, etc.) (Alabau, 1997).

Policies that fit in the last category, regulation, usually have an associated intrinsic legal character and carry the authority of the government behind them. The legal char-

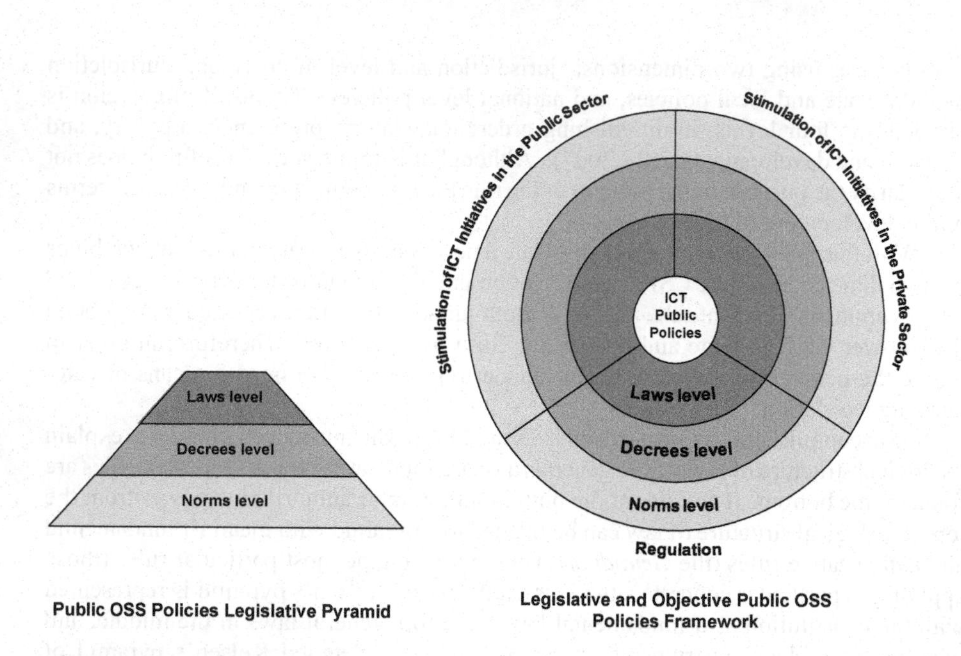

Figure 1. The Public OSS Policies Legislative Pyramid and the Legislative and Objective Framework

acter is the feature that assures national policy will achieve their initial goals. Although most regulations are created hoping that they will be accepted and followed, their legal status oftentimes determines their success. The stronger the legal nature of an ICT policy, the stronger the sanctions against noncompliance, the more likely the policy will have the desired effect.

Drawing from both the Kelsen and Alabau categorization schemes, we have the ability to analyze and categorize OSS policies on two axes: amount of authority/coercive power and form of policy tool toward economic development. The legislative and objective framework is represented in Figure 1.

The legislative and objective framework provides a practical tool to graphically interpret the reach and purpose of a specific ICT policy. We have taken the pyramid and expanded it to encompass a full circle. This circle represents three domains: public, private, and regulatory. This provides us with a mechanism to illustrate the form and domain for each OSS policy initiative.

3 METHODOLOGY

The principal tool used in this research is policy analysis. Policy analysis is defined by Nagel (1980) as "methods associated with determining the nature, causes, and effects of government decisions or policies designed to cope with specific social problems" (p. 3). Putt and Springer (1989) identified five major types of policy research, including exploration, description, causation, estimation, and choice. In the present work, we primarily engage in description and estimation activities. As a form of description policy research, we describe the policies of Venezuela (1) as they exist as physical documents and (2) as they are interpreted and carried out by authorities Policy research estimating activities, in this particular case, are limited given the lifespan of the phenomenon, but in that respect we look at the effects of the public OSS policies in Venezuela on enhancing the ICT sector of the country.

Our data comes from several official Venezuelan channels: the Ministry of Science and Technology website (http://www.mct.gov.ve), the National Center of Information Technology website (http://www.cnti.gob.ve), the National Assembly website (http://www.asamblenacional.gov.ve), and FUNDACITE Merida website (http://www.fundacite-merida.gob.ve). Those websites make available all legal documents, including decrees, resolutions, law proposals, and minutes of legal meetings. There are four legal documents and two plans that constitute the core of the data used in this study (see Table 1). The legal documents are Resolution #237, Resolution #238, Decree 3,390, and the Telecommunications Law (renamed Infogobierno Law). The plans are the National Plan for Innovation, Ccience, and Technology 2005-2030 and the National Plan of Telecommunications, Informatics, and Postal Service.

This paper represents a small portion of a larger research project into public open source initiatives in Venezuela. These research methods complement other forms of qualitative data collection ongoing in Venezuela.

4 PRELIMINARY FINDINGS

The central policy motivating OSS change in Venezuela is Decree 3,390. This decree establishes the mandatory use of OSS in all National Public Administration (see Table 1). Using the legislative and objective framework, this policy falls in the middle ring on the public initiatives wedge. Venezuela also created open source academies (OSA) (Resolutions #237; see Table 1) as an indirect method of stimulating the development and use of open source. The OSAs provide training to government employees and the general public on open source software from an end-user level to a developer-administrator level. In addition, there is an open source factory (OSF) initiative, created simultaneously with the OSAs. The OSFs are developing OSS to be used in government bodies (municipalities, hospitals, and elementary schools) and other forms of entrepreneurship sponsored by the government (small tourism-related businesses). Both initiatives can be categorized in the external ring of the framework on the public initiatives wedge.

The *Infogobierno* Law would provide a stronger authoritative-coercive support for the adoption of OSS since it would have more significant power than the Decree 3,390. This proposal has not yet passed. The delay may be due to the current crafting of the *National Plan of Telecommunications, Informatics, and Postal Service 2007-2013*. The preceding *National Plan of Innovation, Science and Technology 2005-2030* was published in 2005 and included the adoption of OSS in public offices as one of its strategic goals. The *2007-2013* plan is expected to include this focus, but also take intou account the private sector and foreign investors. Venezuela current policies do not include a direct intervention in the private sector. Although it is possible that this sector is making a move to integrate itself to the migration processes.

Table 1 Documents Used in this Work

Date	Governmental Body, Type of document	Description	Objective
November, 2004	Minister of Science and Technology, *Resolution #237*	For resolution of the Minister of Science and Technology, Marlene Yadira Córdova; created the Academia de software Libre (Open Source Academy).	It is created the scientific-technological program of research, denominated "Academia de software Libre," this program will have as an objective the promotion of research, development, innovation, and formation in the open source software area.
December, 2004	National Executive, *Decree 3,390*	President of the Republic, Hugo Chávez Frías signed Decree 3,390. The decree establishes open source as a mandatory first option in all government systems.	The National Public Administration will use open source software as first priority in its systems, projects, and informatics services. To such aims, all of the institutions and offices of the National Public Administration will initiate the progressive adoption of open source software.
August, 2005	National Assembly, *Technologies of Information Law*; first discussion	Article 75 of this law clearly describes the characteristics of the software that the government should use.	The organs and agencies of the Public Power will have to use, primarily and by preference in their systems of information technologies, the computer science programs and applications whose licenses or contracts guarantee in an irrevocable way access to the source code of the program by the user; to execute it with any intention; to modify it and to redistribute the original program as much as its modifications in the same decided conditions of licensing to the original program, without having to pay exemptions to the previous developers.
October, 2005	Minister of Science and Technology, *National Plan of Innovation, Science and Technology 2005-2030 (NPSTI)*	The Minister of Science and Technology published a 152-page document with the National Plan of Science Technology and Innovation. Open source is used as part of the strategic goals of the plan.	Strategic goal #5: Migration of the systems of the public administration to open source systems until reaching a complete adoption within the technological platforms of the state, in a period no longer than 5 years.

Date	Governmental Body, Type of document	Description	Objective
August, 2006	National Assembly, *Infogobierno Law*	The National Assembly began the second discussion of the Law of Technologies of Information. The law now is renamed *Infogobierno Law*. The new version has kept a good part of the content of the prior version. Article 67 specifies the use of open source software in government offices.	The Public Power will have to guarantee that in their systems of information technologies, the computer science programs and applications fulfill the following characteristics: (1) Access to all source code and the transference of the knowledge associated for its understanding. (2) Freedom of modification. (3) Freedom of use in any area, application, or intention. (4) Freedom of publication of the source code and its modifications
April, 2007	Minister of Telecommunications and Informatics, *National Plan of Telecommunications, Informatics, and Postal Service 2007-2013*; in process	The Venezuelan government announces the *National Plan of Telecommunications, Informatics, and Postal Service 2007-2013.*	Using online surveys and board discussions with stakeholders from different sectors of the society and economy of the country, the Minister of the Public Power of Telecommunications and Informatics is creating a roadmap for the next 5 years in the area. The delay in approving the Infogobierno Law could be due to the need to fit the legal document to the new national plan.

Venezuela's open source software policies are still being crafted. The National Plan of Telecommunications, Informatics, and Postal Service 2007-2013 will be published soon. Along with this plan, a final version of the *Infogobierno* Law will be passed. Given the direction of the legal framework of Venezuelan OSS policies, is likely that the mandatory character of OSS will be kept in the new law. The existing initiatives are portrayed in the legislative and objective framework in Figure 2.

5 RESULTS FORTHCOMING

The goal of this research is to demonstrate our thinking about the role of government in the service economy. We have focused on government interventions through laws that mandate the use of open source software and systems. We have created a framework that looks at these interventions through three lenses providing one picture. The first lens is the amount of coercive authority granted the initiative, the second is the public arena in which the initiative seeks to have an effect, and the third is the role the government has taken in each. We have chosen to highlight three aspects of this role: the government in the role of regulator, the government in the role of consumer, and the government in the role of thought leader.

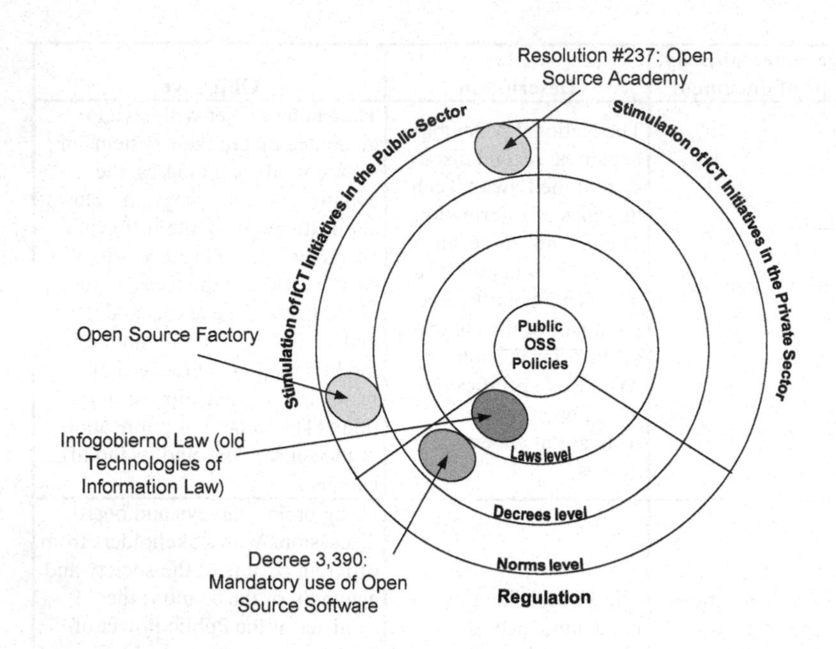

Figure 2. Venezuela's Public OSS Initiatives

As mentioned earlier, this policy analysis work is part of a larger research program investigating public open source. We are currently gathering data in Venezuela. We believe that the analytical framework we have created will not only help us to analyze the policy efforts in Venezuela, but will provide a mechanism to compare efforts across countries that have similar OSS initiatives. Once our data is fully collected and analyzed, we expect to complete the entire picture of the Venezuelan migration using the framework presented in this work. Once the framework is refined and validated, we anticipate it will be possible to apply it to other types of policies influencing the development of the technology-enabled service economy within a nation.

References

Alabau, A. 1997. "Telecommunications and the Information Society in European Regions," *Telecommunications Policy* (21:8), pp. 761-771.

Casiño, T. A. 2007. "Republic of the Philippines House Bill No. 1716," August 1 (http://www.bayanmuna.net/data_files/HB/HB-1716-FOSS.pdf).

CSIS. 2007. "Government Open Source Polices," Center for Strategic and International Studies, Washington, DC, August (http://www.csis.org/tech/it/#oss)

Ghosh, R. A. 2005. "The European Politics of F/OSS Adoption," in *The Politics of Open Source Adoption*, J. Karaganis and R. Latham (eds.), New York: Social Science Research Council.

Kelsen, H. 2002. *Pure Theory of Law* (translation from the second German edition by M. Knight), Berkeley: University of California Press, 1967; reprinted by The Lawbook Exchange, Ltd..

Nagel, S. 1980. *The Policy-Studies Handbook*, Lexington, MA: Lexington Books.

Naranjo, Y. 1982. *Introduccion al derecho*, Caracas: Ediciones Libreria Destino.
Pieterse, J. N. 2001. *Development Theory: Deconstrution/Reconstruction*, London: Sage Publications.
Putt, A. D., and Springer, J. F. 1989. *Policy Research: Concepts, Methods, and Applications*, Englewood Cliffs, NJ: Prentice Hall.
Weber, S. 2003. "Open Source Software in Developing Economies ," unpublished manuscript, University of California, Berkeley (http://www.ssrc.org/programs/itic/publications/ITST_materials/webernote2.pdf).
Weerawarana, S., and Weeratunga, J. 2004. *Open Sources in Developing Countries*, Stockholm: Swedish International Development Cooperation Agency, January (http://www.eldis.org/fulltext/opensource.pdf)..

About the Authors

Edgar Maldonado is a Ph.D. candidate in the College of Information Sciences and Technology at Pennsylvania State University. Edgar will complete his degree in 2008. His research focuses on national-level telecommunication policies specifically directed at open source software. In addition, his work centers on policy in developing countries and their influence on the development of knowledge-based societies and economies. He has an undergraduate degree in Electronic Engineering from Simón Bolívar University, Caracas, Venezuela. Before entering graduate school, Edgar worked as a software support engineer for banking networks, with projects in Venezuela, Trinidad and Tobago, Panamá, Guatemala, and the Cayman Islands. Edgar can be reached at eam264@psu.edu.

Andrea H. Tapia is an assistant professor at the College of Information Sciences and Technology at Pennsylvania State University. Andrea is a sociologist with expertise in social research methods and social theory, applying those to the study of information and communication technologies and their context of development, implementation, and use. Her guiding research question is, "What is the role that technology plays in institutional patterns of power, hierarchy, governance, domination and resistance?" The domains of interest in which she asks this question are government and public institutions, virtual worlds, and social policy. Andrea's work has been funded by the U.S. National Science Foundation, the U.S. Department of Defense, the United Nations, and Penn State's Schreyer's Honors College. Her work has appeared in *The Information Society, Government Information Quarterly, Database for Information Systems Research, Communications of the ACM, Science Technology and Human Values*, and *Information Technology and People*. Andrea can be reached at atapia@ist.psu.edu.

18 CO-ORIENTING THE OBJECT: An Activity-Theoretical Analysis of the UK's National Program for Information Technology

Panos Constantinides
Frank Blackler
Organisation, Work and Technology
Lancaster University
Lancaster, U.K.

Abstract *This paper contributes to research on the success and failure of information and communication technologies (ICT) by focusing on the learning processes associated with the development of new ICT projects and the way they challenge and extend familiar organizational limits. Drawing on recent developments in activity theory, we provide an analysis of oral and written evidence taken before a House of Commons Committee in relation to the UK''s National Program for IT (NPfIT). Our preliminary findings point to the ways in which new objects of activity such as the NPfIT can emerge from the meeting of contrasting forms of discursive activity, as well as how new policy insights can be translated into new organizational practices. We conclude with some implications for further research.*

Keywords ICT success and failure, organizational limits, co-rientation, rhetorical strategies, activity theory

1 INTRODUCTION

This paper contributes to organizational analyses of success and failure in the development and implementation of major information and communication technology (ICT) projects (Brown and Jones 1998; Fincham 2002; Sauer 1999; Wilson and Howcroft 2002). A key theme emerging from this literature is that, given the scale of such projects

Please use the following format when citing this chapter:

Constantinides, P., and Blackler, F., 2008, in IFIP International Federation for Information Processing, Volume 267, Information Technology in the Service Economy: Challenges and Possibilities for the 21st Century, eds. Barrett, M., Davidson, E., Middleton, C., and DeGross, J. (Boston: Springer), pp. 259-270.

and the stakeholders involved, almost any project is potentially vulnerable to the attribution of failure. Despite their flaws (e.g., over-run time scales, over-budget spending, poor consideration of user needs, etc.), however, many ICT projects are still deemed a success (Wilson and Howcroft 2002).

To overcome this paradox, we argue that, rather than focusing on critical success factors and causes of failure, it is more helpful to focus on learning processes associated with the development of new ICT projects and the way they challenge and extend familiar organizational limits (Farjoun and Starbuck 2007; Starbuck and Farjoun 2005). In exploring these dynamics we use the notion of *co-orientation* (Taylor and Robichaud 2004). Co-orientation points to the ways by which, through discursive activity, different groups and individuals work toward new understandings of their shared object of activity and of each other's roles and contributions. Such analysis needs, we suggest, to be located in a broader study of the cultural-historical and material mediations (Engeström 1987, 2004).

We develop this argument by drawing on a recent Committee of Public Accounts (CPA) report on the UK's National Program for IT (NPfIT). The focus is on the ways in which the CPA interrogated project leaders, and the responses of the project leaders to the committee's cross-examination. Our preliminary analysis suggests some initial implications for our understanding of ICT and changes in service organizations which point to further research priorities. Key issues include the ways in which new objects of activity can emerge from the meeting of contrasting forms of discursive activity and ways in which new policy insights can be translated into new organizational practices.

In the following sections, we outline the background to the case study and describe our methodological and analytical directions in more detail. We then provide a brief discussion of our initial analysis of the CPA's and the project leaders' activities of co-orientation, as well as the practical implications of those. We conclude with a discussion of some directions for further research and the expected contributions of the analysis we are developing for large-scale ICT development and changes in service organizations

2 BACKGROUND TO THE CASE STUDY

In 1998, the UK National Health Service (NHS) Executive set a target for all NHS trusts to have electronic patient records in place by 2005. The central vision of the NPfIT was to standardize the previously fragmented ICT delivery in all NHS organizations by introducing an integrated system (Brennan 2005; Currie and Guah 2007). However, the complexity of technological innovation and funding needed for this national program soon made it the biggest public ICT project ever undertaken, with considerable implications for the widespread and timely implementation of the different applications proposed. The complexity of the NPfIT can be summarized into several challenges including the turnover of senior responsible owners (project leaders); doubts about the capability of ICT suppliers to meet the program's demands, leading to some suppliers breaking their contracts; repeated announcements of delays with considerable cost implications and disruption of work to different NHS trusts; implementation problems in hospitals with implications for patient care; poor engagement of clinicians leading to their alienation; restructuring of the program; constant, critical commentary from the press and the CPA;

and, finally, the resignation of the director general of NHS ICT in June 2007 (Currie and Guah 2007; Sauer and Willcocks 2007).

In short, the national program has been moving slower than expected, with deployment plans becoming increasingly unreliable, and responsibility and accountability constantly disputed. The NPfIT has gradually become an object of great dispute and those involved in its development have found themselves struggling to defend their accountability and, in consequence, their project identity, while those responsible for scrutinizing the performance shortcomings demand new approaches to projects of this kind.

3 METHODOLOGICAL AND ANALYTICAL DIRECTIONS

The research draws on data from the oral and written evidence presented before the CPA assigned to examine "the progress made by the Department of Health in implementing the Programme (NPfIT)" (HC 2007, p. 4) on the basis of a National Audit Office report (NAO 2006). The CPA is appointed by the House of Commons to examine "the accounts showing the appropriation of the sums granted by Parliament to meet the public expenditure" (Standing Order No. 148). Despite the general duties of the CPA, however, a closer examination of the evidence in the CPA report suggests that the meeting between the CPA and the NPfIT project leaders was more than a routine activity. Recent articles (Collins 2006; Hoeksma 2006), as well as oral (HC 2007, pp. 30-54) and written evidence in the CPA report (ibid, pp. 136-151), questioned the validity of the NAO report on the grounds that it "failed to ask key questions and to explore crucial evidence regarding [the] NPfIT" (ibid, p. 137); the NAO report was instead found to paint a very positive picture of the project's progress. To this end, the CPA meeting had a distinct agenda in questioning (and redefining, as we will argue later on) the limits of the NPfIT as those were presented in the NAO report.

The participants to the meeting included the CPA and the NPfIT project leaders. The CPA consisted of members of parliament, assisted by the NAO Comptroller and Auditor General and the Treasury Officer of Accounts from the HM Treasury. NPfIT project leaders included the acting chief executive of the NHS; the director general of ICT; the director of ICT and Service Implementation for the NPfIT; the director of Clinical Safety for the NPfIT; a primary care medical director for one of the NPfIT applications; a national clinical lead in the NPfIT; and three representatives from the Department of Health.

To develop an understanding of the evidence presented at the meeting, we asked the following questions: (1) How do different groups and individuals co-orient themselves both to an ICT project and to each other through their discursive activities? (2) What are the implications (products) of this co-orientation?

Our analysis, thus, builds on two themes. First, we undertake an analysis of the oral exchanges at the meeting. This work starts with the contention that discourse is structured in a way that not only describes the world in which it is part, but also categorizes it by bringing phenomena into sight (Parker 1992). We have concentrated on an analysis of the intentions by which the NPfIT has been and is being constructed, exploring ways in which the CPA and those presenting evidence to it co-orient themselves both to each other and to the project under investigation (Taylor and Robichaud 2004).

Second, we are assessing the likely implications of the CPA's analysis for the way such projects should be managed in the future. This work involves analysis of how those involved in the organization and execution of such projects come to collectively produce implications for those projects through their discursive activities. In a recent article, Engeström (2004, p. 97) argued that work meetings "typically not only reflect on the particular issue or case, they also include consequential decision-making. In other words, they are both reflective and practical." He added that, "such settings offer opportunities to capture how history is made in situated discursive actions" (p. 97). The meeting between the CPA and NPfIT project leaders was less dialogic and more adversarial in nature than the meetings to which Engeström was referring. Their aim, however, was the same: to be both reflective and consequential.

In summary, in analyzing large-scale ICT development and changes in service organizations, we focus on the whole activity system within which different groups and individuals try to co-orient themselves. The notion of the activity system is a way of conceptualizing distributed agency around particular objects of activity. These objects of activity, such as the NPfIT, are understood to be collective projects that are stabilized in particular communities of practice through a negotiated set of tools, as well as recognized procedures and a division of labor (Engeström 1987). Thus, by focusing on the whole activity system within which different groups and individuals try to co-orient themselves, we approach micro–macro instances of change as two sides of the same coin. For example, the (micro) discursive exchanges between the CPA and the project leaders (their co-orientation activities) later produce (macro) implications for policy on the organization, implementation, and management of the NPfIT, the object of activity. In other words, scale is enacted. By focusing on the object of activity we can develop an informed understanding of how this enactment unfolds and with what implications for both the object and the individuals participating in its co-orientation.

4 PRELIMINARY ANALYSIS OF ACTIVITIES OF CO-ORIENTATION

Our preliminary analysis explores the activities of co-orientation of the CPA and NPfIT project leaders, as well as the practical implications of the reforms they have produced. Table 1 provides a summary of key themes in the oral evidence presented to the committee. As shown in the table, questions raised by committee members included:

- How is the central procurement of the project affected by local needs?
- Is the project linked to best practice?
- How competent are contractors and suppliers?
- Is the confidentiality of patient data being addressed?
- Are key users being engaged?

Project leaders responses to these questions included assertions that "procurement is national but implementation is local," "best practice elements have been incorporated in the efforts," "participants' competence is monitored." "patient confidentiality is addressed," and "clinicians have plenty of opportunity to get involved." For further details on key themes discussed at the meeting, see Table 1.

Table 1. Key Themes from the Oral Evidence on the NPfIT

Key Questions Raised From Committee Members	Responses (From Project Leaders)	Co-Orientation of the NPfIR*	Practical Implications (New Forms of Discursive Action)*
1. **How is the central procurement of the project affecting local needs?** "Do you believe that there is one standard UK system that can deliver what the project is trying to deliver?"(MP)	**Procurement is national but implementation is local.** "Unless you test whether it will fit, how do you actually know" (DoH representative).	**Unclear responsibility and accountability of local implementation**	**Chief executives and senior management in the NHS to be given authority and resources for local implementation**
2. **Is the project linked to best practices?** "Are you doing something that no other country apparently is attempting?" (Chairman)	**Best practice elements have been incorporated in the efforts.** "It is at times uncomfortable being in a leadership position" (Director general of IT).	**Signs of possible inhibition on innovation, including the implementation of best practices**	**The Department of Health to modify the procurement process so that NHS trusts can develop their own strategy on system selection and implementation**
3. **How competent are the contractors/ suppliers?** "Is the network of suppliers robust enough to withstand the pressure that you are putting on them?" (MP)	**Their competence is monitored.** "There is a balance to strike between the inefficiency of having lots of suppliers and the efficiency of single supply and we are 3 years in a 10 year program" (Director general of IT).	**Suppliers' capacity to deliver unclear**	**The DoH to commission an urgent independent review of the suppliers' performance**
4. **Questions around (a) whether the project is meeting proposed time scale; (b) the total cost of the project; (c) the return on investments (ROI) made on the project.** "…the risk is in the timescale rather than the costs" (MP)	**(a) Some systems have already been installed; suppliers are challenged by the time scale; (b) largest and most ambitious program in the world; (c) currently unable to comment on ROI.** "Yes, because we have transferred finance and completion risk for the most part to the suppliers" (Director general of IT)	**(a) Electronic patient records module delayed extensively; (b) overall expenditure vague; (c) unclear ROI**	**(a) The DoH to develop more robust timetable with suppliers; (b) the DoH to establish more detailed estimations of the total costs and benefits of the project; (c) the DoH to commission an independent assessment of the business case so far**
5. **Is the confidentiality of patient data being addressed?** "How are you going to make sure that staff follow the rules so the security and confidentiality of patients' records is protected?" (MP)	**Patient confidentiality is addressed in the system.** "We deal with that at the design stage…and then we shall be monitoring when things are in practice" (Director of Clinical Safety)	**No concluding co-orientations drawn****	**No explicit discursive actions formulated, but possibilities for emerging trust and governance issues****

Key Questions Raised From Committee Members	Responses (From Project Leaders)	Co-Orientation of the NPfIR*	Practical Implications (New Forms of Discursive Action)*
6. **Are the clinicians (key users) being engaged?** "Do you believe you do have a buy-in from clinicians?" (MP)	**Clinicians had plenty of opportunities to get involved.** "I will just say that GPs are very, very shrewd customers" (Director general of IT)	**Possible resistance from clinicians due to poor engagement**	**The Chief Medical Officers within the DoH to review whether the project has met the needs of the clinicians**
7. **How effective is the project's leadership?** "[There have been] changes, all of which took place in a very short time at the leadership level of this project. Why on earth was so much mobility and lack of continuity permitted?" (MP)	**Leadership has been effective in places but due to changes in the DoH and the scale of the project there have been some shortcomings.** "There was continuity through [the current Director general of IT] and his team on the procurement....[However] you are well aware of the changes which have taken place in the DoH over time." (Acting Chief Executive of the NHS)	**Focus too narrowly on delivery of ICT systems at the expense of broader processes of business and organizational change**	**The DoH to avoid further changes in leadership to improve links with clinicians and efforts to improve NHS services that the project intends to support**
8. **Is the report from the National Audit Office valid/true?** "Sources suggest that the NAO was ground down in a war of attrition with [project leaders] who fought a dogged rearguard action to keep back criticisms." (MP quoting a media article)	**The NAO Report is factual.** "I was not ground down...I bring my work to Parliament and I am satisfied that what I have brought to you is work of high quality, done by my staff." (NAO Auditor General)	**No concluding co-orientations drawn*** **	**No explicit discursive actions formulated*** **

KEY: Bold text represents general themes, whereas normal text represents individual responses.
*These are drawn from the formal conclusions and recommendations made in the report (HC 2007, pp. 5-7) and explored in more detail in the oral evidence.
**Patient confidentiality was brought up in the oral evidence (HC 2007, pp. 31-32) but not mentioned in the formal conclusions and recommendations.
***As with the issue of patient confidentiality, the validity of the NAO report was brought up in the oral evidence by all CPA members (HP 2007, pp. 30-54) but not mentioned in the formal conclusions and recommendations. This interest came out of considerable written evidence provided by various NHS staff, who questioned the findings of the NAO report, as well as the methods by which those findings were drawn (HC 2007, pp. 136-151).

In an initial analysis of the data, we found the project leaders' discourses to be implicated in two particular rhetorical strategies, namely, *synecdoche* and *metonymy* (Corbett 1990; Putnam 2004). Synecdoche is a strategy of representation, a means of expanding the meaning from a part to a larger whole or vice versa. Metonymy is a strategy of reduction, a means of presenting events, situations, problems, etc., in a way that requires alternative meanings and new patterns of association.

For example, using *synecdoche*, project leaders argued that responsibility as to the delays in the delivery of the NPfIT could not fall solely to individual actors, but to teams and their interconnected activities. Using *metonymy*, project leaders argued that, despite some shortcomings in parts of the project, other parts were extremely successful.

Further, these two rhetorical strategies were often used in an intertwined fashion, which made the task of the CPA to go after the evidence even harder. In one exchange, for example, the chairman asked why there had been no deployments of patient administration systems in the 172 hospital trusts, only to be provided with an answer by the director general of IT about the deployment of other systems in 33 trusts in the last 24 months (HC 2007, p. 29). This response points to a combination of synecdoche and metonymy as the director general of IT tried to convince the CPA that not only did they need to evaluate developments in the NPfIT as parts of a greater whole, but also that delays to the program were only relative in the context of the overall time frame of the program.

Certainly, the CPA itself employed its own rhetorical strategies in an effort to align their task to the NPfIT with the broader policy context of the NHS. In this sense, the CPA's preferred rhetorical strategy was that of persuasion toward compliance to outcome-based management. In classical rhetoric, resource deficient efforts to persuasion (i.e., when facts are unclear as in the meeting under focus) are thought to be enhanced by invoking common topics of argumentation, namely, definition, comparison, relationship, circumstance, and testimony (Corbett 1990). These topics represent a system upon which rhetoric builds in order to rise above the lack of clarity caused by resource deficiencies in classification, analysis, and synthesis of meaning.

First, *definition* is an effort to ascertain the specific issue to be discussed (i.e., to define the key terms in a thematic proposition) so that a given audience clearly understands what is being discussed. For instance, a repeated question posed by CPA members was around the costs of the overall program leading to the inference that these costs were wrongly defined, at least for some NHS trusts.

Second, *comparison* between things refers to arguments about similarity and difference, including the degree to which they differ. For instance, CPA members questioned the decision to go for a national, centralized solution when other countries are explicitly avoiding such a strategy.

Third, *relationship* refers to arguments around cause and effect. CPA members asked, for example, whether funding for the NPfIT would have implications for other services.

Fourth, *circumstance* concerns the subdivision of arguments into the possible and impossible. Some CPA members, for example, asked whether the NPfIT was perhaps "too ambitious," showing strong determination on one hand, but being aggressively self-seeking and perhaps self-centered on the other (HC 2007, p. 35). In other words, a program that is not just facing risks, but more so, facing the very likely possibility of failure.

Finally, as in all public committee inquiries, the CPA employed the topic of *testimony*, which, unlike the other topics, does not derive its material from the nature of the question under discussion but from external sources, such as different types of evidence, maxims (i.e., statements about universal matters such as moral and ethical codes) and precedents (i.e., previous decisions and actions taken in similar cases in the past).

In the end, the rhetoric of the CPA in conjunction with the amassing evidence on the different issues raised resulted in a blaming exchange between the project leaders. As noted in Table 1 in their final recommendations, the CPA members not only required that the project leaders follow a completely different strategy (e.g., empower local chief executives for a bottom-up approach), but also asked the NAO to report on the progress of those implementations in another CPA meeting. It remains to be seen, of course, whether or not such general policy dictates will be realized.

5 PRELIMINARY ANALYSIS OF THE PRACTICAL IMPLICATIONS OF CO-ORIENTATION

Concerns expressed by members of the CPA that the project was unmanageable resonate with a recent discussion in the organization studies literature about projects which may push an organization beyond its limits, a concept first explored in Starbuck and Farjoun's (2005) collection of papers reviewing the causes of the 2003 Columbia space shuttle catastrophe and elaborated in their later work (Farjoun and Starbuck 2007). Reviewing lessons from the disaster, Starbuck and Farjoun emphasized the need for NASA to navigate mindfully in conditions of ambiguity and uncertainty, to improve organizational learning and unlearning, to manage complexity systematically, and to keep the organization within its limits. They emphasized too that the Columbia episode needed to be understood in the context of a range of historical, social, political, organizational, and technological factors affecting NASA; nonetheless, lessons from the Columbia episode are, they suggested, directly relevant to other large organizations that are based on distributed knowledge systems, face complex dilemmas, serve multiple constituencies, seek to meet contradictory demands, and that need to innovate, use risky technologies, and are confronted with extreme time and resource pressures.

Elaborating on the idea of organizational limits, Farjoun and Starbuck defined them as "the range, amount, duration, and quality of things they [organizations] can do with their current capabilities, and these limits may originate in their members' perceptions, in their policies, in the technologies they adopt, or in their environments" (p. 543).

What emerges from our initial analysis of the CPA's review is the possibility that the NPfIT is an impossible project because it is pushing the whole activity system to its limits, but that by pushing the limits it holds possibilities for change and improvement. What is interesting in the CPA meeting is that it is in the efforts of the project leaders to exceed previous limits that the limits of the NPfIT are constituted. In other words, while being assigned to organize and execute a project that by definition is bound to exceed any previous limits, because of its scale and possible impact, any mishap or error by the project leaders is immediately interpreted as another step outside the limits of the project.

Certainly, some limits were defined in advance such as the budget and the timescale of the project. These limits were set to serve certain functions; however, they were ill-defined and their implications not well understood. These were limits set at an early stage by people that are currently distanced from the organization and execution of the project itself. Most importantly, these limits were set in isolation from key communities of practice such as the clinicians and the chief executives of the hospitals and primary health centers that represent the key users of the technologies being introduced.

Previous research has pointed to the ways in which complex technologies like the NPfIT compel organizations like the NHS to develop complex structures and complex management processes (Perrow 1999). However, these complex structures and processes involve gaps in coordination and communication with the consequence that organizations have more difficulty managing those technologies effectively. Farjoun and Starbuck discuss the example of NASA's space shuttle, which, because of its multiple and diverse technologies, has inhibited NASA from replacing unreliable and legacy components. Instead, this technological complexity has pushed NASA to add "hierarchical layers and occupational specialties that have narrowly defined functions," which have, in turn, resulted in poor and inappropriate responses to even small, incremental changes (Farjoun and Starbuck 2007, p. 551). This example illustrates the learning limitations that many organizations experience as a consequence of their technological choices, management structures, and processes. Learning limitations may involve a narrow understanding of environmental and organizational changes, as well as difficulties in analyzing those, but also difficulties in adapting to changes in their personnel and technologies. Very often, these learning limitations are a direct consequence of contested political agendas.

From its inception in 1998, the NPfIT has been characterized by unclear and confused priorities because the politicians that controlled its funding demanded that the senior responsible owners (project leaders) pursue performance targets that took for granted the limitations of the project in relation to the complex structures and processes of the NHS, as well as its existing legacy information systems (see Brennan 2005). These politicians had to face reelection campaigns and, therefore, had little patience for goals that required long time horizons. In contrast to this activity, the project leaders were driven to try to exceed those limitations and pursue unrealistic goals out of both their insecurity of potentially losing their job and their ambition to become heros (cf. Hayward and Hambrick 1997). In consequence, the limits of the NPfIT were being exceeded at multiple levels in the activity system, and "even talented and intelligent people with plentiful resources and laudable goals… [found] themselves incompetent to deal with their challenges" (Farjoun and Starbuck 2007, p. 542).

The NPfIT has, thus, become an increasingly contested object both in the ways in which it is constructed and on the possibilities it holds for change. In this view, the NPfIT is neither a failure nor a success; rather, its limits are constantly being reexamined, reconstituted, and exceeded.

6 FURTHER RESEARCH AND EXPECTED CONTRIBUTIONS

A key area for further research becomes how this, presumably hesitant, uncertain and conflicted process can best be understood and managed.

Our preliminary analysis suggests that it may be necessary to open up the activity system around the NPfIT as the appropriate unit of analysis. In addition to examining how participants reconceive of their priorities, this would involve paying attention to emerging technological and organizational arrangements, rules and performance targets, and the extant division of labor (Engeström 1987). We argue that such an analysis would support the analysis of the process of co-orientation and enable a more in-depth analysis of the tensions and conflicts in the discursive actions of the two groups (i.e., the CPA and the project leaders) than has hitherto been possible, while also shedding light on the possible trajectories of change and continuity around the NPfIT.

In this effort, we will pay attention to the limits of the NPfIT as they are defined in the co-orientation of the object by the politicians and project leaders. As evident from our analysis of the discursive actions of the CPA and project leaders, rhetoric and organizations mutually co-orient one another; by engaging in various rhetorical strategies, these key actors not only question issues of responsibility and accountability about the current state of the object, but also, proactively, focus on the creation of new opportunities and capabilities. In this sense, new ICTs and organizational changes are enacted in terms of the expectations and visions that have shaped their potential. Such expectations and visions can be seen to be fundamentally generative in that they give definition to roles and duties, while offering some direction of what to expect and how to prepare for opportunities and risks (Brown et al. 2000). However, these generative outcomes give rise to a number of unintended consequences, which may potentially lead to new meetings and reports to evaluate what went wrong, resulting in new activities of co-orientation. Further research would need to dig deeper into the history behind these rhetorical strategies to unearth the ways by which patterns form to shape organizational roles and relationships, but also to enable or constrain the process of organizing the object.

In conclusion, our research is intended to contribute to organizational analyses of success and failure in the development and implementation of large-scale ICT projects (Brown and Jones 1998; Fincham 2002; Sauer 1999; Wilson and Howcroft 2002) by examining the historical trajectories of such projects, and featuring the importance of collective development as a result of them. Research reported here examined one critical aspect of this process, the interrogation by a House of Commons Committee of Public Accounts of project leaders responsible for a very large and apparently failing ICT project. By focusing on activities of co-orientation and its practical implications it is possible, we suggest, to begin to understand the limits and the limit violations of large-scale ICT and change in organizations. This approach would overcome the analytical paradox of ICT projects as objects vulnerable to attributions of both failure and success. Instead, objects of activity and the limits of the activity systems through which they are enacted would be understood to be constantly open to redefinition and reconstitution. The research challenge, we propose, is to recognize examples of when limits are being extended, to analyze associated processes, and to translate such insights into practical support for those involved.

References

Brennan, S. 2005. *The NHS IT Project: The Biggest Computer Programme in the World Ever!*, Oxford, UK: Radcliffe Publishing.

Brown, A. D., and Jones, M. 1998. "Doomed to Failure: Narratives of Inevitability and Conspiracy in a Failed IS Project," *Organization Studies* (19:1),pp. 73-88.

Brown, N., Rappert, B., Webster, A. (eds.). 2000. *Contested Futures: A Sociology of Prospective Techno-Science*, Aldershot, UK: Ashgate.

Collins, T. 2006. "NAO: The Challenge of Objectivity," *Computer Weekly*, June 26.

Corbett, E. P. J. 1990. *Classical Rhetoric for the Modern Student*, Oxford, UK: Oxford University Press.

Currie, W. L., and Guah, M. 2007. "Conflicting Institutional Logics: A National Program for IT in the Organisational Field of Healthcare," *Journal of Information Technology* (22), pp. 235-247.

Engeström, Y. 1987. *Learning by Expanding*, Helsinki: Orienta-konsultit.

Engeström, Y. 2004. "Managing as Argumentative History-Making," in *Managing as Designing*, R. Boland and F. Collopy (eds.), Stanford, CA: Stanford University Press, pp. 96-101.

Farjoun, M., and Starbuck, W. H. 2007. "Organizing At and Beyond the Limits," *Organization Studies* (28:4), pp. 541-566.

Fincham, R. 2002. "Narratives of Success and Failure in Systems Development," *British Journal of Management* (13), pp. 1-14.

Hayward, L. A., and Hambrick, D. C. 1997. "Explaining the Premiums Paid for Large Acquisitions: Evidence of CEO Hubris," *Administrative Science Quarterly* (42), pp. 103-127.

Hoeksma, J. 2006. "Healthy Optimism?," *E-Health Insider* (Online Newspaper), June 22.

HC. 2007. *Department of Health: The National Programme for IT in the NHS*, House of Commons, Committee of Public Accounts, HC 390, London.

NAO. 2006. *Department of Health: The National Programme for IT in the NHS*, House of Commons, National Audit Office, Comptroller and Auditor General's Report , HC 1173, London.

Parker, I. 1992. *Discourse Dynamics: Critical Analysis for Social and Individual Psychology*, London: Routledge.

Perrow, C. 1999. *Normal Accidents: Living with Highrisk Rechnologies*, Princeton, NJ: Princeton University Press.

Putnam, L. 2004. "Dialectical Tensions and Rhetorical Tropes in Negotiations," *Organization Studies* (25:1), pp. 35-53

Sauer, C. 1999. "Deciding the Future for IS Failures: Not the Choice You Might Think," in *Rethinking Management Information Systems*, B. Galliers and W. L. Currie (eds.), Oxford, UK: Oxford University Press, pp. 279-309.

Sauer, C., and Willcocks, L. 2007. "Unreasonable Expectations—NHS IT, Greek Choruses and the Games Institutions Play Around Mega-Programmes," *Journal of Information Technology* (22), pp. 195-201

Starbuck, W. H., and Farjoun, M. 2005. *Organization at the Limit: Lessons from the Columbia Disaster*, Oxford, UK: Blackwell Publishing

Taylor, J., and Robichaud, D. 2004. "Finding the Organization in the Communication: Discourse as Action and Sensemaking," *Organization* (11:3), pp. 395-413

Wilson, M., and Howcroft, D. 2002. "Re-conceptualising Failure: Social Shaping Meets IS Research," *European Journal of Information Systems* (11), pp. 236-250.

About the Authors

Panos Constantinides is a lecturer in Organization, Work and Technology at Lancaster University Management School (LUMS). Before joining LUMS, Panos was a research associate at the Judge Business School at the University of Cambridge where he also earned his Ph.D. His research interests focus on the relations between organizations, technology and everyday work practice and their implications for new technological innovations and organizational change. His latest research is in the area of quality systems for infection control, management systems for emergency response, and decision making within multidisciplinary teams. Panos can be reached at p.constantinides@lancaster.ac.uk.

Frank Blackler has worked as Professor of Organisational Behaviour in the Department of Organisation, Work and Technology at Lancaster University for the past 20 years. He has a special interest in the management of change and in theories of practice. In recent years he has concentrated on organizational issues in the public sector, working on top management development in the UK's National Health Service and on the reorganization of services for vulnerable children and families. Frank can be reached at f.blackler@lancaster.ac.uk.

19 A MULTIVOCAL AND MULTILEVEL INSTITUTIONALIST PERSPECTIVE TO ANALYZE INFORMATION TECHNOLOGY-ENABLED CHANGE IN THE PUBLIC SERVICE IN AFRICA

Roberta Bernardi
Warwick Business School
University of Warwick
Coventry, UK

Abstract *The research adopts a multivocal and multilevel institutionalist perspective to analyze information technology-enabled change into the structures of the public service in Africa as reflected in changes of practices around information processing. Information systems scripts and guidelines are considered as vocal to new logics of public service (e.g., new public management) imported into the local setting through international public sector reforms. The research will focus on the micro or agent level as the locus of institutional change. Here, formal structures planned at the policy (macro) and organizational level (meso) are modified through sensemaking as users change there is and information processing practices in order to seek realignment between competing logics embedded in new and old public administration models. The analysis will be undertaken based on a case study of the Ministry of Health in Kenya. The research will provide new insights into the implications of institutional mechanisms for the integration of new IT-enabled service models in the public sector of developing countries.*

Keywords Public service, IT-enabled change, Africa, institution theory

Please use the following format when citing this chapter:

Bernardi, R., 2008, in IFIP International Federation for Information Processing, Volume 267, Information Technology in the Service Economy: Challenges and Possibilities for the 21st Century, eds. Barrett, M., Davidson, E., Middleton, C., and DeGross, J. (Boston: Springer), pp. 271-280.

1 RESEARCH AREA AND QUESTION

The research adopts a multivocal and multilevel institutionalist perspective to better understand the implications of information systems for the re-engineering and quality of the public service in Africa.

The value of the research is twofold. First, it is related to the importance of information technology in the restructuring of the public administration for improved accountability and good governance (Ciborra 2005; Fountain 2002; Heeks 2001; World Bank 1997). Actually, under the aegis of the new public management (NPM), most public sector reforms in Africa and in the rest of the developing world leverage IT to reduce hierarchical structures into flatter, more information-efficient organizational forms (Bellamy and Taylor 1992; Cresswell et al. 2006; Lucas and Baroudi 1994; Osborne and Gaebler 1992).

However, divergent opinions on the extent and implications of the IT impact on the institutional arrangements and values of the public administration (e.g., Ciborra 2005; Diamond and Khemani 2006) have highlighted the lack of a linear causality between IT and specific organizational outcomes. This has inspired an analytical approach that is more grounded in the institutional context of IT users (Orlikowski and Robey 1991). Under this perspective, the way users make sense of the information processed by consciously choosing to enact and ignore institutions' scripts encoded in information systems influences the impact of IT on organization structures.

Finally, the value of the research is linked to the peculiarity of the public sector context in Africa, cauterized by specific institutional properties. It is believed that a deeper consideration of the institutional setting may enhance the analysis of the effects of IT-enabled public sector reforms.

In Africa, as in other developing countries, the logics embedded in imported reform models such as the new public management (e.g., Hood 1991) overlap with the divergent and, somehow, competing logics of the old public administration (e.g., Lynn 2006) nested in the existing local public service models (see Table 1).

The former is at the core of development initiatives to enhance *governance* and *accountability* (Bangura and Larbi 2006; Djelic and Sahlin 2006) in order to fight corruption, mismanagement, and inefficient bureaucracies (Drori 2006; Hood 2000; Lynn 2006). It embodies "developmentalist and neoliberal logics" (Drori, 2006) advocating economic values, increased efficiency, and equity through managerialism and market-like mechanisms such as competition (Dunleavy et al. 2006; Heeks 2001; World Bank 2002).

The latter represents the traditional model of state bureaucracies (Lynn 2006) inherited from the colonial and post-independence period (Haruna 2001; Saxena and Aly 1995). It is mainly characterized by logics of politicization and bureaucratization underpinning patrimonial, clientelistic, and rent-seeking behaviors (Batley and Larbi 2006) of authoritarian political regimes. Western countries have identified in it the main causes of the inefficient and ineffective implementation of development policies (Kiragu and Mutahaba 2006; Larbi 2006) opening the door to the NPM movement.

However, value and legal systems pertaining to the traditional bureaucratic institutions (Higgo 2003; Marikanis 1994; Russell et al. 1999) and the "rhetoric of reforms" (Therkildsen 2006) instilled by the pressures of donor countries (Kimaro and Nhampossa, 2005; Kiragu and Mutahaba 2006) have posed no few challenges to the implementation

Table 1. Institutions of the New Public Management and Old Public Administration Models

PA Models	Logics	Institutions	Literature
NPM	Managerialism	• Increased responsibility and decision-making power at managerial level • Depoliticization of implementing structures and functions	Drori 2006 Hope 2001 Olowu 2006
OPA	Bureaucracy and politicization	• External political control (budgets, recruitment policies, political change, donor assistance) • Decision-making is concentrated at top of hierarchy • Input controls, rules and procedures	Bajjaly 1999 Bozeman & Bretschneider 1986 Kraemer & Dedrick 1997
NPM	Accountability	• Result and performance-oriented management system • Personnel and salary reforms and incentive schemes	Hope 2001
OPA	Meritocracy	• Political rewarding system	Grindle 1997 Owusu 2006 Peterson 1998
NPM	Market	• Competition • Externalization of the public service to free market • Disaggregation and agencification	Hope 2001 Grindle 1997
OPA	Monopoly logic	• Internalization of service delivery • Lack of competition • Weak market economy • Centralized control over financial and human resources	Ciborra 2005 Ciborra & Navarra 2005 Heeks 2001
NPM	Customer service	• Responsive, diversified and exclusive service • Customer identity of public service beneficiaries	Ciborra 2005 Drori 2006 Hope 2001
OPA	Politicization of service	• Public service complies with international/national policy priorities • Impersonalized and bureaucratic/administrative public service delivery	Grindle 1997 Owusu 2006 Peterson 1998

of NPM reforms (Batley and Larbi 2006; Kiragu and Mutahaba 2006). This has also been evidenced in the analysis of the failure of IT initiatives at the core of NPM reforms (Bellamy and Taylor 1994; Cordella 2007; Heeks 2001). Actually, the rationalization and decentralization of bureaucratic structures and the inconsistency of aid programs have led to the fragmentation of information systems increasing policy complexity (Dunleavy et al. 2006; Kimaro and Nhamposса 2005).

Given the rationale and modalities of diffusion of the new logics and the way they clash with the old ones, we might expect that the logics of the NPM are either resisted or translated (Czarniawska and Joerges 1996) into new localized models of public service (Hood 2000; Lynn 2006).

This research posits that the way IT users make sense and enact new and old logics is key to the understanding of the influence of IT on organizational change in the public sector of developing countries. Hence, informed by institutional entrepreneurship (Dacin et al. 2002), the objective of the research is to increase the understanding of how the practices of IT users fulfil, ignore, or reinvent norms and meanings underpinning different sets of logics (Schneiberg and Clemens, 2006) through patterns of use and non-use of IT (Orlikowski and Robey 1991). This is based on the argument that a way to analyze how IT-enabled public sector reforms have effectively impacted organizational structures is to focus on changes of practices of IT users (Orlikowski 1992; Orlikowski and Robey 1991). As the latter are embedded into multivocal institutional contexts, it is assumed that users try to realign the institutional order by legitimating new meanings and practices (Johnson et al. 2000; Lounsbury 2007). In their choices, they are driven by different institutional forces underpinning different types of legitimacies.

Hence, the research will adopt a multilevel institutionalist perspective (Chreim et al. 2007) to uncover in which way and under what circumstances IT users react to different legitimating pressures. More specifically, by taking the micro or agency level as an analytical focus, the research will seek to view how IT users either conform to or change *prescribed* or *normative* behaviors embodied in rules and norms at the macro or policy level (e.g., public sector reforms) and meso or organizational level (e.g., professional norms). Institutional mechanisms at the macro and meso levels are connected to each other (Chreim et al. 2007) and are seen as supportive of both NPM and OPA logics. Hence, users can either conform to NPM logics or resist them by reproducing the logics of the old public administration model. However, new opportunities for change might also arise as IT users legitimate their actions as they seek realignment between competing logics.

The proposed research question will be addressed by adopting information behavior as an empirical lens of the practices of IT users. It is argued that the information behavior of individuals depends on how they legitimize their information needs and, in turn, on their choice to either enact or ignore institutions' *scripts* encoded in formal IS designs, and, more broadly, in organizational and regulative models. The focus on information behavior links back to the view of organizations as information processing systems, whereby the structuring of organizations is associated with their information needs (e.g., Galbraith 1977) and the information flows between the different parts of an organization (Mintzberg 1979). The main assumption here is that changes in information processing practices reflect changes into the structures of organizations.

2 MULTILEVEL INSTITUTIONALIST FRAMEWORK

The proposed multilevel institutionalist framework (Figure 1) is meant to analyze how the information processing practices of IT users reproduce, adapt, or combine patterns of meanings embodied in the NPM and OPA logics under the pressure of institutional elements at the macro and meso levels. In particular, it provides a more comprehensive view of how regulative and normative institutional mechanisms at these

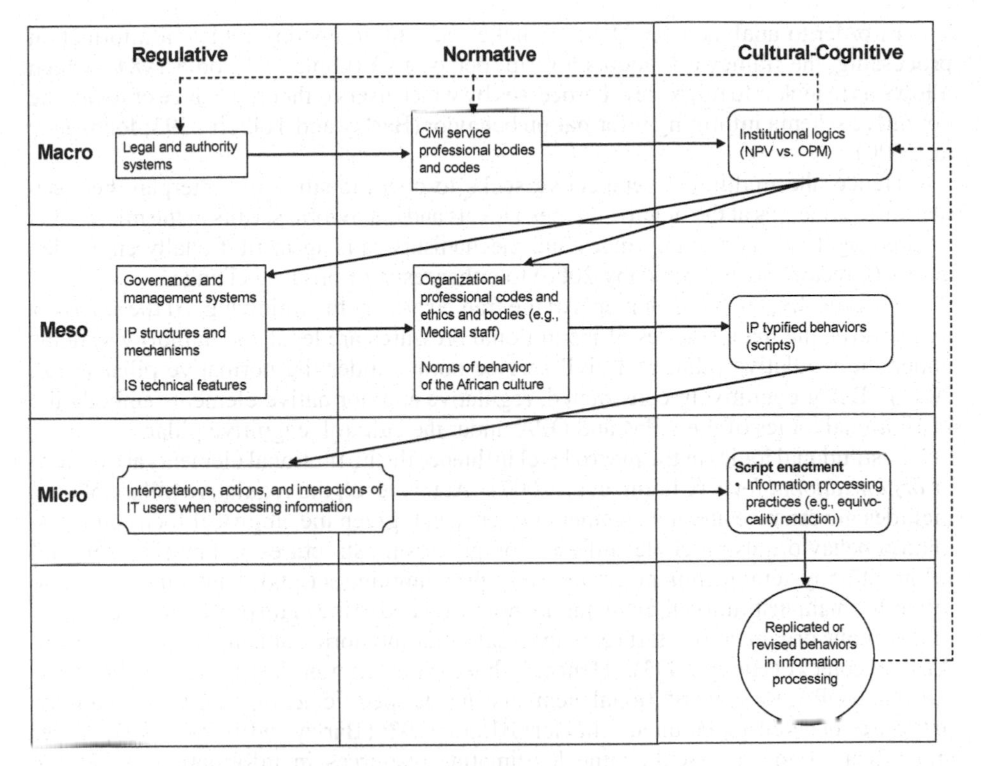

Figure 1. Multilevel Institutionalist Framework

two levels influence the encoding of cultural–cognitive institutions (or institutional logics) at the micro level. It thus accounts for the difference in salience between pillars (Schneiberg and Clemens 2006; Scott 2001) in conditioning the legitimacy-seeking behavior of human actors as they enact different logics.

Drawing on institutional entrepreneurship, the proposed framework seeks to address the limitations of institution theory in taking into account the microfoundations of institutional change (Chreim et al. 2007). By focusing on agency at the microlevel as a locus of change, the framework is meant to analyze the intersubjective and interaction processes through which human agents interpret and enact institutional logics (DiMaggio 1988; Greenwood and Suddaby 2006). It does so by integrating microlevel institution theory with a structuration (Barley and Tolbert 1997) and sensemaking (Weber and Glynn 2006) perspective.

Structuration theory extends the limitations of institution theory by including agency into the definition of structure. By adding this perspective, institutions both constrain and enable action. Hence, institutional logics are not only socially constructed meaning systems (Friedland and Alford 1991) framing the action of human agents, but also a source of interpretive, legitimating and material resources that individual actors can reinterpret and mobilize toward a new institutional order (Sewell 1992). The ability to mobilize resources comes from a sensemaking effort by human actors. The impact of institutional logics depends on how these are sensed by human actors (Johnson et al. 2000).

In order to analyze how IT users make sense of competing logics in information processing, the framework adopts a script perspective (Gioia and Poole 1984), which allows us to look into micro-level processes by which users either reproduce or revise the cognitive schema informing information behavior (Barley and Tolbert 1997; Johnson et al. 2000).

Hence, the multilevel perspective seeks to operationalize the interplay between institutional forces at the macro and meso levels and behavioral scripts at the micro level to shed light on the circumstances and mechanisms causing institutionally embedded actors (Greenwood and Suddaby 2006) to either resist or produce change.

Specifically, the main elements of the framework are the following. At the *macro* or *policy level*, the main sources of institutional pressures are legal and authority systems under the regulative pillar and civil service norms under the normative pillar (Scott 2001). Being cognitively constructed, regulative and normative elements embody the institutional logics of the NPM and OPA under the cultural–cognitive pillar.

Institutional logics at the macro level influence the institutional elements at the *meso* or *organizational level* (Chreim et al. 2007). At this level, the regulative pillar not only includes governance and management systems, but, given the empirical focus on information behavior, also includes information processing structures and mechanisms and information systems technical features (e.g., programming scripts). Under the normative pillar, the main institutional elements are norms related to the main professional category of the public organization staff (e.g., medical staff) and norms of behavior related to the African context (Higgo 2003). Hence, imbued with different institutional logics from NPM and OPA, these institutional elements encode specific *action scripts* (i.e., patterns of typified or taken-for-granted behaviors) (Barley 1986; Barley and Tolbert 1997; Weber and Glynn 2006) representing the legitimating resources in information processing behavior.

Finally, at the *micro level*, IT users draw from this set of institutionalized patterns of information behavior. Such patterns of behavior or scripts are assumed to encode different institutional logics both from the NPM and the OPA. Depending on how these are sensed, IT users build their perceptions and assumptions on information needs and their choice of the best actions, processes, and structures to meet them. Through their actions and interactions, users enact information behaviors that may either replicate or revise the scripts. In the second case, they give rise to new patterns of information behavior underlying changes in the structures of the public administration. Such changes can become institutionalized provided that these new scripts are commonly shared and become taken for granted among public employees (Berger and Luckmann 2004).

3 DATA COLLECTION

A case study is being conducted at the Ministry of Health in Kenya. The case study comprises three units of analysis: the central unit of the health management information systems, the HIV/AIDS division, and the immunization division.

Data collection is based on semi-structured interviews and documentary material. The sample of informants consists of program officers and health information officers. Interviews are meant to gather historical perception of change in information processing for the execution of their tasks following the automation of their health management information systems and the introduction of public sector reforms.

The content analysis of official documents (e.g., reports) aims to spot changes in meanings underpinning new practices in information processing.

4 DATA ANALYSIS

1. The main institutional and technological changes that have occurred over the last 20 years will be allocated to a time period.
2. For each time period, data will be coded into typified behaviors (or *scripts*) in information processing in order to identify variation in patterns of information behavior across time.
3. Patterns of behavior will be confronted with the main institutional logics (NPM or OPA) for each period to highlight commonalities in the case of conforming behaviors or variation in the case of revised behaviors and logics.
4. Finally, the analysis will be focused on the main institutional elements and mechanisms (regulative, normative, and cognitive) that, across the three levels (macro, meso, micro), have contributed to the shaping and shifting of patterns of information behavior encoding specific to institutional logics across time.

5 CONCLUDING REMARKS

The research will provide new directions on how multivocal rationalities at different institutional levels influence the enactment of a technology. By better understanding the relationship between the three mechanisms of institutionalization across multiple institutional levels and logics (Ruef and Scott 1998; Schneiberg and Clemens 2006), the research will provide new insights on how variation, contradiction, ambiguity among different policy, organizational and IT models trigger the sensemaking activity of human agents in shaping the institutional environment at the micro level (Colyvas 2007; Colyvas and Powell 2006; Weber and Glynn 2006).

It addresses the need to better understand how formal changes, envisaged in public sector policies, are translated and reflected onto the performance of IT users. Focusing on the practices of human actors can yield uncovered aspects of IT-driven public sector reforms, such as the emergence of informal and unplanned practices that are not envisaged in public sector reforms. The formal–informal perspective provides a deeper insight into the link between IT and *actual existing development* as opposed to official normative frameworks (Ciborra 2005) and the process by which alternative patterns of actions embedded in IT-led organizational change are created. This can improve the understanding on how new public service delivery models should be regulated and adapted to a specific context.

References

Bajjally, S. T. 1999. "Managing Emerging Information Systems in the Public Sector," *Public Productivity & Management Review* (23:1), p. 40-47.

Bangura, Y, and Larbi, G. A. 2006. "Introduction: Globalization and Public Sector Reform," in *Public Sector Reform in Developing Countries,* Y. Bangura and G. A. Larbi (eds.), New York: Palgrave Macmillan, pp. 1-12.

Barley, S. R. 1986. "Technology as an Occasion for Structuring: Evidence from Observations of CT Scanners and the Social Order of Radiology Departments," *Administrative Science Quarterly* (31:1), pp. 78-108.

Barley, S. R.., and Tolbert, P. S. 1997. "Institutionalization and Structuration: Studying the Links Between Action and Institution," *Organization Studies* (18:1), pp. 93-117.

Batley, R., and Larbi, G. A. 2006. "Capacity to Deliver? Management, Institutions and Public Services in Developing Countries," in *Public Sector Reform in Developing Countries*, Y. Bangura and G. A. Larbi (eds.), New York: Palgrave Macmillan, pp. 99-127.

Bellamy, C. A., and Taylor, J. A. 1992. "Informatization and New Public Management: An Alternative Agenda for Public Administration," *Public Policy and Administration* (7:3), pp. 29-41.

Bellamy, C., and Taylor, J. A. 1994. "Reinventing Government in the Information Age," *Public Money and Management* (14:3), pp. 59-62.

Berger, P. L., and Luckmann, T. 2004. "From the Social Construction of Reality: A Treatise on the Sociology of Knowledge," in *The New Economic Sociology: A Reader*, F. Dobbin (ed.), Princeton, NJ: Princeton University Press, pp. 296-317.

Bozeman, B., and Bretschneider, S. 1986. "Public Management Information Systems," *Public Administration Review* (46:6), pp. 473-73.

Chreim, S., Williams, B. E., and Hinings, C. R. 2007. "Interlevel Influences on the Reconstruction of Professional Role Identity," *Academy of Management Journal* (50:6), pp. 1515–1539.

Ciborra, C. 2005. "Interpreting E-government and Development: Efficiency, Transparency or Governance at a Distance?," *Information Technology & People* (18:3), pp. 260-279.

Ciborra, C., and Navarra, D. D. 2005. "Good Governance, Development Theory, and Aid Policy: Risks and Challenges of E-Government in Jordan," *Information Technology for Development* (11:2), pp. 141-59.

Colyvas, J. A. 2007. "From Divergent Meanings to Common Practices: The Early Institutionalization of Technology Transfer in the Life Sciences at Stanford University," *Research Policy* (36), pp. 456-476.

Colyvas, J. A., and Powell, W. 2006. "Roads to Institutionalization: The Remaking of Boundaries Between Public and Private Science," *Research in Organizational Behavior*, B. Staw (ed.), Greenwich, CT: JAI Press, pp. 305-353.

Cordella, A. 2007. "E-Eovernment: Towards the E-Bureaucratic Form?," *Journal of Information Technology* (22:3), pp. 265-274.

Cresswell, A. M., Burke, G. B., and Pardo, T. A. 2006. "Advancing Return on Investment: Analysis for Government IT, a Public Framework." White Paper. Center for Technology in Government, State University of New York, Albany.

Czarniawska, B., and Joerges, B. 1996. "The Travel of Ideas," in *Translating Organizational Change*, B. Czarniawska, B. Joerges, and G. Sevon (eds.), Berlin: De Gruyter, pp. 13-48.

Dacin, T., Goldstein, J., and Scott, W. R. 2002. "Institutional Theory and Institutional Change: Introduction to the Special Research Forum," *Academy of Management Journal* (45:1), pp. 45-57.

Diamond, J., and Khemani, P. 2006. "Introducing Financial Management Information Systems in Developing Countries," *OECD Journal on Budgeting* (5:3), pp. 97-132.

DiMaggio, P. 1988. "Interest and Agency in Institutional Theory," in *Institutional Patterns and Organizations: Culture and Environment*, L. Zucker (ed.), Cambridge, MA: Ballinger, pp. 173-188.

Djelic, M-L., and Sahlin, K. 2006. "A World of Governance: The Rise of Transnational Regulation," *Transnational Governance: Institutional Dynamics of Regulation*, M-L. Djelik and K. Sahlin (eds.), Cambridge, UK: Cambridge University Press, pp. 32-52.

Drori, G. S. 2006. "Governed by Governance:the Institutionalization of Governance as a Prism for Organizational Change," in *Globalization and Organization: World Society and*

Organizational Change, G. S. Drori, J. W. Meyer, and H. Hwang (eds.), Oxford, UK: Oxford University Press. pp. 91-118.

Dunleavy, P., Margetts, H., Bastow, S., and Tinkler, J. 2006. "New Public Management Is Dead—Long Live Digital-Era Governance," *Journal of Public Administration Research and Theory* (16:3), pp. 467-94.

Fountain, J. E. 2002. "Developing a Basic Social Science Research Program for Digital Government: Information, Organizations and Governance," report of a national workshop, John Fitzgerald Kennedy School of Government. Harvard University.

Friedland, R., and Alford, R. R. 1991. "Bringing Society Back In: Symbols, Practices, and Institutional Contradictions," in *The New Institutionalism in Organisational Analysis*, W. W. Powell and P. DiMaggio (eds.), Chicago: University of Chicago Press, pp. 232-264.

Galbraith, J. R. 1977. *Organization Design,* Reading, MA: Addison-Wesley.

Gioia, D. A., and Poole, P. A. 1984. "Scripts in Organizational Behavior," *The Academy of Management Review* (9:3), pp. 449-459.

Greenwood, R., and Suddaby, R. 2006. "Institutional Entrepreneurship in Mature Fields: The Big Five Accounting Firms," *Academy of Management Journal* (49:1), pp. 27-48.

Grindle, M. S. 1997. "Divergent Cultures? When Public Organizations Perform Well in Developing Countries," *World Development* (25:4), pp. 481-95.

Haruna, P. F. 2001. "Reflective Public Administration Reform: Building Relationships, Bridging Gaps in Ghana," *African Studies Review* (44:1), pp. 37-57.

Heeks, R. 2001. "Understanding E-Governance for Development," i-Government Working Paper Series, Paper No. 11, Institute for Developmnt Policy and Management, University of Manchester, Manchester, UK (http://unpan1.un.org/intradoc/groups/public/documents/NISPAcee/UNPAN015484.pdf).

Higgo, H. A. 2003. "Implementing an Information System in a Large LDC Bureaucracy: The Case of the Sudanese Ministry of Finance," *Electronic Journal of Information Systems in Developing Countries* (14:3) (http://www.ejisdc.org/ojs2/index.php/ejisdc/article/viewFile/88/88).

Hood, C. 1991. "A Public Management for All Seasons," *Public Administration* (69:1), pp. 3-19.

Hood, C. 2000. "Paradoxes of Public-Sector Managerialism, Old Public Management and Public Service Bargains," *International Public Management Journal* (3:1), pp. 1-22.

Hope, K. R. 2001. "The New Public Management: Context and Practice in Africa," *International Public Management Journal* (4:2), pp. 119-134.

Johnson, G., Smith, S., and Codling, B. 2000. "Microprocesses of Institutional Change in the Context of Privatization," *Academy of Management Journal* (25:3), pp. 572-580.

Kimaro, H. C., and Nhampossa, J. L. 2005. "Analyzing the Problem of Unsustainable Health Information Systems in Less-Developed Economies: Case Studies from Tanzania and Mozambique," *Information Technology for Development* (11:3), pp. 273-298.

Kiragu, K., and Mutahaba, G. 2006. "Public Service Reform in Eastern Africa and Southern Africa: Issues and Challenges," a report of the proceedings of the Third Regional Consultative Workshop on Public Service Reform in Eastern and Southern Africa, 2005, Dar Es Salaam: Mkuki Na Nyota Publishers.

Kraemer, K. L., and Dedrick, J. 1997. "Computing and Public Organizations," *Journal of Public Administration Research and Theory* (7:1), pp. 89-112.

Larbi, G. A. 2006. "Applying the New Public Management in Developing Countries," in *Public Sector Reform in Developing Countries*, Y. Bangura and G. A. Larbi (eds.), New York: Palgrave Macmillan, pp. 25-52.

Lounsbury, M. 2007. "A Tale of Two Cities: Competing Logics and Practice Variation in the Professionalization of Mutual Funds," *Academy of Management Journal* (5:2), pp. 289-307.

Lucas Jr., H. C., and Baroudi, J. 1994. "The Role of Information Technology in Organization Design," *Journal of Management Information Systems* (10:4), pp. 9-23.

Lynn, L. E. 2006. *Public Management: Old and New,* New York: Routledge.

Marikanis, A. E. 1994. "Public Sector Employment in Developing Countries: An Overview of Past and Present Trends," *International Journal of Public Sector Management* (7:2), pp. 50-68.
Mintzberg, H. 1979. *The Structuring of Organizations: A Synthesis of the Research,* Englewood Cliffs, NJ: Prentice Hall.
Olowu, D. 2006. "Decentralization Policies and Practices under Structural Adjustment and Democratization in Africa," in *Public Sector Reform in Developing Countries,* Y. Bangura and G. A. Larbi (eds.), New York: Palgrave Macmillan, pp. 228-254.
Orlikowski, W. J. 1992. "The Duality of Technology: Rethinking the Concept of Technology in Organizations," *Organization Science* (3:3), pp. 398-427.
Orlikowski, W. J., and Robey, D. 1991. "Information Technology and the Structuring of Organizations." *Information Systems Research* (2:2), pp. 143-169.
Osborne, D., and Gaebler, T. 1992. *Reinventing Government: How the Entrepreneurial Spirit Is Transforming the Public Sector,* Reading, MA: Addison-Wesley.
Owusu, F. 2006. "Differences in the Performance of Public Organisations in Ghana: Implications for Public-Sector Reform Policy," *Development Policy Review* (24:6), pp. 693-705.
Peterson, S. B. 1998. "Saints, Demons, Wizards and Systems: Why Information Technology Reforms Fail or Underperform in Public Bureaucracies in Africa," *Public Administration and Development* (18:1), pp. 37-60.
Ruef, M., and Scott, W. R. 1998. "A Multidimensional Model of Organizational Legitimacy: Hospital Survival in Changing Institutional Environments," *Administrative Science Quarterly* (43:4), pp. 877-904.
Russell, S., Bennett, S., and Mills, A. 1999. "Reforming the Health Sector: Towards a Healthy New Public Management," *Journal of International Development* (11:5), pp. 767-775.
Saxena, B., and Aly, A. 1995. "Information Technology Support for Re-engineering Public Administration: A Conceptual Framework," *International Journal of Information Management* (15:4), pp. 271-293.
Schneiberg, M., and Clemens, E. 2006. "The Typical Tools for the Job: Research Strategies in Institutional Analysis," *Sociological Theory* (3), pp. 195-227.
Scott, W. R. 2001. *Institutions and Organizations,* London: Sage Publications.
Sewell, W. H. 1992. "A Theory of Structure: Duality, Agency., and Transformation," *American Journal of Sociology* (98:1), pp. 1-29.
Therkildsen, O. 2006. "Elusive Public Sector Reforms in East and Southern Africa," in *Public Sector Reform in Developing Countries,* Y, Bangura and G. A. Larbi (eds.), New York: Palgrave Macmillan, pp. 53-81.
Weber, K., and Glynn, M. A. 2006. "Making Sense with Institutions: Context, Thought and Action in Karl Weick's Theory," *Organization Studies* (27:11), pp. 1639-1660.
World Bank. 1997. "World Development Report 1997: The State in a Changing World," New York: World Bank.
World Bank. 2002. "World Development Report 2002: Building Institutions for Markets," Washington, DC: World Bank.

About the Author

Roberta Bernardi is currently a Ph.D. student in the Information and Management Department of the Warwick Business School (UK). Her doctoral research takes an institution theory perspective in order to analyze how the usage of information among IS users within a determined policy, organizational, and technological setting may affect the restructuring of public service in Africa. As part of her research she has conducted a case study at the Ministry of Health in Kenya, where she also worked in a parliamentary information system project sponsored by the United Nations.

Part 4:

Outsourcing and Globalization of IT Services

20 LEGITIMACY MANAGEMENT AND TRUST IN OFFSHORING INFORMATION TECHNOLOGY SERVICES

Michael Barrett
Judge Business School
Cambridge University
Cambridge, UK

C. R. Hinings
School of Business
University of Alberta
Edmonton, AB Canada

Eivor Oborn
Judge Business School
Cambridge University
Cambridge, UK

Abstract *Our study examines the evolution of offshoring of information technology services between a multinational telecommunications firm and Indian vendors. The firm's strategy sought to access critical resources of talented software professionals at low cost and to facilitate growth in a highly competitive telecommunications sector. Legitimacy management proved critical in explaining the evolution of this offshoring arrangement between a client firm and its four Indian vendors. Our findings surface the strategies and activities adopted by proponents in their challenge of gaining, maintaining, building, and repairing legitimacy. We examine how the subsequent reactions and interactions of other audiences iteratively influenced the legitimacy dynamics of offshoring. We also contribute an understanding of the role of trust in the challenges of managing legitimacy, and conclude with practical implications for institutionalizing offshoring arrangements.*

Please use the following format when citing this chapter:

Barrett, M., Hinings, C. R., and Oborn, E., 2008, in IFIP International Federation for Information Processing, Volume 267, Information Technology in the Service Economy: Challenges and Possibilities for the 21st Century, eds. Barrett, M., Davidson, E., Middleton, C., and DeGross, J. (Boston: Springer), pp. 283-299.

Keywords Offshoring, legitimacy, institutionalization, trust, IT services

1 INTRODUCTION

The Indian information technology enabled services industry grew at more than 50 percent per annum between 2000 and 2003, with revenue growth of 43 percent to U.S.$5.1 billion in 2004-05, up from U.S.$3.6 billion the previous year (NASSCOM 2005). By 2008, the expected potential of exports is projected to be in the range of $21 billion to $24 billion (the Indian government's Department of IT, Annual Report 2003-04), employing upwards of 1.2 million people.

Rapid changes in technology and global competition over the last decade have radically increased the needs of high technology firms to develop innovative relationships with vendor firms for software development (Nicholson and Sahay 2001). These vendor–client relationships are often emergent and ambiguous, evolving from cooperative relationships such as recurrent contracting (Ring and Van de Ven 1992) to relational contracting or strategic alliances. The evolution of these relationships may involve closer collaboration as vendors become partners, and an "extension of the client firm" (Humphreys 1998).

While recent research has recognized the importance of trust and control in offshoring relationships (Sabherwal 1999; Sahay et al. much less is known about the actual process of evolution of the relationship at the level of work practice (Barley and Kunda 2001; Orlikowski 2000). In our case study, offshoring was referred to as externalization or the extension of the client firm through outsourcing relationships with Indian vendors. We focus on the challenges of managing legitimacy and highlight the different dimensions of trust at play in the evolution of offshoring relationships.

Surprisingly little empirical research has uncovered the challenges of legitimacy management (Suchman 1995) or how practices can be more or less legitimated over time in the process of institutionalization (Colyvas and Powell 2006). In our study on the evolution of offshoring practices involving multiple organizations, network legitimacy is particularly important (Human and Provan 2000; Provan et al. Doyle 2004). We build on this emerging stream of literature by empirically analyzing how legitimacy of a network of multiple organizations changes over time. In so doing, we examine a hitherto neglected but important area: the role of trust in these legitimacy processes, which is particularly important at this network level. We ask the following questions: What are the various strategies and activities adopted by proponents in managing the challenges of legitimacy during the institutionalization of offshoring IT services? What role does trust play in the ensuing legitimacy dynamics?

In the next section, we discuss our key concepts, namely legitimacy and trust. We then describe the research process used in our longitudinal case study of an offshoring relationship between a North American telecommunications firm, Globalco, and its Indian vendors. Our case analysis illuminates the strategies and activities deployed by actors of the clients' outsourcing department in managing legitimacy challenges, and the reactions of other audiences within the firm and across vendors. We follow with a discussion of the role of trust in managing these legitimacy challenges at Globalco, and conclude with practical implications for institutionalizing offshoring IT services.

2 LEGITIMACY AND TRUST

2.1 Legitimacy

Legitimacy is commonly referred to being largely taken for granted, and accepted as appropriate and right by external constituencies (Aldrich and Fiol 1994). There are both evaluative and cognitive dimensions of legitimacy, which can be holistically incorporated in defining legitimacy as "a generalized perception or assumption that the actions of an entity are desirable, proper, or appropriate within some socially constructed system of norms, values, beliefs, and definitions" (Suchman 1995, p. 574).

A key feature of legitimacy is that practices, beliefs, or rules are reproduced without significant mobilization and are resistant to contestation (Colyvas and Powell 2006; Jepperson 1991). Legitimacy is conferred when influential constituencies or audiences perceive actors as being in alignment with a shared social reality, which can be an organization, or a network of multiple organizations that seek to operate from a common meaning system, or society at large. Thus, audiences perceive the legitimate practice as not only more worthy, but more meaningful and *trustworthy* (Suchman 1995).

Somewhat ironically, despite its centrality, legitimacy has been largely taken for granted and is in need of conceptual development (Colyvas and Powell 2006; Suchman 1995). In particular, there is little work on the process by which legitimacy is gained, maintained, and lost (Baum and Powell 1995; Suchman 1995; Suddaby and Greenwood 2005).

Suchman (1995) synthesized the role actors play in skillful legitimacy management, drawing on a selection of strategies aimed at gaining, maintaining, and repairing legitimacy. For example, actors may gain legitimacy by manipulating the environment through the enactment of their claims in carrying out a successful institutionalization project (DiMaggio and Powell 1991). Strategies for maintaining legitimacy broadly include perceiving future changes and protecting past accomplishments. The former focuses on facilitating the organization's ability to recognize audience reactions and foresee emerging yet unforeseen challenges while the latter can involve managers actively policing internal operations for reliability and responsibility, stockpiling trust.

Actors can also creatively reestablish legitimacy or repair legitimacy following a crisis or significant disruption. Managers may decide to deny, excuse, or justify the threat, although they do so at the risk of depleting long-term legitimacy reserves. Alternatively, managers may accept these threats and attempt to facilitate relegitimation (Suchman 1995).

Recent research on network legitimacy theory focuses on the processes used by interorganizational networks to establish legitimacy (Provan et al. 2004). Human and Provan (2000), in their study of manufacturing networks, highlight that achieving legitimacy requires that both outside constituents/audiences and inside member organizations view the network concept as an acceptable *form* for organizing multiple organizations and providing services. The network also has to be viewed as a distinct *entity* with its own identity. They also identified a third dimension of legitimacy, *interaction*, which recognizes the need for cooperation among organizations.

In our study, we focus on the evolution of the offshoring department and the efforts of its staff to establish and build legitimacy for offshoring. We propose that the successful evolution of offshoring arrangements requires legitimacy management strategies by key actors to establish network legitimacy at the *form*, *entity*, and *interaction* levels. Furthermore, we suggest that ongoing active trust is central to the development of legitimacy at each of these levels and we, therefore, turn to the literature on trust.

2.2 Trust

While a definition of trust remains elusive (Das and Teng 2001), it is widely recognized as a key element in collaborative interorganizational relationships (Newell and Swan 2000; Ring 1996). Trust recognizes the need for risk-taking in the context of dependent interpersonal relations. For example, Mayer et al. (1995) suggest that trust must include a willingness to take a risk in the relationship and to be vulnerable. Dimensions of trust have been categorized as cognition based, affect based and system trust, which we now discuss in turn.

Cognition-based trust, such as competence trust, depends on the predictability of one party to forecast another party's behavior. This dimension of trust is emergent, with the trustor granting trust based on prior experiences that demonstrate predictable target behavior. As such, cognition-based trust building requires information or knowledge about past actions or knowledge-based trust (Shapiro et al. 1992). Goodwill trust, another element of cognitive trust, focuses on one party's perception of the intentions of the other party to demonstrate a special concern for others' interests above their own (Barber 1983; Doney et al. 1998). Parties exhibiting goodwill trust in a supplier client relationship will be prepared to make investments to extend and support a range of tasks and to continue developing the relationship (Humphreys 1998).

Another form of cognitive-based trust, calculative trust, suggests that developing trust involves a calculative, rational process whereby one party calculates the costs and/or rewards of another party cheating or cooperating in a relationship (e.g., Dasgupta 1988; Williamson 1985). Trust in business relationships often develops first on a calculative basis (Child 1998, 2001), as parties assess what they might get from the relationship, and what their risks and their vulnerabilities are likely to be (Doney et al. 1998). One type of calculative trust (Shapiro et al. 1992), deterrence-based trust, suggests that people will do what they say they will do to the extent to which the deterrent, whether it be contractual penalties or increased monitoring of behavior (Humphreys 1998), is clear, possible, and likely to occur (Doney et al. 1998).

In contrast to cognitive-based trust, affect-based trust recognizes attitudes, behaviors, (Lewicki and Bunker 1995), and their importance in the evolution of trust. At each exchange point, affect-based trust influences the ongoing experience and meaning of the relationship. Two types of affect-based trust have been distinguished in the literature. First, characteristic-based trust relies on ascribed characteristics such as ethnic backgrounds as good reasons to trust (Zucker 1986) while identification-based trust (Shapiro et al. 1992) suggests that trust develops when partners identify with shared values and norms at a bonding phase of the relationship (Child 2001).

System trust (Luhmann 1979) is an important complement to the above-mentioned dimensions of interpersonal trust. Giddens (1990) suggests that the sustaining of trust is challenging in globalization as the proliferation of systems of expertise (e.g., standardized company procedures and practices) leads to the disembedding (or stretching) of social relations whereby social relations are lifted out from local contexts of interaction. At the same time, there are reembedding processes which facilitate trust maintenance at the access points, which, Giddens suggests, provide the link between personal and system trust. These access points are typically found in boundary spanning roles such as relationship managers (Sydow and Windeler 1998). These individuals often serve as representatives to reinforce trust in systems such as standards, rules, and procedures comprising the system.

Our theoretical position starts with and builds on the above literature on trust and legitimacy. We suggest that the different dimensions of trust built up through interaction are vital in supporting legitimacy management strategies throughout the evolution of the offshoring arrangement.

3 RESEARCH PROCESS

We conducted a qualitative longitudinal case study of an offshoring relationship between a multinational company we call Globalco and its four Indian partner firms to examine the complexities and dynamics of collaboration (Arino and De la Torre 1998; de Rond and Bouckikhi 2004). The field study sought to understand Globalco's initial formation and subsequent development of offshoring arrangements.

Our real-time study took place over 2 years with prior historical reconstruction over 10 years of the offshoring arrangement. We sought to develop an in-depth understanding of the actions and perceptions of different actors within the firms who influenced the formation and development of the offshoring arrangements. We approached the concept of process as a sequence of events and actions that describes how things change over time (Giddens 1984; Pettigrew 1990).

3.1 Globalco and its Four Indian Vendors

Globalco is a North American multinational firm in the telecommunications industry providing a wide range of networking and infrastructure products and services to customers in 150 countries. At the turn of the 21st century, it had revenues of approximately U.S.$20 billion and employed over 60,000 people. During this period, Globalco faced intense competitive pressure within its dynamic deregulated industry. As such, there were significant demands for technological expertise in the development of new products along with the maintenance of existing large systems. To this end, they developed an externalization (offshoring) vision in 1991, which sought to develop contractor relationships with firms in India, China, Russia, Vietnam, and the Philippines to support their work.

We carried out a historical reconstruction of the development of the offshoring strategy with a particular focus on Globalco's relationship with its Indian vendors, three of whom were located in Bangalore and one in Mumbai. Vendors varied in size and expertise and offered a variety of services to meet the needs of Globalco. For example, the largest vendor was the IT consultancy of an Indian multinational, and one of the pioneers of offshore outsourcing. This firm was the least dependent on Globalco for business, and was able to provide large numbers of coders and developers predominantly for software development activities. In contrast, the smallest vendor depended on Globalco for over 60 percent of its revenues. This firm provided highly trained personnel to develop niche digital signal processing solutions for Globalco. The other vendors were rapidly growing mid-sized vendors offering a mix of services, which were successfully developing an international reputation in IT services offshoring.

3.2 Data Collection and Analysis

Our first phase of interviews took place in Globalco's Office in North America. We interviewed four directors of the newly established offshoring office, ORD, responsible for the implementation of the outsourcing programs and three senior research and development managers working with teams in the Indian partner firms. A second phase took place 6 months later in Bangalore and Mumbai where a total of 45 semi-structured interviews were conducted with a range of individuals across the 4 Indian vendor firms including CEO's, project managers, software engineers, and relationship managers. In keeping with our longitudinal approach to the study of these offshoring relationships, we conducted a third round of 30 interviews a year later in North America and a final set with the Indian partners shortly thereafter. We also collected and analyzed documents concerning the development of the outsourcing arrangements and strategies.

Our subsequent analysis and theory building drew on three processes highlighted by Langley (1999): induction, deduction, and inspiration. Our inductive process to analysis drew on the open coding technique (Strauss and Corbin 1990, 1998). Initial coding and conceptualization involved careful reading of, and reflection on, empirical data through which key concepts, themes, and issues were identified and subsequently sorted into categories emerging from the data. We started by organizing the data by client and vendor participants. On the client side, we focused on senior managers, offshoring managers, line managers, and expatriate managers. On the vendors' side, we separated out senior managers, project managers, software engineers/developers, and vendor relationship managers. We examined interview transcripts to identify participants' statements that reflected interpretations surrounding the initial formation and subsequent development of the offshoring relationship. We also examined the changing contexts within which the relationship developed over time, and identified key events in the evolution of the offshoring arrangements (see Arino and De la Torre 1998; Langley 1999). Two researchers independently used the qualitative software program, Nudist, as an aid in the analysis process. Among others, trust and control were dominant themes identified as critical in the on-going development of the offshoring relationship. We subsequently went back to the literature to explore theories of institutionalization and legitimacy as there was a good fit between the underlying themes and this literature (see Boudreau and Robey 2005).

4 TRUST AND STRATEGIC ACTION IN MANAGING LEGITIMACY

In this section, we present our case findings highlighting the strategic action adopted by Globalco's top managers and ORD directors in managing legitimacy of the offshoring relationship. Four main challenges were identified, namely: gaining legitimacy, maintaining legitimacy, building legitimacy, and repairing legitimacy.

First, early on in the evolution of the relationship, we identified strategies and activities introduced to gain legitimacy for offshoring. In keeping with our theoretical approach, we also identified the reactions of other audiences within Globalco and the vendor firms to the concept of offshoring. Second, we identified actors' strategies to maintain legitimacy and audience reactions. Third, despite challenges, top management sought to further build legitimacy toward offshoring, which was now viewed as a competitive necessity to support their hi-tech strategy. Fourth, as delegitimation set in due to persistent performance and customer risk, ORD and line managers sought to repair legitimacy.

Below, we discuss the strategic action adopted by proponents, the reactions and interactions with other audiences, and the role of trust in managing these legitimacy challenges.

4.1 Gaining Legitimacy

Table 1 summarizes the strategic actions adopted in gaining legitimacy for offshoring as a discursive concept and the role of trust in this endeavor. Below, we elaborate on the strategic actions and subsequently synthesize the role of different dimensions of trust.

Two senior Globalco managers of Indian origin led a taskforce to develop an offshoring feasibility project, which would examine how Globalco could increase R&D productivity and reduce time to market. A successful offshoring strategy was expected to provide Globalco with the flexibility to focus their resources on strategic areas such as the development of new products for new markets.

The CEO at the time, along with a top manager lent crucial support for offshoring to many countries including India. As a founding member of ORD explained,

Table 1. Role of Trust and Strategic Action in Gaining Legitimacy

Legitimacy Management Challenge	Strategic Action	Role of Trust
Gaining Legitimacy	Offshoring feasibility project	Overcome distrust of offshoring
	Budget structures and ORD office set up	Characteristic-based trust enables bonding
	ORD as brokers develop knowledge of vendors capabilities and sell to line managers	No deterrence-based trust to signal long-term relationships building identification based trust
	Line managers offer independent low risk tasks to Indian vendors	Competence-based trust of vendor capabilities

The CEO made the corporation look outside...we were to stay lean and mean and go out and get the best from the world... [we] check[ed] out the viability of India...as the company's brain power, problem solver for research...not replacing MIS.

The initial rationale for offshoring strategy was access to resources, in particular "*people with good minds and excellent software skills and low cost.*" Globalco's vision was recognized as distinctive in the Indian software industry.Indian vendors readily accepted offshoring as legitimate and supportive of their goals of going beyond being mere suppliers toward true partnership. As an Indian vendor senior manager affirmed,

In a nutshell I have seen how they have had (from the beginning) essentially a bigger vision, a bigger goal in mind rather than use India as a cheap resource. They [Globalco] have never projected India as a place for cheap resource.

Gaining legitimacy for offshoring among Globalco line managers and staff was challenging and slow at the beginning. Offshoring was neither desirable nor appropriate to Globalco staff, who were protectionist of local jobs. The senior manager of Indian origin within Globalco who founded the internal offshoring unit explained, "*The [key] issue in the early days was that of acceptance of the concept of offshoring by the organization.*" As important a challenge was the "not invented here" attitude that staff held as a result of their perceived technical superiority. The CEO tried to shift these attitudes, as an ORD director commented, "*He tried to get rid of poor attitudes and promoted the idea that just because it is a poor country we shouldn't expect to lower [our] standards.*"

By addressing directly the negative bias held by Globalco staff, the CEO and top management worked hard at bringing discursive legitimacy to offshoring and promoting the importance of looking for resources outside the firm. The CEO also provided financial support for a feasibility study on offshoring to India, budgeted funds specifically for offshoring, and required managers to declare their offshoring goals and objectives in their budgeting process.

ORD was started up by four directors, two of whom were of Indian origin. ORD was charged with the responsibility to implement and develop offshoring at Globalco. From the beginning, their broker role was crucial to gain legitimacy for offshoring in India: "*We had to achieve credibility with internal line managers and be trusted by Indian vendors.*" Establishing their own credibility with managers was necessary to develop trustworthiness of ORD as an entity. Their role as "*lubricant and shock absorber*" also focused on "*taking the awkardness out of working in India.*" They made frequent visits to India, gaining an in-depth knowledge of the Indian vendors' skills and capabilities. In their broker role, they subsequently promoted and matched different vendor capabilities to Globalco labs around the world.

The Indian champions setting up ORD at Globalco were keen to contribute to the development of "*mother India,*" and established trust and commitment with the Indian vendors. As a senior manager of an Indian vendor confirmed, "*There was a commitment to the relationships with the four partners and a focus on raising the standards as to what is achievable in India, not just undercutting costs.*"

Standard vendor contracts billed on time and materials were developed between Globalco and the vendor firms. As a founding Indian director of offshoring noted, the emphasis of the contracts was on rewards, as penalties signaled mistrust: "*We didn't put penalty clauses in the contract....we drove it by rewards.*"

After the initial introduction of Indian vendors by a member of the ORD team, champion line managers of Globalco labs initiated offshoring relationships with those vendor firms who met the necessary criteria and best supported their business strategy. These contracts were initially for independent stand-alone tasks such as program testing and bug fixing. Such tasks were perceived to be of relatively low risk and required low levels of direct, face-to-face communication with the Globalco lab. Indian vendors would send the completed tasks electronically to Globalco for testing and integration.

Within a year, the strategy of taking advantage of the economic brain power in India was heralded as a good one. Indian software developers had proved their raw competence and gained a reputation for superior technical talent. The Globalco line managers trusted Indian developers to perform independent tasks, and this early success was critical to establish legitimacy for offshoring and the ORD entity. Line managers were keen to further develop the relationship and gave Indian partners increased responsibility to perform complex and interdependent tasks as part of global project teams. The additional work tasks now included a broader range of activities, including coding and low level design, which required in-depth domain knowledge of telecommunications and Globalco's proprietary products and languages.

Globalco facilitated these new working relationships by improving connectivity and communication, upgrading the infrastructure to two 64 KB links to enable voice and data paths. Vendors also increased their investment and risk in the relationship, for example, by expanding their premises and ramping up their workforce in anticipation of increased business with Globalco, as explained by an vendor CEO, "*They take you into their plans, and then on the basis of these plans, they ask you to bid what you can do for their plans....The whole organization is very open, once you are taken into their club.*" The successful efforts by ORD to match vendors with labs and to facilitate growth in the nature and scope of offshoring activities were significant in further gaining legitimacy of vendors for offshoring.

A number of dimensions of trust played an important role in facilitating the above-mentioned strategic actions towards gaining legitimacy. First, top managers and ORD sought to overcome initial mistrust and gain legitimacy for offshoring through discursive strategies and mobilizing resources. Second, characteristic-based trust between Globalco managers of Indian origin and their fellow countrymen in Indian vendor firms enabled bonding between these organizational units. Further, a deliberate strategy by the ORD managers to avoid deterrents-based trust, in particular not to include penalty clauses in contracts, was designed to build long-term relationships and build identification-based trust toward Indian vendors' values and goals of true partnership and moving up the value chain. Fourth, competence-based trust between line managers and Indian vendors furthered the gaining of legitimacy as line managers developed knowledge about the capabilities and technical prowess of vendors.

4.2 Maintaining Legitimacy

Table 2 summarizes the strategic actions adopted in maintaining legitimacy during subsequent interactions, and highlights the role of trust in this process. Below, we start by elaborating the strategic actions deployed and then discuss the role of different dimensions of trust.

Table 2. Role of Trust and Strategic Action in Maintaining Legitimacy

Legitimacy Management Challenge	Strategic Action	Role of Trust
Maintaining Legitimacy	ORD implement management practices and personnel to support expanded form of offshoring	Goodwill trust by both parties to extend tasks and develop relationship
	ORD introduce/impose Western human resources practices to manage attrition	Systems trust to further offshoring activities Micromanage to make up for lack of competence based trust Control through systems of expertise and people at access points

Following the change in the nature and type of offshoring activities involving new interactions, there was a new challenge to maintaining legitimacy. These activities involved new interactions, which now had to be deemed trustworthy and legitimate in their own right. ORD responded by inculcating a number of their management practices and processes to support Indian vendors in working effectively as global team members. This included technical training on particular Globalco technologies and management practices.

Despite the introduction of a number of coordinating mechanisms, signs of delegitimation soon became apparent. Line managers started to complain about a perceived loss of control over project timetables and delays from the vendors, which were threatening customer satisfaction. Globalco line managers accused the Indian managers of poor project management practices involving late deliverables, which impacted the perceived quality of their work:

> *One of the things we like are no surprises....If you [are] not going to deliver on a certain time—we want to know it as soon as you know it. Not the day before a project is to be delivered...it's not that...the development partners aren't good...they don't like disappointing.*

A recurrent problem that plagued projects was the attrition of talented Indian software developers. Three different types of attrition were identified as persisting in the relationship. The first type was the emigration of developers from partner firms to the United States. Indian firms planned for this attrition by creating a buffer through over staffing projects. The second type of attrition stemmed from developers moving across different Globalco projects within the vendor firm, while the third type involved the movement of developers from Globalco projects to other client work within the firm. Vendor managers explained these actions of attrition by emphasizing the need for them as a company to meet the career development of their staff.

In the eyes of line managers and ORD, these performance and customer risk problems were starting to delegitimize offshoring as desirable and appropriate. Globalco's line managers reacted by being cautious as to the level and type of work they provided to Indian vendors: "*The clients are afraid of giving too much of the critical deliveries to India because if the key guy goes away they are behind the ball, there is no recovery plan.*" Line managers also attempted to "de-risk" by micromanaging vendor's work. By monitoring and controlling local developments at a distance, they hoped to ensure that project deadlines and key deliverables were met.

In further efforts to maintain legitimacy of offshoring, ORD mediated between the line managers and Indian vendors and attempted to introduce a range of human resources practices to manage attrition, even though vendors felt attrition was a part of the Indian work landscape with which they had to learn to live. ORD also required vendors to use "gold awards" for retention of key resources, and requested vendors to carry out an employee satisfaction survey they had designed, which the largest vendor flatly rejected as undermining local forms of knowledge and control. In an attempt by ORD to gain compliance with procedures, senior managers were posted to each of the vendor's sites to influence their systems.

Vendors were willing to endure these systems of control as they trusted Globalco to help them meet their goals and interests in moving up the value chain. Furthermore, Globalco also paid very well initially and was genuinely interested in them doing high-end strategic work, which contributed to the mainstream of Globalco's work as opposed to the industry norm of replacing business MIS.

However, over time, vendors were adversely affected by the significant "churn" of new product development experienced by Globalco and the industry at large. They experienced significant manpower planning challenges involving the ramp up and disbanding of staff with the birth and death of projects. These dynamic cycles increased vendor overheads in "shadowing" key resources on projects, and this adversely affected the once-attractive profit potential of Globalco contracts.

A number of dimensions of trust played an important role in facilitating the strategic actions adopted in responding to the challenge of maintaining legitimacy. First , the successful legitimacy building efforts stimulated significant goodwill trust, demonstrated by the increasing level of infrastructure investment to extend offshoring activities. From Globalco's perspective, goodwill trust in the relationship was evident due to the fact that, while the partnership remained a legal contract it was much more similar to a joint venture in its operationalization. Second, trust in systems of expertise was introduced to support the expanded set of offshoring activities. Despite these efforts by ORD to lubricate the continued growth of offshoring activities, line managers did not build systems trust in vendors' interactive offshoring activities, often complaining of unacceptable performance and customer risk and responded by micromanaging vendors' efforts. Third, ORD managers tried to build vendors' trust in different systems of expertise (e.g., HR mechanisms) in attempts to control and manage attrition. For example, ORD introduced expatriate postings within vendor firms to be access points to facilitate or control the reembedding of systems of expertise.

4.3 Building Legitimacy

Table 3 summarizes the strategic actions adopted by top management in building legitimacy during subsequent interactions, and highlights the associated role of trust. Below, we start by elaborating strategic actions and then synthesize the role of different dimensions of trust at play in building legitimacy.

Table 3. Role of Trust and Strategic Action in Building Legitimacy

Legitimacy Management Strategy	Strategic Action	Role of Trust
Building Legitimacy	Top management encourage vendor ownership of mature technologies	Building and reciprocating identification based trust
	Discursive legitimacy of offshoring by bestowing partners with "sister lab" and "global lab" status	Top management control requires increased risk-taking

Despite the above-mentioned strains of maintaining legitimacy, opportunities emerged to further build offshoring. First, the dramatic contextual changes in the telecommunications sector led Globalco's top management to strategically undertake a right-angle turn toward Internet-based development, and this provided a new impetus for the future potential of offshoring activities. Top managers provided further investments in infrastructure to allow vendors to "almost" function as a "global lab," and rhetorically demonstrated commitment and increased legitimacy for offshoring by conferring "sister lab" status to Indian vendors.

In response to these new opportunities and challenges, line managers started to rely on vendors to take responsibility for code ownership, new product releases, and market ownership of mature technologies so as to free themselves up to work on leading edge technology developments. The discursive legitimacy of sister lab status stirred significant optimism of a true partnership among Indian vendors as suggested, by a vice president: "*But I believe that the world is moving towards...partnering as something which is the call of the day, in the true sense of partnering....I believe that Globalco is definitely a partner of choice to us.*" Vendors readily embraced this shift toward partnership and also the move away from "*tool work to include products and markets.*" This evolution of the relationship gained legitimacy with Indian vendors, for whom the meaning of offshoring was aligned with their goals and interests.

In building legitimacy, two key dimensions of trust were salient. First, identification-based trust was reciprocated and built in response to the strategic actions and discursive legitimacy, which furthered the offshoring relationship. Second, top management displayed more trust through an increased appetite for risk taking.

4.4 Repairing Legitimacy

Table 4 summarizes the strategic actions adopted in supporting the right-angle turn and the role of trust in surmounting this legitimacy challenge. Below, we start by elaborating strategic actions and then synthesize the role of different dimensions of trust.

Globalco line managers continued to complain about attrition and its effects on project management and customer satisfaction. Their concerns were accentuated in light of the movement from code to product ownership. ORD directors remained unhappy with the efforts of vendors in addressing this ongoing problem. Even more seriously for the relationship, they became highly suspicious of vendors' low incentive levels to deal with

Table 4. Role of Trust and Strategic Action in Repairing Legitimacy

Legitimacy Management Challenge	Strategic Action	Role of Trust
Repairing Legitimacy	ORD implement an experience based pay model	Perceived lack of proactivity stymies competence trust
	Senior Globalco managers instill Western culture and work practices in vendor firms	Goodwill trust declines over new charging model
	ORD introduces an Indian offshore development center	Introduce new systems of control to reduce risk and repair competence trust

attrition but rather a curious willingness to merely live with the problem. ORD perceived a zero-sum game and alleged that highly experienced staff leaving projects were being replaced by fresh, inexperienced recruits at full replacement cost to Globalco. An offshoring director hinted at the potential financial gains the vendors could accrue by adopting such strategies:

> *The current strategy for controlling attrition doesn't manage productivity but is merely a simple substitution principle. A new person is substituted for an old, experienced worker in Globalco technology....So in that respect the attrition and losses are good for the contractors, the partners.*

This view of vendors' charging policy led to mistrust and challenged the desirability and appropriateness of offshoring to provide low costs and provide further revenue to support Globalco's right angle turn. These concerns of moral and pragmatic legitimacy brought into question the appropriateness of the entity responsible for organizing offshoring across the network of client–vendor organizations.

While vendors sought to justify the performance threats of attrition, ORD directors disassociated Globalco from procedures that were depleting long-term legitimacy reserves. ORD allowed the introduction of penalty clauses in contracts and restructured vendor rates. Instead of being calculated as a linear model based on number of resources, an experience-based pay model was adopted that used tiered rates based on years of experience.

A second challenge Globalco experienced was the perceived lack of proactivity by Indian vendors, which was deemed critical if they were effectively to support Globalco's right-angle turn and access to regional markets.

Repairing legitimacy of offshoring, therefore, involved a rethinking as to the legitimate form required to adequately support the right angle turn. ORD sought professional advice from a consultant and subsequently restructured around an offshore development center (INDCO). It was proposed as an extension of ORD to be located in India, which would leverage and integrate Indian vendor resources. INDCO would specifically be responsible for ownership of leading-edge products that had high customer impact as well as mature products with high customer impact.

INDCO's strategy for repairing legitimacy included the introduction of subject matter experts to reduce the need for micromanagement by line managers and to eventually facilitate new opportunities for vendors to ultimately move up the value chain. The focus on building an experience base within the vendors was cast as a critical success factor in the de-risking strategy and building legitimacy of offshoring at the interaction level. Globalco theorized that line managers would develop competence trust in Indian vendors, which would lead to less control or micromanagement on their part. As a senior offshoring Director noted,

> *How do I de-risk? Some companies do this by bodyshopping. This is best done by raising the experience basis, which involves the use of subject matter experts. When the experience basis is low, there is a tendency for Globalco managers to micromanage. In either case, costs are high.*

However, vendors hotly contested the INDCO center proposal, perceiving it as a significant loss of legitimacy for offshoring. They believed the INDCO center would sustain the arm's length vendor relationship, thwarting vendors' desire for true partnership and delaying their ability to move up the value chain.

A number of dimensions of trust played an important role in facilitating the strategic actions adopted in responding to the challenge of repairing legitimacy.

First, a number of perceived risks led to an erosion of trust relations, which challenged legitimacy and led ORD to develop repair strategies. ORD and line managers continued to perceive high levels of performance and customer risk as a result of attrition, and this was only accentuated as they started to transfer product and market ownership to vendors. These concerns, along with beliefs that Indian vendors lacked proactivity, adversely affected the building of competence trust in supporting this expanded set of offshoring activities. Even more seriously, perhaps, was the loss of goodwill trust by ORD over suspicions of the existing charging model, which they deemed to be scandalously inappropriate and a challenge to moral legitimacy (Suchman 1995). The new experience based pay model, however, simultaneously impacted on vendors' goodwill trust in the relationship. Third, ORD attempted to repair legitimacy through restructuring both the concept and entity of offshoring in support of the right-angle turn. INDCO sought to de-risk and build competence trust through the aid of locally based subject matter experts. Vendors interpreted INDCO as a lack of identification trust by Globalco, adversely affecting vendors' goodwill trust. Efforts to rebuild goodwill trust through the joint task force sought to restore legitimacy of offshoring through a process of developing identification-based trust between Globalco and its vendors. These efforts highlighted the tight coupling of trust and legitimacy that pervaded the relationship between vendors and Globalco.

5 CONCLUSION

This paper introduces theoretical developments on legitimacy and trust to examine the evolution of offshoring of IT services around global software development. In so doing, we contribute a first step in understanding the role of trust in legitimacy dynamics.

Our research highlights the different challenges of managing legitimacy, strategic actions that proponents can adopt to respond to these challenges, and the role of trust in the unfolding legitimacy dynamics. For example, even when the concept of offshoring had become legitimated, mistrust that developed over recurrent interactions (e.g., how attrition was managed and the introduction of the experience-based pay model) can call into question the legitimacy of the entity. Repairing legitimacy in this situation led proponents to dramatically restructure, which ultimately led to efforts at enhancing cognitive legitimacy through the joint client–vendor taskforce.

As our case shows, dimensions of trust are developed through interactions between various sets of actors across the different legitimacy management strategies. For example, *gaining legitimacy* was achieved through interactions between top management, ORD, and line managers who successfully built affect-based trust with vendors. Furthermore, line managers relied heavily on cognition-based trust (in particular, competence-based trust) throughout the evolution of the offshoring arrangement. *Maintaining legitimacy* was dependent on further building competence-based trust through interactions between line managers and vendors. In addition, system trust (and control) was relied on heavily by ORD during interactions with vendors. Top management's *building of legitimacy* evoked affect-based trust (in particular identification-based trust) with vendors and highlighted the way groups of actors with various interests depend on different dimensions of trust. ORD's efforts at *repairing legitimacy* relied on control through restructuring and introducing systems of expertise into vendor firms, which they subsequently resisted and instead sought to establish mutual identification-based trust.

Finally, our case illuminates that legitimacy building is an iterative process (Provan et al. 2004), whose success or failure at any interaction can affect other dimensions of legitimacy, such as calling into question the meaning of the concept of offshoring or restructuring the entity. Further research is needed to validate and build on our findings, and deepen our understanding of the evolution of offshoring arrangements. Nonetheless, we believe our conceptual developments on legitimacy management and trust are a useful starting point in this direction and may be a helpful guide for managers of both client and vendor firms involved with institutionalizing offshoring of IT services.

References

Aldrich, H. E., and Fiol, C. M. 1994. "Fools Rush In? The Institutional Context of Industry Creation," Academy *of Management Review* (19), pp. 645-670.

Arino, A., and De la Torre, J. 1998. "Learning from Failure: Towards an Evolutionary Model of Collaborative Ventures," *Organization Science* (9), pp. 306-325.

Barber, B. 1983. *The Logic and Limits of Trust,* New Brunswick, NJ: Rutgers University Press.

Barley, S., and Kunda, G. 2001. "Bringing Work Back In," *Organization Science* (12)1), pp. 76-96.

Baum, J., and Powell, W. 1995. "Cultivating an Institutional Ecology of Organizations: Comment on Hannan, Carroll, Dundon and Torres," *American Sociological Review* (60), pp. 529-538.

Boudreau, M.-C., and Robey, D. 2005. "Enacting Integrated Information Theory: A Human Agency Perspective," *Organization Science* (16:1), pp. 3-18.

Child, J. 1998. "Trust and International Strategic Alliances: The Case of Sino-Foreign Joint Ventures," in *Trust Within and Between Organizations: Conceptual Issues and Empirical Applications*, C. Lane and R. Bachmann (eds.),Oxford, UK: Oxford University Press.

Child, J. 2001. "Trust: The Fundamental Bond in Global Collaboration," *Organizational Dynamics* (29:4), pp. 274-288.

Colyvas, J., and Powell, W. 2006. "Roads to Institutionalization: The Remaking of Boundaries Between Public and Private Science," *Research in Organizational Behavior* (25), pp. 305-353.

Das, T. K., and Teng, B. 2001. "Instabilities of Strategic Alliances: An International Tensions Perspective," *Organization Science* (11:1), pp. 77-101.

Dasgupta, P. 1988. 'Trust as a Commodity," in *Trust: Making and Breaking Cooperative Relations*, D.Gambetta (ed.), Oxford, UK: Blackwell, pp. 49-72.

de Rond, M., and Bouchikhi, H. 2004. "Dialectics of Strategic Alliances," *Organization Science* (15), pp. 56-69.

DiMaggio, P. J., and Powell, W. W. 1991. *The New Institutionalism in Organizational Analysis*, Chicago: University of Chicago Press.

Doney, P. M., Cannon, J. P., and Mullen, M. R. 1998. "Understanding the Influence of National Culture on the Development of Trust," *Academy of Management Review* (23:3), pp. 601-620.

Giddens, A. 1984. *The Constitution of Society*, Cambridge, UK: Polity Press.

Giddens, A. 1990. *The Consequences of Modernity*, Cambridge, UK: Polity Press.

Human, S. E., and Provan, K. G. 2000. "Legitimacy Building in the Evolution of Small Firm Multilateral Networks: A Comparative Study of Success and Demise," *Administrative Science Quarterly* (45), pp. 327-365.

Humphreys, L. 1998. "Trust and the Transformation of Supplier Relations in Indian Industry," in *Trust Within and Between Organizations, Conceptual Issues and Empirical Applications*, C. Lane and R. Bachmann(eds.), Oxford, UK: Oxford University Press, pp. 214-272.

Jepperson, R. L. 1991. "Institutions, Institutional Effects, and Institutionalism," in *The New Institutionalism in Organizational Analysis*, W. W. Powell and P. J. DiMaggio (eds.), Chicago: University of Chicago Press, 1991, pp. 143-163.

Langley, A. 1999. "Strategies for Theorizing from Process Data," *Academy Management Review* (24:4), pp. 691-710.

Lewicki, R. J., and Bunker, B. B. 1995. "Trust in Relationships: A Model of Trust Development and Decline," in *Conflict, Cooperation, and Justice,* B. B. Bunker and J. Z. Rubin (eds.), San Francisco: Jossey-Bass, pp. 133-173.

Luhmann, N. 1979. *Trust and Power,* Chichester, UK: John Wiley & Sons.

Mayer, R. C., Davis, J. H., and Shoorman, D. 1995. "An Integrative Model of Organizational Trust," *Academy of Management Review* (20:3), pp. 709-734.

NASSCOM. 2005. "Indian ITES-BPO Industry: NASSCOM Analysis," (http://www.creativebpo.com/IndianITES-BPOFactsheet.pdf).

Newell, S., and Swan, J. 2000. "Trust and Inter-Organizational Networking," *Human Relations*, (53:10), pp. 1287-1328.

Nicholson, B., and Sahay, S. 2001. "Some Political and Cultural Issues in the Globalization of Software Development: Case Experience from Britain and India" *Information and Organization* (11), pp. 25-43.

Orlikowski, W. J. 2000. "Using Technology and Constituting Structures: A Practice Lens for Studying Technology in Organizations," *Organization Science* (11:4), pp. 404-428.

Pettigrew, A. 1990. "Longitudinal Field Research on Change: Theory and Practice," *Organization Science* (1:3), pp. 267-292.

Provan, K. G., Lamb, G., and Doyle, M. 2004. "Building Legitimacy and the Early Growth of Health Networks for the Uninsured," *Health Care Management Review* (29:2), pp. 117-128.

Ring, P. S. 1996. "Fragile and Resilient Trust and Their Roles in Economic Exchange," *Business & Society* (35:2), pp. 148-175.

Ring, P. S., and Van de Ven, A. H. 2001. "Structuring Cooperative Relationships Between Organizations," *Strategic Management Journal* (13), pp. 483-498.

Sabherwal, R. 1999. "The Role of Trust in Outsourced IS Development Projects," *Communications of the ACM* (42:2), pp. 80-86.

Sahay, S., Nicholson, B., and Krishna, S. 2003. *Global IT Outsourcing: Software Development Across Borders*, Cambridge, UK: Cambridge University Press.

Shapiro, D., Sheppard, B. H., and Cheraskin, L. 1992. "Business on a Handshake," *Negotiation Journal* (8), pp. 365-377.

Strauss, A. L., and Corbin, J. 1990. *Basics of Qualitative Research: Grounded Theory, Procedures, and Techniques,* Newbury Park, CA: Sage Publications.

Strauss, A., and Corbin, J. 1998. *Basics of Qualitative Research Techniques and Procedures for Developing Grounded Theory* (2nd ed.), London: Sage Publications.

Suchman, M. 1998. "Managing Legitimacy: Strategic and Institutional Approaches," *Academy of Management Review* (20:3), pp. 571-610.

Suddaby, R., and Greenwood, R. 2005. "Rhetorical Strategies of Legitimacy," *Administrative Science Quarterly* (50), pp. 35-67.

Sydow, J., and Windeler, A. 1998. "Organizing and Evaluating Interfirm Networks: A Structurationist Perspective on Network Processes," *Organization Science* (9:3), 265-284.

Williamson, O. E. 1985. *The Economic Institutions of Capitalism,* New York: Free Press.

Zucker, L. G. 1986. "Production of Trust: Institutional Sources of Economic Structure 1840-1920," *Research in Organizational Behaviour* (8), pp. 53-111.

About the Authors

Michael Barrett is the director of the M.Phil. program on Innovation, Strategy, and Organization at the Judge Business School, University of Cambridge. His research interests include Information technology and service innovation; IT implementation and organizational change; and IS services outsourcing and globalization. Michael is an associate editor for *MIS Quarterly* and *Information Systems Research* and a member of the editorial boards of *Organization Science, Information and Organization,* and *Journal of the Association of Information Systems.* He has worked as an IS consultant with Oracle Canada Corp, and in production management with Colgate Palmolive Jamaica Ltd. Michael can be reached at m.barrett@jbs.cam.ac.uk.

Bob Hinings is Professor Emeritus in the Department of Strategic Management and Organization, School of Business, University of Alberta, and Senior Research Fellow in the Centre for Entrepreneurship and Family Enterprise. He is currently carrying out research on strategic organizational change in professional service firms, healthcare, and the Canadian wine industry. He is a Fellow of the Royal Society of Canada, a Fellow of the U.S. Academy of Management, and an Honorary Member of the European Group for Organizational Studies. He has been a recipient of the Distinguished Scholar Award from the Organization and Management Theory Division of the U.S. Academy of Management. Bob can be reached at chinings@ualberta.ca.

Eivor Oborn is a senior research associate at the Judge Business School where she earned her Ph.D. Her research interests are broadly concerned with knowledge development in multidisciplinary teams for service innovation, and have centered on multi-professional healthcare and software development teams. She is also studying new technology adoption and diffusion within the UK National Health Service and the processes of legitimising change within institutions. Eivor can be reached at emdo2@hermes.cam.ac.uk.

About the Authors

21 INFORMATION TECHNOLOGY OUTSOURCING IN THE SERVICE ECONOMY: Client Maturity and Knowledge/Power Asymmetries

Aris Komporozos-Athanasiou
Judge Business School
University of Cambridge
Cambridge, U.K.

Abstract *The service economy calls for a new, interdisciplinary methodology for defining and valuing information technology services outsourcing needs. Parameters such as ill-informed provider selection and poor contract management have hitherto dominated the IT consulting literature, yet have offered inadequate explanations to the high failure rates in global outsourcing arrangements. This paper takes a different approach in examining the causes of the problem; we discuss the knowledge and power asymmetries that appear to prevent both parties from realizing potential benefits in the market. The concept of* ***self-knowledge*** *as opposed to* ***relationship management*** *is suggested. We posit that knowledge/power asymmetries can be better comprehended when the two parties are considered as interacting entities that influence each other in a dynamic way. Under this spectrum, we discuss the value of a client-focused maturity assessment in realizing potential outsourcing benefits.*

Keywords Outsourcing, knowledge and power, client maturity, asymmetries, coevolution

1 IT OUTSOURCING: RELATIONSHIP MANAGEMENT VERSUS SELF-KNOWLEDGE

The scale of outsourcing arrangements has grown exponentially over the last 10 years. It has come to include a significant transfer of assets and staff to a vendor who

Please use the following format when citing this chapter:

Komporozos-Athanasiou, A., 2008, in IFIP International Federation for Information Processing, Volume 267, Information Technology in the Service Economy: Challenges and Possibilities for the 21st Century, eds. Barrett, M., Davidson, E., Middleton, C., and DeGross, J. (Boston: Springer), pp. 301-310.

assumes profit and loss responsibility. The resulting $200 billion market offers possibilities for significant cost saving, improvement in agility, control, and risk exposure. However, a large number of outsourcing arrangements fail to secure the agreed deliverables, often resulting in a lose-lose situation for both the service provider and the service recipient. It is estimated that the United Kingdom alone has wasted $18 billion on failed information technology outsourcing projects.[1]

What is more, outsourcing has come to constitute a much more dynamic and complex activity in the service economy. Knowledge creation, sharing and diffusion are of key value to understanding and managing an organization's outsourcing needs. Hence, sourcing decisions over business processes cannot be considered in abstraction from knowledge issues throughout the lifecycle of an outsourcing arrangement.

It is true that knowledge facets of outsourcing have generally been acknowledged in the extant literature; however, they tend to be considered mainly in relation to the transfer of IT staff and specialized know-how to the outsourcing provider. Devolution of knowledge is thought to be caused by the outsourcing arrangement itself. The prescription for the problem typically includes strategic pathways to successful outsourcing through efficient relationship management. Managing the relationship with the service provider pertains to the power side of these knowledge issues. It is also argued to be a lever for gaining power over the relationship and thus controlling the flux of knowledge capital established by service level agreement (SLA) interorganizational channels.

Nevertheless, our research points at another milieu wherein power and knowledge are manifested: the client's own organization. One could argue that it is from within the organizational boundaries that knowledge-related hydras commence. It is, in other words, the lack of *self-knowledge* on the client's side that prevents it from capitalizing on the (assumed) increased control over the outsourcing counterpart (provider). Vital issues such as the inability to define outsourcing needs clearly, the failure to understand and evaluate the role of different business processes, and the absence of a comprehensive view of the intra-organizational linkages, commonly result in a "rule of thumb" approach of sourcing strategy. What we will call *self-knowledge* is hence a *conditio sine qua non* for an effective engagement in an outsourcing relationship. Moreover, the *informed power* generated from a self-education process is key to building a mutually beneficial SLA.

Before we discuss these points in more depth, it is useful to attempt an analysis of the sourcing relationship context, the outsourcing market, and consider some interesting trends pertaining to the behavior of the involved parties.

2 KNOWLEDGE/POWER ASYMMETRIES

A typical outsourcing contract includes a maturity certification requirement for the service provider. Such third party validation, forced by the customer and often paid by the provider, typically involves capability maturity model (CMM) type assessments offered by organizations such as Accenture, TPI, and PricewaterhouseCoopers. Major service providers hold certifications that appear to assure risk reduction for clients and,

---

[1]"Outsourcing Overruns Cost UK Taxpayers £9bn," *The Register*, December 27, 2007, in Taylor and Toft 2008 (originally published at *Kablenet* (Kable's Government Computing), "Report Says Outsourcing Overuns Cost Billions," December 24, 2007.

therefore, increase the probability of a correct solution being delivered. Furthermore, through the (repeated) process of third party certification, and as long as the validations offered are not content-free, they force an increase of knowledge on the part of the suppliers. Ergo, the client is given various options of different suppliers who insist on self-knowledge and are able to keep costs for the customer at a minimum in response time. However, looking at the outsourcing service recipient side, there is currently no form of third party validation of skills, self-knowledge, and maturity, that has achieved a degree of market penetration similar to the aforementioned provider-focused CMM-type assessments.[2]

Nonetheless, in an increasingly commoditized market, an apparent shift of power toward the buyer is taking place, a tendency predicted to grow in the next few years (Willcocks and Lacity 2006). The client is seemingly becoming more capable of managing the outsourcing relationship and dictating the terms and conditions of the SLA, but does this increase of customer power have an equivalence in increased (self) knowledge and, if not, does this matter? We suggest that it does not and that this leeching of understanding can be detrimental to the long-term future of an organization.

An asymmetry of power and knowledge appears to exist in the outsourcing market. The conventional view on this asymmetry suggests that it is in favor of the client. It is thus proposed that the client's control in driving and designing outsourcing deals will result in increased benefits on the service recipient side.[3] From our observation of commercial and public organizations, it is arguable that the leeching of internal expertise and lack of self-understanding will at best reduce, and at worst preclude, long-term benefits. The client organizations are likely to devolve themselves of real power in the course of the outsourcing contract. In effect the asymmetries function in favor of the provider who assimilates knowledge and expertise.

This asymmetry has also resulted in a gradual saturation of the existing assessment methodologies and weakened their credibility. What is more, they have imperilled their very functionality by favoring a homogeneous, highly mature environment on the provider's side, which allows for little or no differentiation among them. Thus, even the marketing-justified purposes of the produced models are severely undermined as the client is presented with a palette of seemingly hyper-mature service providers.

3 THE OUTSOURCING LITERATURE GAP

Academic and best practice literature has generally assumed this conventional power view and it has thus focused on developing frameworks that address the contract signing and post-contract period of the outsourcing relationship.

---

[2]For instance, the Software Engineering Institute at Carnegie Mellon has only recently developed client CMM, which we discuss in section 3 of this paper. See also the discussion of the "peacock's tail" client–provider relationship in section 6.

[3]Indicative of this view is Willcocks and Lacity's (2006) study. They present their estimations on the future of the outsourcing market, suggesting that "clients are taking more control in driving and designing outsourcing deals" (p. 279) and that "they will move en masse from 'hype and fear' into maturity" by 2011 (p. 281). They subsequently suggest that providers are currently overselling (their own) capability to the outsourcing recipient and that they would need to follow a similar to the client's core competences model to match the client practices.

Various high-level models outline strategic visions related to IT outsourcing. Scholars such as Venkatraman (1997) describe an IT organization structured according to the "value center" concept. This concept divides IT activities into four categories (service center, investment center, cost center, and profit center) that differ with regard to their purpose (business capability versus operational efficiency) and their risk propensity (minimize risk versus maximize opportunity). Such a structure of the IT organization could allow for more informed and rational IT outsourcing decisions. Moreover, decision-making frameworks have typically provided the customer with a road map *to successfully establish and manage the IT outsourcing relationships*. This includes actual sequences of steps or more general best practice guides such as the eight building blocks model proposed by Cullen and Willcocks (2003). Di Romualdo and Gurbaxani (1998) define three different strategic intents for IT outsourcing (IS improvement, commercial exploitation, and business impact) and outline how organizations should pursue these different intents when approaching outsourcing. Also, decision-making models such as McFarlan and Nolan's (1995) strategic grid and Kernet al.'s (2002) winner's curse provide organizations with a way of structuring the rationale underlying their outsourcing decisions. The Carnegie Mellon "eSourcing Capability Model for Client Organizations," the latest, and most coherent to date, approach in measuring outsourcing capabilities, also takes a relationship management perspective. It focuses on a set of best practice guidance throughout the life of an outsourcing arrangement, but similarly to the aforementioned models, it does not consider the organization's maturity prior to the sourcing lifecycle.

There seems to be a gap in the literature; the precontract phase of an outsourcing agreement is not covered by any of these models. In effect, *no causal relationship has been probed between the organization's in-house capabilities and a successful outsourcing project*. The unasked question is, what are the necessary organizational skills within the client's IT organization, before it undertakes an outsourcing project?

It can be argued that an outsourcing arrangement is more likely to deliver the expected benefits when the client already has a level of self-awareness with regard to its outsourcing capabilities. That can only be achieved by rigorous engagement in a self-education process, through introspection and reconsideration of the organization's IT and business capabilities.

4 THE VALUE OF A CLIENT-FOCUSED MATURITY ASSESSMENT MECHANISM

As has become clear from our analysis, a client-focused maturity assessment mechanism can be greatly facilitative in tackling the outsourcing market asymmetries. A third party assessment of the client's outsourcing capabilities could be developed around the information technology infrastructure library (ITIL) platform. One such attempt is the client outsourcing maturity model (COMM) developed by Komporozos-Athanasiou (2006) and Perez-Hallerbach (2007). COMM attempts to encompass a more coherent and detailed method for the examination of an organization's process maturity. Moreover, it extends the process of a focused and unidimensional ITIL in order to include softer dimensions such as *governance* and *people*. Hence, it enhances ITIL's applicability on IT services outsourcing and suggests that COMM-mature client organizations are more likely to engage in successful and mutually beneficial outsourcing agreements with a service provider (Komorozos-Athanasiou 2007).

At the process level, the model looks at issues such as the use of best practice models in IT management, the comprehension of and adherence to the service and service lifecycle concepts[4] and the use of quantitative models and outsourcing frameworks in the IT organization, prior to an outsourcing engagement. The governance dimension assesses the mechanisms through which IT strategy is articulated and communicated as well as its linkages with the organization's business processes and it draws from a wide set of criteria covering, among others, an assessment of its IT investment evaluation behavior.

One should note that the COMM methodology differs from the one that process-focused models (i.e., Carnegie-Melon's CMM) commonly adopt, in that it *provides a wider definition of maturity*. A knowledge-based definition of maturity is adopted and hence the key notion of business and IT alignment is being examined under the spectrum of cognitive congruence (Merali 2002). COMM recognizes the importance of the *service lifecycle* for organizing business processes and orchestrating them around what the final product is, the service; however, it incorporates a *people dimension* that assesses the degree of intra-organizational understanding (among different stakeholders) of basic assumptions, concepts, values, and needs. Following the concept of *Ba* (Nonaka and Noboru 1998), which describes the shared space where knowledge is created through a process of socialization (including interfunctional communication), COMM attempts to capture the diverse interpretations throughout the organizational interactions. Furthermore, it proposes a platform for bringing together key stakeholders (user groups, finance, operations, research and development) and employs a methodology for developing analytic models that can be shared between stakeholder groups in order to enable joint exploration of design, implementation, and management issues (Taylor et al. 2004).

5 THE ROLE OF THE SERVICE PROVIDER

Having considered the importance of client maturity and the necessity of self-understanding prior to the engagement in an outsourcing relationship, we have generally assumed that high maturity characterizes the provider side. One should note that the service provider is exposed to more specialized knowledge and benefits significantly from the continuous self-assessment process (through its various certification and assessment mechanisms). However, our research indicates that this suggested maturity *does not necessarily translate into mature behavior toward the client.*

Often, the need for a "big bid" forces the provider to engage in a unsophisticated, commodity-based approach in its relationship with the client because the client is unprepared to assimilate a more mature level of services and services support due to its lack

[4]Service concept: By packaging business needs into well-defined products that are delivered by the IT organization, a basic object of interest is created that the processes within the IT organization can handle, modify, and, perhaps more importantly, talk about in a well-understood language.

of needs understanding and internal capabilities. In such an immature business environment, any attempt to raise issues of self-awareness and knowledge diffusion within the service recipient's organization is fruitless. Consequently, providers are presented with ill-defined outsourcing needs, which are then poorly addressed. No attempt is made to draw the client's attention to its inability to understand and thereby define its needs in a comprehensive fashion, not least for fear of losing the deal; a "quick fix" solution is preferred instead. We hence observe the development of a vicious circle of low maturity throughout the client–provider relationship, *which commences well before the signing of the outsourcing contract*, in the bidding/selection period. This attitude has come to be embedded in both parties' culture, preventing the solutions offered from providing substantial benefits and the outsourcing market as a whole from realizing potential profits.

It becomes evident from the above discussion that *a service provider bears the primary responsibility in triggering a change in the mentality of both parties*. An "open box" philosophy in the services provision offers the ground for such a transition to a more mature relationship context. The tools, and most importantly the mindset, this paper attempts to advocate suggest an "intelligent engagement" methodology that can (1) potentially increase the "fit for purpose" of a solution, and (2) offer a means of understanding the impact of outsourcing investment on the client's business objectives.

6 OUTSOURCING AS A COEVOLUTIONARY RELATIONSHIP

The asymmetries discussed above can be better comprehended when the two parties (provider and client) are considered as interacting entities that influence each other in a dynamic way. Under this spectrum, we can examine different formations of the outsourcing relationship using examples of coevolutionary contexts exhibited by biological systems.

6.1 The Competitive Predator–Prey Relationship[5]

The outsourcing relationship can take the form of a competitive predator–prey interaction, which may lead to an *arms race*.[6] Both outsourcing counterparts are confined in a game of control over the terms and conditions of the SLA in order to maximize their own benefit, disregarding knowledge issues that pertain to the relationship. To illustrate a typical situation resulting from this relationship pattern, we cite the case of the U.S. Navy IT outsourcing deal mega-failure. EDS assumed control of all information systems and software provision for 690,000 users and 4 network operation centers at the Navy

---

[5]Here, both parties are effectively preys, as both are faced with the threat of failure. The case of "pollination" is the other form of a predator–prey pattern, where both parties engage in a symbiotic relationship.

[6]An evolutionary struggle between competing sets of coevolving genes that develop adaptations and counter adaptations against each other, resembling an arms race.

and the Marine Corps.[7] As of 2006, EDS had made a total provision of $875 million against the contract, albeit failing to meet the stringent performance targets that had been included in the SLA. As a result, EDS suffered a major loss due to their failure to fully deploy their intended solutions and charge under the contract. The client seemingly managed to achieve a victory over the provider, first by imposing strict SLA terms and subsequently by financially punishing the errant supplier (Taylor and Tofts 2005). Nonetheless, neither of the two parties won in the end, insofar as the client was not receiving the services and systems needed, presumably continuing to operate with equipment they intended to replace (equipment that one can assume was not delivering the capabilities they desired).

6.2 Pollination

In this relationship, the outsourcing provider acts as a pollinator, transmitting the client's tacit knowledge and self-understanding to a know-how repository. This repository is continuously enriched via interaction with more clients and subsequently communicated back to the clients in the form of IT outsourcing consulting. In this case, the provider recognizes the need for dealing with an "educated" client with whom they do not solely attempt to close a deal, but rather engage in a mutually beneficial relationship (at a later stage). This type of pattern necessitates the adoption of long-term considerations and planning toward enriching the engagement by *establishing knowledge sharing channels and promoting open box practices.*

6.3 The Peacock's Tail

The peacock's tail phenomenon constitutes perhaps the most accurate analogy with the dominant form of outsourcing relationship, observed in the market: the provider adjusts its behavior to the client's selection criteria. It, therefore, seeks external maturity assessment, turning to the third party certification market to add more such qualifications to its certification arsenal. This behavior, while the client side remains out of the certification process, seemingly increases the provider's chances to be selected in an outsourcing bid. Nonetheless, as previously discussed, it is likely that this unilaterally performed behavior will effectively result in limited provider manoeuverability, and ultimately, in reduced benefits from the outsourcing arrangement. Likewise, in evolutionary biology, male peacocks develop spectacular tails, which facilitate sexual selection and are therefore considered to be an advantage in reproduction. However, the peacock's tail almost certainly reduces the male's survival: the tail reduces manoeuverability, powers of flight, and makes the bird more conspicuous (Burgess 2001). Interestingly, certain studies have moreover indicated that the peahen has little or no interest in the appearance of the peacock, insofar as there is no evidence that it can recognize subtle aesthetic features (Cronin 1991). In the outsourcing milieu, it is arguable that an immature client is not able to distinguish between providers of different maturity levels and thus the respective *maturity certification often becomes a mere aesthetic feature.*

---

[7]See http://www.eds.com/sites/nmci/, "Navy Marine Corp Intranet (NMCI)."

These metaphors shed more light on a number of key issues that we have hitherto identified. In the EDI–U.S. Navy arms race case, the evident lack of self-knowledge at the pre-engagement stage on the client (U.S. Navy) side, "resulted in outsourcing a mess" (Willcocks and Lacity 2006), a state wherein the client mitigates responsibility for a business process, over which it has already lost control, and hence in contracting for services, which the organization did not have the mechanisms in place to evaluate. However, as we earlier discussed, one should not assume that the provider bears no responsibility for this lose-lose situation. Although CMM certified and mature, the provider did not appear willing to invest (and be a pollinator) in raising the client's self-awareness with regard to its own outsourcing needs. The role of the third party certifications is crucial here, since a significant flux of provider investment toward them is generated, fueling a fierce competition in the IT services market. Nonetheless, as we saw in the peacock's tail example, *insofar as a similar third party validation for the client is neither available nor bilaterally accepted,* these certifications will often function merely as "tranquilizers" for client companies when dealing with their outsourcing conundrums. In this context, the resulting knowledge/power asymmetries are likely to trigger a vicious circle of low maturity, which will characterize the outsourcing relationship in its later stages.

7 CONTRIBUTION AND FUTURE RESEARCH

The value of a maturity assessment mechanism that specifically addresses the client organization's business environment can be significant in triggering a change in the mindset of IT managers. The aim of this paper was to open a new trail in the outsourcing literature for research that includes

(1) Power and knowledge issues considered under the same analytical lens
(2) Introspection and self-education viewed as a part of the outsourcing decision-making lifecycle
(3) Knowledge asymmetries utilized as an alternative epistemological approach in explaining the unrealized outsourcing benefits

This paper has attempted to raise the managers' awareness on aspects of IT services outsourcing that have been under-considered in the extant literature. We deployed a knowledge/power lens to analyze the inter- and intra-organizational pathologies. At the market level, we attempted to reposition the current maturity debate at the client side and we underscored the *inseparability of power and knowledge in the formation of the client–provider relationship.* We discussed the value of a client-focused framework (such as COMM) that points at neglected soft organizational characteristics and places issues of individual and shared understanding at the heart of its methodological platform. We then utilized the biological metaphor to illustrate how different patterns of dynamic outsourcing relationships may be observed in an asymmetrical market.

Nevertheless, the power side was only superficially considered in our intra-organizational analysis and it is undoubtedly a question that merits further investigation. Specifically, the Foucaultian epistemology has a great deal to contribute to the study of

IS phenomena (Willcocks 2006) and the knowledge-related challenges posed within different socio-political contexts. Furthermore, we discussed the asymmetrical nature of the outsourcing market focusing on the resulting increased leverage of the supplier. However, in the long term, one should consider the consequences of what ultimately constitutes a *knowledge mismatch for the outsourcing market as a whole.* Under this spectrum, it is arguable that the expected long-term benefits cannot be realized by either party.

It can be argued that what is needed in the outsourcing market is *a shift of the focus from a unilateral to a reciprocal approach in considering outsourcing needs, services, and outsourcing maturity.* Developing the meta-capabilities of the client must involve organizational learning that imports knowledge from beyond the organizational boundaries. In this context, the value of third arty validation mechanism for the client lies not only in pinpointing the organization's outsourcing capabilities, but moreover in engaging it in a process of self-assessment that will ultimately stimulate organizational learning.

Acknowledgements

The author wishes to acknowledge the contribution and support of Chris Tofts and Richard Taylor, both from HP Labs Bristol.

References

Burgess, S. 2001. "The Beauty of the Peacock Tail and the Problems with the Theory of Sexual Selection," *TJ Magazine* (15:2), pp. 94-102.

Cronin, H. 1991. *The Ant and the Peacock*, Cambridge, UK: Cambridge University Press.

Cullen, S., and Willcocks, L. P. 2003. *Intelligent IT Outsourcing: Eight Building Blocks to Success*, Amsterdam: Elsevier.

DiRomualdo, A., and Gurbaxani, V. 1998. "Strategic Intent for IT Outsourcing," *Sloan Management Review* (39:4), pp. 67-80.

Kern, T., Willcocks, L. P., and van Heck, E. 2002. "The Winner's Curse in IT Outsourcing: Strategies for Avoiding Relational Trauma," *California Management Review* (44:2), pp. 47-69.

Komporozos-Athanasiou, A. 2006. "The Client Outsourcing Maturity Model as a Mechanism for Investigating IT Outsourcing Relationships in the Public Sector," HP Lab Tech Report (HPL-2006-169).

Komporozos-Athanasiou, A. 2007. "Assessing Client Maturity: A Key to Successful Outsourcing," HP Labs Technical Report (HPL-2007-124).

McFarlan, W. F., and Nolan, R. L. 1995. "How to Manage an IT Outsourcing Alliance," *Sloan Management Review* (39:4), pp. 67-80.

Merali, Y. 2002. "The Role of Boundaries in Knowledge Processes," *European Journal of Information Systems* (11:1), pp. 47-60.

Nonaka, I., and Noboru, K. 1998. "The Concept of 'Ba': Building a Foundation for Knowledge Creation," *California Management Review* (43:3), pp. 40-54.

Perez-Hallerbach, I. "Assessing the IT Outsourcing Maturity of Organizations: The Case of a Large Consumer Products Company," HP Labs Technical Report (HPL-2007-36).

Taylor, R., and Tofts, C. 2005. "Death by a Thousand SLAs: A Short Study of Commercial Suicide Pacts," HP Lab Tech Report (HPL-2005-11).

Taylor, R., and Tofts, C. 2008. "Model Based Services Discovery and Management Industrial Research and Practice Paper," HP Lab Technical Report (HPL-2008-19).

Taylor, R., Tofts, C., and Yearworth, M. 2004. "Open Analytics," HP Labs Tech Report (HPL-2004-138).

Venkatraman, N. 1997. "Beyond Outsourcing: Managing IT Resources as a Value Center," *Sloan Management Review* (38:3), pp. 51-54.

Willcocks, L. P. 2006. "Michel Foucault in the Social Study of ICTs: Critique and Reappraisal," Department of Information Systems Working Paper Series #138, London School of Economics and Political Science.

Willcocks, L. P., and Lacity, M. 2006. *Global Sourcing of Business and IT Services*, London: Palgrave Macillan 2006.

About the Author

Aris Komporozos-Athanasiou is an M.Phil. candidate at Judge Business School, University of Cambridge. He holds a B.Sc. in Economics and an M.Sc. in Information Systems from the University of Warwick. Prior to joining Cambridge, he worked as a services science researcher at Hewlett Packard Labs. His research integrates social theory and information systems and his dissertation explores different concepts of agency in the development of collaborative innovation. Aris can be reached at ariskomp@yahoo.gr.

22 HOW INFORMATION SYSTEMS PROVIDERS DEVELOP AND MANAGE EXPERTISE AND LEVERAGE THEIR CLIENT RELATIONSHIPS FOR COMPETITIVE ADVANTAGE

Robert Gregory
Michael Prifling
Institute for Information Systems
University of Frankfurt, E-Finance Lab
Frankfurt am Main, Germany

Abstract *Information technology vendors are continuously growing into global service providers. To reap the benefits of the changing global economy and gain a competitive advantage, providers need to see their client relationships as strategic assets and leverage them for expertise development and knowledge integration. How do IS service providers absorb knowledge from client relationships over multiple projects at multiple levels? How do IS service providers transfer and disseminate knowledge internally at and across multiple levels and integrate it to generate value-creating competencies? These questions are investigated in a longitudinal qualitative study employing interpretive case-study methods. The case included in the analysis is a 4-year IS outsourcing project between a large European bank and one of the largest Asian service providers. Services were delivered through a global virtual team, including offshore and onshore locations, and a particular strategy was employed to transfer knowledge cross-functionally, integrate this knowledge internally within the organization, and utilize it effectively in the global service delivery system. However, our initial analysis of the first interviews reveals that there was a discrepancy between the expected and actual performance of the service provider. Accordingly, the knowledge transfer and management processes need to be analyzed in more detail. With this research study, we aim at contributing to the domain of IS offshore outsourcing and services science as well as to the theory on vendor capabilities and knowledge management.*

Please use the following format when citing this chapter:

Gregory, R., and Prifling, M., 2008, in IFIP International Federation for Information Processing, Volume 267, Information Technology in the Service Economy: Challenges and Possibilities for the 21st Century, eds. Barrett, M., Davidson, E., Middleton, C., and DeGross, J. (Boston: Springer), pp. 311-319.

Keywords Strategic client–vendor relationships, IS outsourcing, IS service providers, expertise development, knowledge integration, knowledge transfer, global delivery of services, globalization of knowledge work, global virtual teams, organizational learning

1 INTRODUCTION, MOTIVATION, AND RESEARCH QUESTION

While information systems vendors aspire transforming themselves into global service providers and the economy evolves toward a global service economy, new strategies are needed by firms to reap the benefits of the emerging opportunities. Firms are continuously focusing on their extended enterprise to enhance their customer relationships, their product and service offerings, and their revenue growth in order to gain a competitive advantage (Krishnan et al. 2007). The challenge for firms in the new globalized economy is to leverage resources, exploit competencies, manage partner relationships, and explore opportunities (Krishnan et al. 2007). Currently, we have a limited understanding of how organizations develop and deploy the necessary capabilities to position themselves in the global service economy in order to harvest the value-adding benefits, being the product of successfully orchestrated interorganizational relationships (Krishnan et al. 2007). Therefore, further research is needed that addresses these issues and helps explain how interorganizational relationships—buyer–supplier relations are one possible form (Uzzi 1997)—can be leveraged for value creation and be converted into competitive advantage. Empirical investigations are needed to examine the value of strong customer ties and how they can be realized across multiple relationships (Saraf et al. 2007).

Buyer–supplier relations must be studied at multiple levels (Hitt et al. 2007; Uzzi 1997). The same is true for organizational knowledge management which occurs at and across different organizational levels (Nonaka and Takeuchi 1995). However, multilevel research in management is scarce, which holds particularly true for IS research (Hitt et al. 2007). Therefore, this research study will make a theoretical contribution by analyzing knowledge-sharing in client–vendor outsourcing relationships as well as knowledge management and dissemination at multiple levels.

While the client's strategic motives for IS outsourcing and the client–vendor relationship as such have been examined, there are only a few studies on the vendor's perspective (Dibbern et al. 2004). More empirical studies are needed to explain precisely how IS service providers interchange knowledge with their clients and generate value-creating competencies (Dibbern et al. 2008; Ethiraj et al. 2005; Levina and Ross 2003). Therefore, our research questions are

(1) How do IS service providers absorb knowledge from client relationships over multiple projects at multiple levels?
(2) How do IS service providers transfer and disseminate knowledge internally at and across multiple levels and integrate it to create value-creating competencies?

2 RESEARCH METHODOLOGY

The research methodology chosen for this research project is an exploratory and interpretive case study design (Walsham 1993). The epistemological position we draw upon is the interpretive research paradigm implicating, among other things, that we did not start our investigations with any predefined propositions or hypotheses as would have been the case for a positivist research methodology (Dubé and Paré 2003). Rather, the theoretical basis of our study evolves over time as we gain a deeper understanding of the relevant issues that play a role in our research context (Walsham and Sahay 1999). Qualitative research methodologies seem especially opportune for revealing processes and mechanisms of how service providers develop and manage expertise gained from their client relationships, as they avoid the distance employed by their counterparts through quantitative measures and the like (Walsham et al. 2007). In order to analyze processes in detail, investigators have to immerse deeply into the phenomenon and capture information from the object of study, the process, as well as the context (Cappelli and Sherer 1991).

Interviews so far have been conducted with 12 people from the client and 3 people from the vendor organization. In addition to the primary data, secondary data has been collected for the purposes of data triangulation. Also, with the active participation of at least two researchers in the collection and analysis of the data, the requirements for investigator triangulation were met (Yin 2003). In the second interview round (April to July 2008), we will conduct an additional 10 interviews with onshore project workers from the Indian service provider as well as 10 interviews with the client organization. Interviews with offshore project members in India are scheduled for the second half of 2008.

Data collection and analysis for this study follows a two-step approach. In the first data collection phase, following the recommendations of Glaser and Strauss (1967) interviews were conducted in an open-ended fashion for the identification of categories and patterns. In the second data collection phase, we will elaborate in more detail upon the identified categories. Also, for the following interview rounds, the extant literature (i.e., client–vendor relationships, relational capital, absorptive capacity, organizational and individual learning, knowledge management, IS outsourcing) will be included as additional data to compare with the empirical data (Glaser 1998) and guide the theory-building process (Eisenhardt 1989).

3 THEORETICAL FOUNDATIONS

As the world is becoming "flat," the nature of competition is changing and leveraged on a global level (Friedman 2005). More services are being disaggregated globally and delivered from multiple places around the world (Apte and Mason 1995; Mithas and Whitaker 2007). Firms are competing through their extended enterprise and according, to Kanter (1999), a firm's strategic relationships—nurtured by collaboration—are one of the main sources for competitiveness in the 21st century. According to the relational view of the firm, which is an extension of the resource-based theory, interorganizational relationships can be a source of competitive advantage, arising from interfirm business processes, routines, or interfirm specialization (Dyer and Singh 1998). Applied to the IS context, service providers can generate unique capabilities through knowledge sharing routines with their clients that are enabled by relational capital. Relational capital in this context consists of trust, information transfer, and joint problem-solving in the client–vendor relationship (Uzzi 1997). It enables the formation of interfirm knowledge sharing routines, which are defined as regular patterns of interfirm interactions that permit the transfer, recombination, or creation of specialized knowledge (Dyer and Singh 1998; Grant 1996). These routines are built up over time, as the parties develop a mutual understanding and shared knowledge space through the sharing of information (Hitt et al. 2006). Furthermore, as IS service providers gain more knowledge about their clients they augment their partner-specific absorptive capacity, which refers to the ability acquire knowledge, assimilate it, and use it for commercial ends (Cohen and Levinthal 1990). Partner-specific absorptive capacity is a further enabler of interfirm knowledge-sharing routines (Mowery et al. 2002). Interfirm knowledge-sharing routines and repeated exchange with clients translate into client-specific capabilities (e.g., business knowledge, application domain knowledge) which leverage the service quality and expertise of IS service providers (Ethiraj et al. 2005). An additional benefit for IS service providers resulting from client-specific capabilities and client relationships, characterized through stability, trust, and reciprocity, is the generation of new client relationships through reputation-based mechanisms, which enables the further extension of its knowledge base (Levina and Ross 2003).

The literature review above shows us that IS service providers develop unique capabilities and expertise through knowledge-sharing routines with their clients that are enabled by relational capital (i.e., trust, information transfer, and joint problem-solving) and partner-specific absorptive capacity. Relational capital in the client–vendor relationship emerges and evolves as a consequence of multiple individual activities (Ring and Van de Ven 1994). These individual activities take place on multiple levels upon which the interorganizational relationship develops and emerges (Klein et al. 2001; Koh et al. 2004). Furthermore, various scholars have argued that interorganizational relationships are inherently multilevel (Barden and Mitchell 2007; Brass 2001; Hitt et al. 2007; Klein et al. 2001). To acknowledge the multilevel nature of client–vendor relationships in IS, our analysis focuses on project manager relationships (individual level) and collective exchange experiences on the organizational level (Barden and Mitchell 2007; Koh et al. 2004). Representative for organization-level ties between client and vendor firm are boundary-spanning members with little decision-making authority such as sales personnel (Barden and Mitchell 2007). Additionally, organization-level structures and routines developed during past exchange relationships can give a deeper insight into knowledge-sharing on an organizational level (Barden and Mitchell 2007; Gulati 1995). Similar to relational capital and client–vendor relationships, analysis of absorptive capacity must be conducted on multiple levels. For example, Dibbern et al. (2008) analyzed the IS vendor's absorptive capacity on the team member level, focusing on their prior experiences and creativity skills. Similarly, Cohen and Levinthal (1990) argued that a firm's absorptive capacity depends on the individuals who stand at the interface of the firm or subunits within the firm. They continue to argue that some key individuals may have boundary-spanning roles, similar to Barden and Mitchell (2007).

Ensuring that knowledge is transferred from multiple client relationships is not sufficient for IS service providers to develop a competitive advantage. Besides external knowledge integration where knowledge is absorbed from external sources and blended with internal resources, internal knowledge integration, combination, and configuration is important, too (Grant 1996; Henderson and Clark 1990; Kogut and Zander 1992; Okhuysen and Eisenhardt 2002; Tiwana et al. 2003). The management of knowledge internally needs to occur at and across different organizational levels, similar to the way knowledge is absorbed from client relationships at multiple levels (Nonaka and Takeuchi 1995). Applying a systems view (e.g., Weick 1969) to organizational knowledge management, Garud and Kumaraswamy (2005) found out that there are at least three essential elements to building an organizational knowledge system. First, knowledge is created at the individual level through learning mechanisms. Second, communities of practice are based on informal relationships, shared language, and thought-worlds. Third, repositories enable the codification and central storage of organizational knowledge. A systems perspective further emphasizes the importance of virtuous circles (Masuch 1985; Nonaka and Takeuchi 1995) where the above mentioned elements interact with each other to enable that knowledge spirals up from the individual to the collective levels of the organization (Garud and Kumaraswamy 2005). The authors, however, leave the exploration of the precise interaction between these elements for further research. An interesting issue in the context of knowledge management at large IS service providers is the high rate of personnel turnover (Arora et al. 2001; Fairell, Kaka and Stürze 2005; Oshri et al. 2007), which can cause severe disruptions in the formation of team mental models or shared thought-worlds (Hsu et al. 2007). Another challenge is to assure that individuals in the organization contribute valuable knowledge to repositories and contribute actively to internal knowledge transfer (Garud and Kumaraswamy 2005). Furthermore, if IS service providers develop strategic expertise out of multiple projects with different client organizations (Ethiraj et al. 2005; Levina and Ross 2003), service providers need to assure that knowledge accumulated from one client project is reused for other clients (Ravishankar and Pan 2008). However, the provider also needs to take into account confidentiality agreements with clients.

4 THE CASE STUDY AND PRELIMINARY FINDINGS

The case study on which we base our analysis consists of a large and technically complex IT reengineering and integration project. A large European retail and investment bank had two separate IT systems to handle all of the bank's current accounts up and running for several years. These two systems had to be integrated with each other to create a new, common current-account platform. To master the technically complex reengineering and integration tasks, a large Indian service provider was contracted. Originally, the plan was to outsource approximately 60 percent of the technical work to the Indian provider and keep 40 percent within the retained organization. Due to the difficulties of understanding the customer's business processes and supporting IT functions in detail, the distribution of work had to be changed so that 50 percent was conducted by each of the parties. A client-side project manager said, "Later, we actually had a distribution of 60-60 percent, as there were so many unforeseen expectations, requirements, and hidden functionality, so that every party had to invest extra resources for the project to succeed." Many client project members were unsatisfied with the competence level of the service provider even though the initial vendor selection was carried out in a careful manner according to CMMI certification guidelines. Some of our interview partners explained the lack of knowledge transfer from vendor to client with the high rate of personnel turnover in India. However, our Indian interview partners explained to us the effective their knowledge transfer methodology. According to them, knowledge gained from various client relationships is accumulated and transferred by different means. On the one hand, part of the knowledge is made explicit and stored in a central file sharing database with online access from around the globe. Project team members from the vendor firm working from onshore or offshore locations can access the database and get information about functional, business, and technical issues. However, project documentation on these issues are not always sufficient for effective knowledge transfer. Therefore, on the other hand, knowledge is transferred through so-called knowledge transfer sessions. To initiate such a session, any project team member can contact an author of some documentation from the online database or senior project workers specialized in a particular area and ask for personal advice. A personal meeting will then be arranged so that the implicit knowledge that cannot be transmitted by the means of any documentation can be effectively transferred to the project team members that need advice.

The analysis of the first interview round yields conflicting results. The reasons for a lack of knowledge transfer and the knowledge transfer and integration strategy at the vendor organization need to be investigated in greater detail. The question is how IS service providers accumulate knowledge over multiple client relationships and then integrate this knowledge internally. Furthermore, how do IS service providers cope with high rates of personnel turnover? How do they assure that knowledge accumulated from one client project is reused in a new client project without disrupting the confidentiality of the former client? How do they actively manage their project staff to collectively contribute to the generation of organizational knowledge? How do different knowledge management elements (i.e., individual-level learning, communities of practice, and knowledge repositories) interact with each other to enable that knowledge spirals up from the individual to the collective levels of the organization? These questions will be investigated in the next interview rounds over the following 12 to 15 months in both India and Germany. Qualitative interviews with one of the largest IS service providers in India are scheduled for the second half of 2008. Additionally, further interviews will be conducted with clients from the European banking industry.

5 EXPECTED CONTRIBUTIONS OF THE STUDY

The expected contributions of this study are to find some answers to the questions mentioned above. We therefore aim at contributing to the domain of IS offshore outsourcing as well as to the theory on vendor capabilities and knowledge management. Through an interpretive, exploratory, and longitudinal research design, we aim at generating new concepts in line with the theory-building guidelines offered by the grounded theory development methodology (Glaser and Strauss 1967; Walsham 2006).

References

Apte, U. M., and Mason, R. O. 1995. "Global Disaggregation of Information Intensive Services," *Management Science* (41:7), pp. 1250-1262.

Arora, A., Arunachalam, V. S., Asundi, J., and Fernandes, R. 2001. "The Indian Software Services Industry," *Research Policy* (30:8), pp. 1267-1287.

Barden, J. Q., and Mitchell, W. 2007. "Disentangling the Influences of Leaders' Relational Embeddedness on Interorganizational Exchange," *Academy of Management Journal* (50:6), pp. 1440-1461.

Brass, D. 2001. "Networks and Frog Ponds: Trends in Multilevel Research," in *Multilevel Theory, Research, and Methods in Organizations,* K. Klein and S. Kozlowski (eds.), San Francisco: Jossey-Bass, pp. 557-571.

Cappelli, P., and Sherer, P. D. 1991. "The Missing Role of Context in OB: The Need for a Meso-Level Approach," *Research in Organizational Behavior* (13), pp. 55-110.

Cohen, W. M., and Levinthal, D. A. 1990. "Absorptive Capacity: A New Perspective on Learning and Innovation," *Administrative Science Quarterly* (35:1), pp. 128-152.

Dibbern, J., Goles, T., Hirschheim, R., and Jayatilaka, B. 2004. "Information Systems Outsourcing: A Survey and Analysis of the Literature," *The DATA BASE for Advances in Information Systems* (35:4), pp. 6-102.

Dibbern, J., Winkler, J., and Heinzl, A. 2008. "Explaining Variations in Client Extra Costs Between Software Projects Offshored to India," *MIS Quarterly* (32:2), pp. 333-366.

Dubé, L., and Paré, G. 2003. "Rigor in Information Systems Positivist Case Research: Current Practices, Trends, and Recommendations", *MIS Quarterly* (27:4), pp. 597-635.

Dyer, J. H., and Singh, H. 1998. "The Relational View: Cooperative Strategy and Sources of Interorganizational Competitive Advantage," *The Academy of Management Review* (23:4), pp. 660-679.

Eisenhardt, K. M. 1989. "Building Theories from Case Study Research," *Academy of Management Review* (14:4), pp. 532-550.

Ethiraj, S. K., Kale, P., Krishnan, M. S., and Singh, J. V. 2005. "Where Do Capabilities Come from and How Do They Matter? A Study in the Software Services Industry," *Strategic Management Journal* (26), pp. 25-45.

Fairell, D., Kaka, N., and Stürze, S. 2005. "Ensuring India's Offshoring Future," *McKinsey Quarterly*, Special Edition, pp. 74-83.

Friedman, T. L. 2005. *The World is Flat: A Brief History of the Twenty-First Century*, New York: Farrar, Straus and Giroux,

Garud, R., and Kumaraswamy, A. 2005. "Vicious and Virtuous Circles in the Management of Knowledge: The Case of Infosys Technologies," *MIS Quarterly* (29:1), pp. 9-33.

Glaser, B. G. 1998. *Doing Grounded Theory: Issues and Discussions*, Mill Valley, CA: Sociology Press.

Glaser, B. G., and Strauss, A. L. 1967. *The Discovery of Grounded Theory: Strategies for Qualitative Research*, Chicago: Aldine Publishing Company.

Grant, R. M. 1996. "Prospering in Dynamically-Competitive Environments: Organizational Capability as Knowledge Integration," *Organization Science* (7:4), pp. 375-387.

Gulati, R. 1995. "Social Structure and Alliance Formation Patterns: A Longitudinal Analysis," *Administrative Science Quarterly* (40:4), pp. 619-652.

Henderson, R. M., and Clark, K. B. 1990. "Architectural Innovation: The Reconfiguration of Existing Product Technologies and the Failure of Established Firms," *Administrative Science Quarterly* (30:1), pp. 9-30.

Hitt, M. A., Beamish, P. W., Jackson, S. E., and Mathieu, J. E. 2007. "Building Theoretical and Empirical Bridges Across Levels: Multilevel Research in Management," *Academy of Management Journal* (50:6), pp. 1385-1399.

Hitt, M. A., Bierman, L., Uhlenbruck, K., and Shimizu, K. 2006. "The Importance of Resources in the Internationalization of Professional Service Firms: The Good, the Bad, and the Ugly," *Academy of Management Journal* (49:6), pp. 1137-1157.

Hsu, J. S. C., Parolia, N., Jiang, J. J., and Klein, G. 2007. "The Impact of Team Mental Models on IS Project Teams' Information Processing and Project Performance," in *Proceedings of the Second International Research Workshop on Information Technology Project Management*, Montreal, Quebec, Canada, December 8, pp. 39-49.

Kanter, R. M. 1999. "Change Is Everyone's Job: Managing the Extended Enterprise in a Globally Connected World," *Organizational Dynamics* (28:1), pp. 6-22.

Klein, K., Palmer, S., and Conn, A. 2001. "Interorganizational Relationships: A Multilevel perspective," in *Multilevel Theory, Research, and Methods in Organizations,* K. Klein and S. Kozlowski (eds.), San Francisco: Jossey-Bass, pp. 267-308.

Kogut, B., and Zander, U. 1992. "Knowledge of the Firm, Combinative Capabilities, and the Replication of Technology," *Organization Science* (3:3), pp. 383-397.

Koh, C., Soon, A., and Straub, D. W. 2004. "IT Outsourcing Success: A Psychological Contract Perspective," *Information Systems Research* (15:4), pp. 356-373.

Krishnan, M. S., Rai, A., and Zmud, R. 2007. "The Digitally Enabled Extended Enterprise in a Global Economy," *Information Systems Research* (18:3), pp. 233-236.

Levina, N., and Ross, J. W. 2003. "From the Vendor's Perspective: Exploring the Value Proposition in Information Technology Outsourcing," *MIS Quarterly* (27:3), pp. 331-364.

Masuch, M. 1985. "Vicious Circles in Organizations," *Administrative Science Quarterly* (30:1), pp. 14-33.

Mithas, S., and Whitaker, J. 2007. "Is the World Flat or Spiky? Information Intensity, Skills, and Global Service Disaggregation," *Information Systems Research* (18:3), pp. 237-259.

Mowery, D. C., Oxley, J. E., and Silverman, B. S. 2002. "The Two Faces of Partner-Specific Absorptive Capacity: Learning and Cospecialization in Strategic Alliances," in *Cooperative Strategies and Alliances,* F. J. Contractor and P. Lorange (eds.), Amsterdam: Elsevier Science, pp. 291-319.

Nonaka, I., and Takeuchi, H. 1995. *The Knowledge-Creating Company*, New York: Oxford University Press.

Okhuysen, G. A., and Eisenhardt, K. M. 2002. "Integrating Knowledge in Groups: How Formal Interventions Enable Flexibility," *Organization Science* (13:4), pp. 370-386.

Oshri, I., Kotlarsky, J., and Willcocks, L. 2007. "Managing Dispersed Expertise in IT Offshore Outsourcing: Lessons from Tata Consultancy Services," *MIS Quarterly Executive* (6:2), pp. 53-65.

Ravishankar, M. N., and Pan, S. L. 2008. "The Influence of Organizational Identification on Organizational Knowledge Management (KM)," *Omega* (36:2), pp. 221-234.

Ring, P. S., and Van de Ven, A. H. 1994. "Developmental Processes of Cooperative Interorganizational Relationships," *The Academy of Management Review* (19:1), pp. 90-118.

Saraf, N., Langdon, C. S., and Gosain, S. 2007. "IS Applications Capabilities and Relational Value in Interfirm Partnerships," *Information Systems Research* (18:3), pp. 320-339.

Tiwana, A., Bharadwaj, A., and Sambamurthy, V. 2003. "The Antecedents of Information Systems Development Capability in Firms: A Knowledge Integration Perspective," in *Proceedings of the 24th International Conference on Information Systems*, A. Massey, S. T. March, and J. I. DeGross (eds.), Seattle, Washington, December 14-17, pp. 246-258.

Uzzi, B. 1997. "Social Structure and Competition in Interfirm Networks: The Paradox of Embeddedness,"*Administrative Science Quaterly* (42:1), pp. 35-67.

Walsham, G. 1993. *Interpreting Information Systems in Organizations*, Chichester, UK: Wiley.

Walsham, G. 2006. "Doing Interpretive Research," *European Journal of Information Systems* (15:3), pp. 320-330.

Walsham, G., Robey, D., and Sahay, S. 2007. "Foreword: Special Issue on Information Systems in Developing Countries," *MIS Quarterly* (31:2), pp. 317-326.

Walsham, G., and Sahay, S. 1999. "GIS for District-Level Administration in India: Problems and Opportunities," *MIS Quarterly* (23:1), pp. 39-65.
Weick, K. E. 1969. *The Social Psychology of Organizing*, Reading, MA: Addison-Wesley.
Yin, R. K. 2003. *Case Study Research: Design and Methods* (3rd ed.), Thousand Oaks, CA: Sage Publications.

About the Authors

Robert Gregory is a Ph.D. candidate at the Institute of Information Systems at Johann Wolfgang Goethe University, E-Finance Lab, Frankfurt, Germany. His research interests focus on strategic issues of information systems outsourcing and information technology project management. Robert can be contacted at gregory@wiwi.uni-frankfurt.de.

Michael Prifling is a Ph.D. candidate at the Institute of Information Systems at Johann Wolfgang Goethe University, E-Finance Lab, Frankfurt, Germany. His research interests focus on information technology project management and vendor-related issues in outsourcing. Michael can be contacted at Prifling@wiwi.uni-frankfurt.de.

23 MIND THE GAP! Understanding Knowledge in Global Software Teams

Aini Aman
School of Accounting, Faculty of Economics and Business
Universiti Kebangsaan Malaysia
Bangi Selangor, Malaysia

Brian Nicholson
Manchester Business School
Manchester, United Kingdom

Abstract *This paper presents a conceptual framework and preliminary empirical analysis of knowledge gaps between global software team members in the UK and India. Drawing on episodes from rich case study evidence of a UK software firm based in the UK and software development sites in India, the conceptual framework is used to explore the data and to understand the knowledge gaps encountered. These are in relation to the level of knowledge, educational background, and experience of members. This study unpacks the notion of knowledge of software development into domain, technical, and application knowledge and considers the implications of prior knowledge, experience, and education background. It is anticipated from the preliminary findings that a practical and theoretical contribution will improve our understanding of the complexities of knowledge in such global arrangements.*

Keywords Knowledge management, global software teams

Please use the following format when citing this chapter:

Aman, A., and Nicholson, B., 2008, in IFIP International Federation for Information Processing, Volume 267, Information Technology in the Service Economy: Challenges and Possibilities for the 21st Century, eds. Barrett, M., Davidson, E., Middleton, C., and DeGross, J. (Boston: Springer), pp. 321-330.

1 INTRODUCTION

This paper aims to present a conceptual framework and preliminary analysis of the causes and implications of knowledge gaps in offshore software development work, when teams are separated by time, distance, and culture (Carmel and Tija 2005; Sahay et al. 2003). There has been limited prior research focusing on knowledge in offshore software development. Exceptions include Sarker and Sahay (2004) and Nicholson and Sahay (2004) who have improved our understanding of knowledge transfer issues and embeddedness of knowledge. However, the issue of what constitutes gaps in knowledge when developers are distributed globally and the implications of the gaps remains under-researched.

The theoretical framework draws on three linked concepts of tacit–explicit knowledge, knowledge types, and level of knowledge. Tacit knowledge is unarticulated and can be acquired informally or through practical experience, while explicit knowledge is the codifiable knowledge that can be acquired through formal study (Lam 1997; Polanyi 1966). Understanding explicit knowledge requires the understanding of tacit knowledge within the context of social interaction (Thompson and Walsham 2004). Second, a literature search revealed several authors who have identified similar category types of software development knowledge including domain, application and technical knowledge (Curtis et al. 1988; Waterson et al. 1997). Third, each member in a software development team possesses different levels of prior knowledge derived from their unique education background and set of experiences (Lam 1997; Waterson et al. 1997). The differences in the types and levels of knowledge owned by software development team members are referred to as knowledge gaps. Thus, the research question driving this inquiry is concerned with the causes and implications of knowledge gaps in offshore software development.

The empirical basis for our analysis is a longitudinal case study of a UK-based software firm with a development center in India where projects are split between teams in India and the UK. India is widely regarded as the world's leading venue for offshore software.[1]

The rest of this paper is organized as follows. In the next section, we present our theoretical framework, followed the by research methodology and a case description. Subsequently we present the analysis, and finally, we conclude this paper with anticipated theoretical and practical contributions.

2 THEORETICAL FRAMEWORK

This section presents a theoretical framework which was not developed *a priori* but according to the emerging themes found in the data analysis and intensive literature reviews and was refined during data analysis as events emerged. This inductive approach to theorization has been adopted by Barrett and Walsham (1999), Walsham and Sahay (1999), Nicholson and Sahay (2001), and Barrett et al. (2005).

---

[1]A. T. Kearney produces a location-attractiveness index, available at www.atkearney.com.

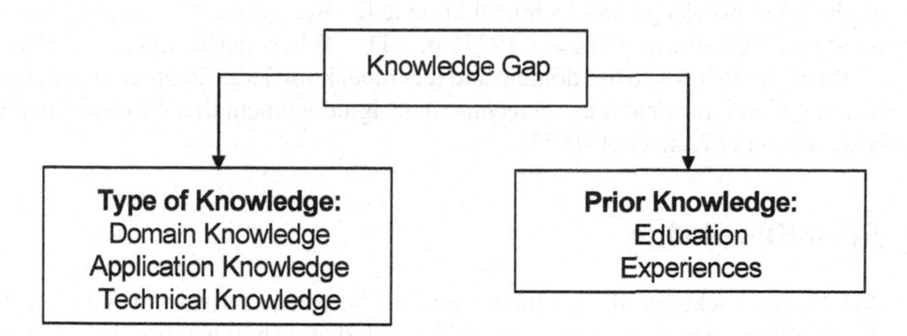

Figure 1. Conceptual Framework

In order to understand the process of managing knowledge gaps in offshore software development, we extend Curtis et al. (1988) on the types of knowledge using the ideas of prior knowledge (Lam 1997) and focusing on the experiences and educational background. We also take into consideration that the process of integrating knowledge situated in a range of sites is difficult because of the differences in the norms of participation and the physical context of work that varies across sites (Sole and Edmondson 2002). Our basic theoretical framework is shown in Figure 1.

2.1 Type of Knowledge

Domain knowledge is "business knowledge" that represents the formal and informal rules and practices of the particular company, which is needed in requirement analysis and design (Curtis et al. 1988, Sahay et al. 2003). Domain knowledge presents challenges when team members are globally distributed. For example, Nicholson (1999) provides a case study of global software development for UK housing benefits. India has no equivalent social security system and thus Indian developers had no immediate intuitive conception of the inputs, outputs, processes, and functions of such a system. By contrast, domain knowledge about the UK social security system was already held by UK-based developers, all of whom had been born and brought up in the UK. Some had even directly engaged with the UK housing benefits system as a user when making housing benefit claims.

Technical knowledge is concerned mainly with coding (Curtis et al. 1988) but may also include use of tools such as compilers, interpreters, and debuggers for generating code or high-level programming languages such as C, C++, Pascal, or Java. A developer should be able to produce and test the source code for software based on his prior knowledge of the relevant requirements, architecture, and design documentation (Curtis et al. 1988; Quintas 1993). For example, a developer needs to have this knowledge in order to understand the design document and translate it into an implementation language or machine-readable form.

Application knowledge is the ability to perform "soft" analysis such as requirements elicitation, relationship management, resource allocation, and tracking in the software development process (Curtis et al. 1988). Application knowledge provides a "mapping" between domain knowledge and technical knowledge that is relevant for the particular software being developed (Guarino 1997, p. 11). When performing requirements analysis, the developer has to use domain and technical knowledge in order to understand the requirement and translate it into a technical design document that will have meaning to other developers (Walz et al. 1993).

2.2 Prior Knowledge

Educational background inevitably differs between software team members depending on how they gain their technical knowledge. They acquire their technical knowledge either through formal education or training (Lam 1997) from various institutions. An illustration of the implications of this is provided by Nicholson and Sahay (2004) in the case of a UK-based software house with a subsidiary in India. The Indian subsidiary recruited graduates from some of the second- and third-tier colleges in India which had a stronger emphasis on technical knowledge than did the first-tier colleges (Nicholson and Sahay 2004). In addition, some of the skills of these lower-tier graduates were found to be outdated, with a majority of them having had their programming course on COBOL, with little or no experience with newer languages like Java and Visual Basic. This technical emphasis clashed with the UK firm's desire for application knowledge involving creative, out-of-the-box thinking, which tend to be taught in higher-tier institutes of technology.

Domain and application knowledge can be acquired through experience (Sahay et al. 2003) that are organized and embedded in organizational routines and procedures (Lam 1997). Each team member in offshore software development work brings disparate experiences into the project that contribute to the differences in the levels of experiential knowledge among team members. The case of UK social security benefits described earlier illustrates how the onshore team in the UK might face difficulties in communicating their knowledge to an offshore team in India, which has no direct, first-hand experience of making a social security claim. Making such tacit knowledge explicit in any requirements specification to be passed to Indian developers represents a major challenge.

3 RESEARCH METHODOLOGY

We adopt an intensive case study approach (Walsham 1995; Yin 1994) carried out in a UK-based company named Alp (a pseudonym) engaged in offshore software development in India. Alp was established in 1997 with 40 staff in the UK and India. During 1999, Alp entered into a partnership with a local company in Mumbai for offshore software development in order to take advantage of low labor costs. The partnership with the small local company called ABS in India lasted for almost two years. Alp (UK) set up a wholly owned offshore software development center in India in July 2002. The type of software development work sent to India depends on the type and nature of the projects. Typically, the onshore team in the UK will perform the requirements elicitation and some high-level technical architecture. The offshore team in India will undertake the design and the rest of the development work such as coding and testing. Communications and discussions between onshore and offshore teams are informal through telephone, e-chat, and e-mail. The focus of the analysis in Alp is on the ATI Co (a pseudonym) project phases one and two. ATI Co currently has 180 staff and decided to set up a data warehouse on construction information and web-based delivery to subscribers. Phase one involved the pay-per-view process as a trial with one portal, while phase two was expanding that for any portal with a subscription engine. Phase one had a four-person team full-time for five months, whereas phase two was extended over a further five months.

The research began with a historical reconstruction starting from Alp's inception of operations offshore and following the subsequent operations during the period of the study between 2001 and 2005. Interview information collected at periodic intervals in India and the UK was cross-checked where possible against other sources of evidence such as documentation and company reports. Data collection used a variety of methods including unstructured and semi-structured interviews, documentation reviews, and observations taken in the UK and during field trips to India. Interviewing, repeat interviewing, and triangulation of sources over time using various techniques of data collection provided multiple perspectives and information on emerging concepts and allowed for cross-checking (Eisenhardt 1989; Pettigrew 1990). Documents such as functional specifications, project plans, project reports, test cases, e-chat between onshore and offshore teams, and other documents related to software projects and work procedures were reviewed. In addition, observations of team interactions and meetings during projects were undertaken during the fieldwork in the UK and India.

There have been a total of 93 hours of interviews with 30 individuals including directors, operation directors, project managers, project leaders, senior developers, developers, quality analyst, administrator, marketing and technical support (Table 1). Interviews were handled in different ways, including face-to-face copresent, conference call, telephone and e-chat. Interviews lasted for at least 1 and a maximum of 2.5 hours. The interviews were transcribed and subsequently summarized. The data, once collected, was analyzed by identifying themes related to knowledge gaps. Transcripts were coded into the categories and a theoretical summary produced. The theoretical framework used was not developed *a priori* but developed according to the emerging themes found in the data analysis and intensive literature reviews. Later, an intermediate report was presented to the operations manager in one of the case companies for further explanation and verification of our interpretations.

Table 1. Details of Interviews

Job Title	UK	India
Director, Operations Director	30 hours	4 hours
Project Manager, Consultant	20 hours	2 hours
Project Leaders	–	5 hours
Senior Developer, Developer, QA Analyst	–	13 hours
Administration, Marketing, Technical Support	2 hours	6 hours
Customer	4 hours	–
Total hours	56 hours	37hrs

4 ANALYSIS AND DISCUSSION

This section presents specific episodes during ATI projects to illustrate the causes and implications of knowledge gap according to the types of knowledge and prior knowledge.

4.1 Type of Knowledge

Episode 1: Domain Knowledge. In ATI phase one, Robert, a project manager in Alp, gathered client requirements and sent a document consisting of a series of flowcharts and text-based system requirements to Samir, a senior developer in India. Later, Robert explained the requirements to Samir through e-chat but Samir had difficulty understanding them. As Samir explains,

> *I did not understand what Robert was trying to say and Robert did not understand what I was trying to say. It was very difficult to get the same understanding in virtual space.*

The misunderstanding was related to the gaps in domain knowledge between Robert and Samir. Robert had a better understanding of the requirement than Samir about the client and their business environment as he lives in the UK and had worked closely with the client previously, while Samir had never met the client, never been to the UK, and never worked with the client. Samir could not understand the users of the application, why they needed the information, what types of information needed to be stored, or how the user would access the information. Again, Samir explains,

> *In one incident, the client wants something to be displayed on the screen. When I looked into it, I had difficulty to understand....the problem is I did not understand the client's business completely and how it runs.*

Without appropriate domain knowledge, Samir had no intuition about the requirement (Winograd 1995) and had to make his own assumptions based on his knowledge about the business environment in India, forming his framework of meaning to make sense of the requirement. However, Samir had no experience of a similar company to ATI in India and had difficulty in imagining the functions, inputs, outputs, and interface of the completed application in use. Domain knowledge is required to develop "a picture in their mind of the context" (Cramton 2001, p. 359). The implication is that he was not able to fill in the gaps with his experiences of similar systems in India. According to Samir,

> *We did not have enough knowledge about the client's business. We had to assume a lot of things... in most of the cases, the client talks in terms of business, which is quite different from the places where we work in India; the working culture, the processes, and the standards. We couldn't imagine.*

Episode 2: Application and Technical Knowledge. A further episode was encountered in ATI phase two. Jeevan, an onshore project manager, has to communicate to Pratap, a project leader in India who often claimed that he did not have sufficient information to perform the tasks. He clarifies,

> *The client tells the onshore team about the requirements, and the onshore team forwards the requirements to us. We have to assume. We didn't have enough information about the system. Questions were not adequately answered, and the information wasn't going down to the developer level.*

The cause of such a claim is the gap in application and technical knowledge between Pratap and Jeevan. According to Pratap, Jeevan did not have appropriate technical knowledge and because of this, despite his application knowledge of the relevant techniques, he was unable to envisage the program code execution and thus explain the requirement in the way that the Indian team could understand:

> *He should understand what the client wants and be able to explain with data, to foresee the possible design or coding and to communicate it to the offshore teams.*

Application knowledge requires imagination of the requirement in the technical knowledge such as the functions, processes, and interface of the system from the requirements. Winograd (1995, p. 69) explains that the programmer needs to be able to "visualize what the program will be like and what can be done with it....[because the] abstract representation, such as written descriptions, flow charts and object class hierarchies cannot provide a grounded understanding." Jeevan's inability to explain with data and foresee the possible design or coding when communicating with Pratap, may result in inaccurate assumptions (Cramton 2001). The implication is that Pratap had to imagine the functions, processes, and interface of the system from the requirements and his imagination is conditioned or filtered by his knowledge, which is derived from the Indian context.

4.2 Prior Knowledge

Episode 1: Education Background. The differences in educational background between Robert and Samir the contribute to knowledge gap. Robert has a degree in Software Engineering from the UK, while Samir only has a technical certificate from a local technical college in India. Samir had to write using nontechnical terms when communicating with Robert. As a director in India explains,

> *He is not going to understand code or technical parts. They have to write in the language that he understands. It's all about giving information in a form that can make sense to the other person.*

Episode 2: Experiences. Experiences between onshore and offshore teams also differ. In ATI phase two, the offshore teams had to produce a design based on a given requirement by the onshore team. Tom, a designer in the UK, tried to explain to Pratap using a design document sent through e-mail. However, he could not include all of the details in the document as some of the knowledge could not be made explicit. Tom explains,

> *We try to explain as many assumptions as possible formally. Every document that we write has assumptions in it but the assumptions do not tell you so much about the business. There will always be implicit assumptions in the design.*

Because of the implicit assumptions, Pratap was always trying to figure out what was happening in the UK without having much experience and knowledge to do that effectively. For example, Pratap suggested a colorful interface design but Tom knew that the client would not accept this as he knew that the client corporate color was black and red. Such information was not formally articulated in the requirements or subsequent documents but constituted part of the tapestry of background domain knowledge or experiences about the client held tacitly by Tom and other UK-based developers. Tom explains,

> *They often suggest different alternatives. Every time, we will go back to the initial design that we gave them. That is because we have experience in UK cultures and business environments or because we probably did not give all the information that we probably could have done. It's those things that are almost intangible; things we know just by living in this country.*

5 CONCLUSION

The research question of this paper is concerned with taking some first steps in improving our understanding of the causes and implications of knowledge gaps in offshore software development. The main anticipated theoretical contribution is in the form of the conceptual framework. We unpack the notion of software development knowledge including domain, application, and technical knowledge (Curtis et al. 1988) and the implications of prior knowledge (Lam 1997). Preliminary findings indicate the importance of the concept of imagination (Winograd 1995), which was demonstrated when Indian software developers used their imagination to understand the domain knowledge and make assumptions on the required design that were sometimes inaccurate. Imagination is required to integrate the system as a whole or to connect diverse things and is linked to prior knowledge.

The practical contribution of the research is in the conceptual framework, which enables analysis of potential for knowledge gaps between team members, and the cognitive "filters" on imagination and interpretation, which domain knowledge presents in particular. Software development teams may use the concepts to diagnose failures and better understand the potential breakdowns. They may benefit from understanding the kinds of knowledge that are necessary to communicate in the projects on which they are working. Formal training in the knowledge types and potential breakdowns would be of benefit and this paper presents a starting point in the form of case studies for discussion.

References

Barrett, M., Cooper, D. J., and Jamal, K. 2005. "Globalization and the Coordinating of Work in Multinational Audits," *Accounting, Organizations and Society* (30:1), pp. 1-24.

Barrett, M., and Walsham, G. 1999. "Electronic Trading and Work Transformation in the London Insurance Market," *Information Systems Research* (10:1), pp. 1-22.

Carmel, E., and Tija, P. 2005. *Offshoring Information Technology: Sourcing and Outsourcing to a Global Workforce*, Cambridge, UK: Cambridge University Press.

Cramton, C. D. 2001. "The Mutual Knowledge Problem and its Consequences for Dispersed Collaboration," *Organization Science* (12:3), pp. 346-371.

Curtis, B., Krasner, H., and Iscoe, N. 1988. "A Field Study of the Software Design Process for Large Systems," *Communications of the ACM* (31:11), pp. 1268-1287.

Eisenhardt, K. M. 1989. "Building Theories from Case Study Research," *Academy of Management Review* (14:4), pp. 532-550.

Guarino, N. 1997. "Understanding, Building, and Using Ontologies," LADSEB-CNR, National Research Council., Padova, Italy (http://ksi.cpsc.ucalgary.ca/KAW/KAW96/guarino/guarino.html).

Lam, A. 1997. "Embedded Firms, Embedded Knowledge: Problems of Collaboration and Knowledge Transfer in Global Cooperative Ventures," *Organization Studies* (18:6), pp. 973-996.

Nicholson, B. 1999. *The Process of Software Development Across Time and Space: The Case of Outsourcing to India*, unpublished Ph.D. thesis, Salford University.

Nicholson, B., and Sahay, S. 2001. "Some Political and Cultural Issues in the Globalisation of Software Development: Case Experience from Britain and India," *Information and Organization* (11:1), pp. 25-43.

Nicholson, B., and Sahay, S. 2004. "Embedded Knowledge and Offshore Software Development," *Information and Organization* (14:4), pp. 329-365.

Pettigrew, A. 1990. "Longitudinal Field Research on Change: Theory and Practice," *Organization Science* (1:3), pp. 267-292.

Polanyi, M. 1966. *The Tacit Dimension*, London: Routledge.

Quintas, P. 1993. "Introduction—Living the Lifecycle: Social Processes in Software and Systems Development," Chapter 1 in *Social Dimensions of Systems Engineering: People, Processes, Policies and Software Development*, P. Quintas (ed.), London: Ellis Horwood Limited, pp. 1-7.

Sahay, S., Nicholson, B., and Krishna, S. 2003. *Global IT Outsourcing*, Cambridge, UK: Cambridge University Press.

Sarker, S., and Sahay, S. 2004. "Implications for Space and Time for Distributed Work: An Interpretive Study of US-Norwegian Systems Development Teams," *European Journal of Information Systems* (13), pp. 3-20.

Sole, D., and Edmondson, A. 2002. "Situated Knowledge and Learning in Dispersed Teams," *British Journal of Management* (13), pp. 17-34.

Thompson, M. P. A., and Walsham, G. 2004. "Placing Knowledge Management in Context," *Journal of Management Studies* (41:5), pp. 725-747.

Walsham, G.,and Sahay, S. 1999. "GIS for District-Level Administration in India: Problems and Opportunities," *MIS Quarterly* (23:1), pp. 39-65.

Walsham, G. 1995. "Interpretative Case Studies in IS Research: Nature and Method," *European Journal of Information Systems* (4), pp. 74-81.

Walz, D., Elam, J., and Curtis, B. 1993. "Inside a Software Design Team: Knowledge Acquisition, Sharing and Integration," *Communications of the ACM* (36:10), pp. 62-77.

Waterson, P. E., Clegg, C. W., and Axtell, C. M. 1997. "The Dynamics of Work Organization, Knowledge and Technology During Software Development," *International Journal Human-Computer Studies* (46), pp. 79-101.

Winograd, T. 1995. "Heidegger and the Design of Computer Systems,." in *Technology and the Politics of Knowledge*, A. Feenberg and A. Hannay (eds), Bloomington, IN: Indiana University Press, pp. 108-127.

Yin, R. K. 1994. *Case Study Research: Design and Methods* (2nd ed.), Thousand Oaks, CA: Sage Publications.

About the Authors

Aini Aman is a senior lecturer at Universiti Kebangsaan Malaysia (UKM), Malaysia. Since 1993, she has been involved in teaching, research, and consultancy projects in the broad area of auditing, accounting, and business. Her Ph.D. involved an in-depth study of the offshore software development process in the UK, India, Bangladesh, and Malaysia. Her post doctoral study involved work with Brian Nicholson on the project "Risk and Control of Offshore Outsourcing of Accounting Services" commissioned by the Institute of Chartered Accountants in England and Wales. Aini's current research is on accounting outsourcing in Malaysia and her current consultancy work is an evaluation of e-government orojects in Malaysia. Aini can be contacted at aini@ukm.my.

Brian Nicholson is a senior lecturer at Manchester Business School. Since 1995, he has been involved in teaching, research, and consultancy projects in the broad area of global outsourcing of software and other business processes. This has involved work in India, China, Costa Rica, Iran, Egypt, Malaysia, and Bangladesh. Brian's research at the firm level has resulted in several influential publications in international journals and a book, *Global IT Outsourcing* (Cambridge University Press, 2003). Policy-level consultancy studies have been undertaken for the governments of Costa Rica and Iran to stimulate software exports. In the case of Costa Rica, this resulted in production of a national level strategy. Recent work has been commissioned by the Institute of Chartered Accountants in England and Wales, "Risk and Control of Offshore Outsourcing of Accounting Services." Brian can be contacted at brian.nicholson@mbs.ac.uk.

24 COMPLICATING UTOPIAN AND DYSTOPIAN VIEWS OF AUTOMATION: An Investigation of the Work and Knowledge Involved in the Call Center Offshoring Industry in India

Paul R. Devadoss
Mike W. Chiasson
Lancaster University Management School
Lancaster University
Lancaster, UK

Abstract *Offshoring is motivated by the relocation and standardization of organizational services to remote locations—typically the so-called developing nations—in order to achieve substantial cost efficiencies. Standardized business practices, aided by information technologies, are assumed to mobilize and recover the service practices in these new contexts. In this paper, we examine the boundary objects and boundary work involved in call center work. Data from several interviews with managers, industry consultants, and agents in the call center industry reveal that the recovery of call center practices in India involves substantial managerial and employee work, in order to manage and stitch together the diverse cultural and practical interests of the various groups. As a result, beneath the automated and simplified appearance of call center work is an underlying complexity of boundary work and boundary objects involved in linking the various participants—both human and nonhuman—into a temporarily stable industry. The result is a complication to both utopian and dystopian views of call center work.*

Keywords Offshoring, call centers, boundary objects, case study

Please use the following format when citing this chapter:

Devadoss, P. R., and Chiasson, M. W., 2008, in IFIP International Federation for Information Processing, Volume 267, Information Technology in the Service Economy: Challenges and Possibilities for the 21st Century, eds. Barrett, M., Davidson, E., Middleton, C., and DeGross, J. (Boston: Springer), pp. 331-339.

1 INTRODUCTION

The effect of the embedded information technology artefact on the transformation and creation of industrial practices is an important subject. An important topic related to this is the various attempts to use IT in order to redistribute industrial work to other regions through offshoring. Examples include the offshoring of advice and software development to other countries. In these particular situations, various IT artefacts and managerial practices are developed and deployed in an attempt to redistribute, replicate, and control service work in a new region.

However, the offshoring industry confronts a number of social and technical challenges in a new geographical region. Differences in work force skills, training, language, culture, institutional contexts, and industrial practices provide numerous challenges to the redistribution and recovery of industrial practices elsewhere (Krishna et al. 2004). In response, numerous people become involved in the construction of IT and industrial systems in order to recover and recreate industrial work practices that emulate the desires of foreign customers within a new region. We refer to these managerial and employee activities as boundary work, and the various logics that tentatively link the exchange of time, money, attention and work across social and technical participants, as boundary objects.

In this paper, we explore the boundary work and objects involved in the call center offshoring industry in India. We examine the challenges of constructing boundary objects that link the heterogeneous interests of diverse groups in this industry—employees, managers, customers, and offshoring clients—with each other. We draw upon the definition of boundary objects from Star and Griesemer (1989) as the things which are "both plastic enough to adapt to local needs and constraints of the several parties employing them, yet robust enough to maintain a common identify across sites….Like a black board, a boundary object 'sits in the middle' of a group of actors with divergent viewpoints" (p. 46).

Boundary objects have been used in information systems research to understand boundary spanning across communities through IT boundary objects (Gasson 2006). Such boundary objects hold important implications for the design and use of IT artefacts (Karsten et al. 2001). The boundary objects we consider in this case move beyond the purely IT to other objects consistent with Star and Griesemer's definition. In our case, these include offshoring logics, process controls, training, and operations management. These techniques and procedures are used to produce boundary objects that are plastic enough to adapt to local needs and constraints, but common enough to produce an exchange that maintains the interests of the supplier and the receiver of the exchange. Without needing to explicitly identify Western customers and clients, we can identify the boundary objects employed across managers and employees in order to translate the heterogeneous group interests in the production of a call center industry in India. In doing so, we address and complicate both utopian and dystopian views of call center work by illustrating the extensive work required by managers and employees to realize the call center service sector in India. However, by considering this as boundary work, we question both utopian and dystopian views of call center work by suggesting that while it may appear to be a one-side relationship, the challenges of employee turnover and the increasing mobility of call center employees is challenging the viability of this industrial system in the long-term.

2 CASE BACKGROUND

A call center is defined "as a dedicated operation in which computer-utilizing employees receive inbound—or make outbound—telephone calls, with those called processed and controlled either by an Automatic Call Distribution (ACD) or predictive dialing system" (Taylor and Bain 1999, p. 102). Call centers use a number of IT tools to manage telephone calls, customer data, workflow processes, and quality control systems.

A vision of a call center is as a flexible, friendly and knowledge-based workplace staffed by cooperative employees, smiling down the phone as they help the customer, who enhance the image of an organization (Bain and Taylor 2000, p. 3). This view of a call center shows employees as empowered in their identification of customer needs, and to provide relevant service through appropriate support from information technology (Frenkel et al. 1998).

The opposite and dystopian view is that call centers are an *electronic panopticon*, where complete control over the employees is possible through the integration of telephony and computing (Fernie and Metcalf 1997). Call centers are characterized as sweatshops, with rows of agents in cubicles, answering call after call, while under constant surveillance and pressure by management (Belt et al. 2000). The process is labeled as "assembly line in the head" (Taylor and Bain 1999, p. 107) or "Taylorisation of the white-collar work in call centers" (Richardson and Howcroft 2006, p. 60).

Call center costs in India can be one-third the cost in Western countries, largely through cheaper labor costs (Dossani and Kenney 2003). Despite the cost savings attraction, Richardson and Howcroft (2006) suggest that the complete routinization and standardization of the call center is very difficult. For example, attempts to recruit and retain certain employees who can speak a language have been problematic (Callaghan and Thompson, 2002). We believe that these attempts to construct this industry reflect a larger problem of managing the quality and quantity in call center settings (Taylor and Bain, 1999), and reflect the boundary work and construction of objects, such as computer systems and training practices, which can translate and enrol the diverse participant interests involved in the outsourced call center.

In summary, call centers are, to a large extent, both a dream and an ever-shifting reality for those involved in building the complex socio-technical logics that will enroll not only outsourcers, but the managers and workers in the call center. The work involved in making this happen is nuanced and complex, and therefore far from the simplicity of automation. We explore this boundary work next.

3 METHODOLOGY

A qualitative case study (Yin 2003) of three call centers in India was conducted between 2003 and 2007. Data was primarily collected through 23 semi-structured interviews with various managers and call center agents. The companies examined were all located in Bangalore, a hotbed of IT off-shoring in India. The interviews were focused on understanding the nature of managerial work, and their struggles to manage the heterogeneous groups and interests involved in the call center industry. In addition, industry consultants and advisors were also interviewed in order to gain general insights into the development of the industry in India. The data was interpreted and organized according to theoretical ideas in boundary work and objects, and are discussed in the findings section (Walsham 2006). In order to understand the nature of the industry, and the role of information technology in relation to call center work, we have chosen to analyze the data at an industrial level, rather than the organizational level (Chiasson and Davidson, 2005).

4 FINDINGS: BOUNDARY OBJECTS

4.1 The Offshoring Logics

Since the mid-1990s, many organizations have offshored various services to developing economies such as India. Consequently, business process outsourcing (BPO) has been one of the fastest growing sectors in India since the mid-1990s. The success of early adopters, such as American Express and General Electric, has convinced others to consider the same.

Despite a common perception that call centers are modern day sweatshops, the managers we interviewed believe that a call center agents' salaries are still relatively good for fresh graduates in a job market with limited opportunities. As one industry human resources manager comments,

> *All call centers are like that—if you look at the population, the profiles of people working in call centers—most of them will be fresh college grads, people with 2 to 3 years of experience, whom the rest of the industry or the rest of the world wouldn't touch 4 to 5 years ago. Today these guys are going in there and making as much money as anybody else did sometime back.*

Combined with this relative salary potential, a rapid growth of the industry in India has lead to new call centers opening every other day. Trained call center staff are now being poached with small salary increments, which has rapidly escalated salaries and recruitment costs. Despite employment opportunities, call center staff also feel their work is a temporary career option, especially since few move from answering calls into managerial roles. As a result, fresh graduates enter the industry in order to earn money before moving on to other educational or career options.

Beyond these economic and workforce issues, managers also suggest that despite a perception of offshoring driven by only cost-efficiencies, they claim that the specificity of making call centers work in India involves numerous business and cultural logics in order to satisfy the diverse interests of the various groups involved in this industry. Given this, we turn toward the nature of the boundary work and objects used to support exchange across these diverse participants.

4.2 Process Controls

Within the call center, organizational processes are facilitated by numerous IT systems, such as shared databases of customer information, call routing, load balancing, monitoring etc. The systems mediate the exchange of information, advice, outsourcer scripts, and call center work across the heterogeneous interests involved in the call center industry.

Standardized scripts prompted by information systems are meant to preserve and manage key managerial objectives in call center conversations, including the necessity to deal with calls both quickly and effectively. In organizations where such scripts are not provided, other "quality" practices are used, such as the checking of 30 or 40 conversational elements in recorded calls. The parameters include greeting, proper addressing of customer, identifying customer queries, providing relevant advice, identifying sales opportunities, etc. The analysis is then fed back to the agent, through their coach or team leader. As the head of quality at a call center commented,

> *It allows teams of people who listen to calls to identify systems aspect problems and various processes and then give feedback back to management. So the focus here is on very specific processes.*

The work involved in designing the systems to measure every second of an agent's time are extensive and almost panopticon. However, even the most stringent boundary object requires adjustment. Although the management is generally interested in the quantity of calls, they have also increased their focus on quality because too much quantity can be detrimental to quality. In order to counteract the growing quantity but decreased quality of calls, small teams have organized weekly competitions in order to encourage both higher quality and quantity targets, recognizing those who manage to produce minimal quality triggers. The result is a revised system of boundary objects in order to manage the complexity of *quality* that involves considerable effort from all groups.

4.3 Agent Training

Quality considerations also moves us into the numerous and complex practices and logics required to achieve it. Outsourcers typically relocate call centers based on the simple premise that achieving a certain quality of interaction among Western native English speakers and English speaking Indians is possible with a huge cost savings to the organization. However, the call is affected by numerous cultural, language, and accent differences with the customers. With only process knowledge, agent conversations with Western customers involve two complex and competing objectives. They have to work at keeping the conversation natural, with a neutral accent, while also exchanging relevant knowledge within a certain time period.

To do so, significant work is required in call center training. Agent training generally lasts six weeks, and spans general as well as call and conversational skills. Organizations began by training agents with American accents, but many now use a neutral accent as a result of difficulties in realizing a pure regional accent. Instead, combined with a neutral accent, they train agents to choose and use common phrases familiar to target customers, rather than mimicking a specific customer accent.

Through emphasizing these softer skills, the training involves difficult and complex boundary work to capture and represent a culture. One head of quality suggests that as the industry matures, the responses of Indian call center agents will begin to match Western expectations:

> *[Quality] doesn't have to do with the [call] process. But it has to do with the confidence level, it has to do with a lot of software issues, confidence with which the majority of the people in the U.S. will approach the call or take the callers is much higher than a consultant [agent] over here.*

Despite the apparent dominating and controlling logic of call center training, the agents believe the training prepares them to talk to people from diverse cultures, and provides an opportunity for better work and career prospects. They are trained to speak in certain ways, look for cues in the customers' conversation, and familiarize themselves, where possible, with customer's local information such as sports, weather, etc. As a result, both the managers and the employees, through boundary work and objects involved in cultural and linguistic training, achieve their separate interests: standardized quality of service, the management of call quantities, and career mobility.

4.4 Operations Management

Despite the extensive training and boundary work in the organization, the industry has created new demands for societal and governmental work. For example, the new call centers were affected by poor public transport in large cities, which created difficulties in achieving operational stability during agent shift changes. As one industry analyst noted,

> *We had to arrange transport, and even organize lunch to ensure employees were available when their shifts began.*

A general manager at a call center commented,

> *People are picked up and dropped from their residence—door-to-door pickup. No other industry is doing that today. They get a free duty meal. This duty meal is checked on a regular basis; the dieticians control the amount of calories, the food committee which comprises of people in this organization who lay down what they like. So we kind of align ourselves, that we are here for you, right? And comfort is certainly very high.*

With rapid expansion in the industry, experienced employees are often poached, fueling competitive attrition and instability in workforce expertise. Management has found it difficult to comprehend the attrition despite the "comfortable" work environment in call centers. A coach in a call center commented that he does not understand why agents leave, since they are earning more than most fresh graduates in other jobs, and are employed by Fortune 500 companies. An agent, however, saw attrition as a result of the stress of routine work:

> *Basically call centers have thousand employees; you can make people take calls for a year, two years, [then] there should be a lateral shift...that is why you have this attrition rate....The stress level is also very high.*

In response, significant managerial work is now focused on addressing the high attrition rates through increased incentives and modifications to organizational practices and labeling. For example, new job titles were added to create the illusion of career progression, along with small but regular increments to pay. Perhaps not surprising, neither has generally solved the attrition problem.

Recently, emphasis has been placed on doing what call centers have rarely done in the past: creating small teams with team leads, in order to mentor and foster agents' needs. One coach commented about his relationship with his team:

> *The coach plays a very important role in keeping the team intact. Half of them who stay in the company, they love their coaches' work....My team...they just look forward to the week offs...we hire a transport or a cab and go to Kaveri fishing camps, overnight stay in forests, and things like that.*

Despite management perceptions of call center work as comfortable, with good salaries and a good position for surplus graduates, the attraction and retention of agents remains the biggest source of boundary work in the call center industry's ability to continue and expand. With increasing salaries as one of the few remaining options, the industry may eventually undermine the original reasons for the industry's creation—low cost—and either disappear or mutate into another industrial form (Caldwell 2002).

5 CONTRIBUTIONS AND CONCLUSION

This discussion raises a number of contributions. The call center industry in India is a contested space where the various and diverse interests among the groups are negotiated and temporarily connected through boundary work and boundary objects. These boundary objects often include embedded IT artefacts in an attempt to replicate and stabilize work practices and their effects. The result is more of a negotiated truce than a stable order, rendered possible through the temporary use of boundary objects and work.

Our work illustrates that the easy replication of call center work to other low-cost regions hides the extensive boundary work and boundary objects required to navigate across intergroup ties within the industry. The case illustrates a shifting set of boundary "fronts," where managers, outsourcers, outsourcees, and employees are involved in the shifting features of this service industry.

Boundary work and object perspectives on call center work also provide a new perspective on a purely dystopian and utopian view of call center work. In the dystopian case, workers are restricted and imprisoned individuals with few options, while in the utopian case, workers have pure and unrestricted agency to pursue a knowledge-based career. Our case shows that mechanistic attempts to render the dystopian sweatshop by management have produced high employee turnover and the poaching of employees to other firms. This suggests that the employee is not a complete prisoner of circumstances. At the same time, the utopian views, which suggest that call centers are an important part of the new Indian economy and a stepping-stone for development, need to experience the hard and monotonous conditions of call center work, driven by a need to satisfy quantity (i.e., standardization) and cost-related interests. The boundary work and object view of call centers illustrates many emergent and competing outcomes for the heterogeneous interests of numerous groups, and the attempts to reconcile these diverse interests through boundary work and boundary objects. In this case, information technology has increased the reach of remote practices and logics from Western companies, which both affect and are affected and transformed by the social and technical settings in which they are recovered.

Here, the IT artefact plays a political role in a complex game of attraction and separation, in the post-structural possibilities of various "productive" engagements, made possible by the engagement of various groups. The work here is immense and complex, and the cost-centered hopes of Western companies and the career aspirations of Indian call center workers confront a complex reconciliation of their diverse interests. The boundary work depends and affects the emergent and somewhat unpredictable mixing and translation of diverse group interests, both disciplined and undisciplined, by the exchange across boundary object systems, so that their interests are perceived to be furthered by their continued relationships. This provides a revised direction and focus for call center, service sector, and information systems research and practice.

References

Belt, V., Richardson, R., and Webster, J. 2000. "Women's Work in the Information Economy: The Case of Telephone Call Centres," *Information, Communication and Society* (3:3), pp. 366-385.

Caldwell, B. 2002. "Outsourcing Cost Reduction Creates Paradox: How to Still Make a Profit," Gartner Dataquest, Stamford, CT.

Callaghan, G., and Thompson, P. 2002. "We Recruit Attitude: The Selection and Shaping of Routine Call Centre Labour," *Journal of Management Studies* (39:2), pp. 233-254.

Chiasson, M. W., and Davidson, E. 2005. "Taking Industry Seriously in Information Systems Research," MIS Quarterly. 29 (4), pp. 591-605

Dossani, R., and Kenney, M. 2003. "Went for Cost, Stayed for Quality? Moving the Back Office to India," *Berkeley Roundtable on the International Economy*, Paper BRIEWP156, August 7 (http://repositories.cdlib.org/brie/BRIEWP156).

Fernie, S., and Metcalf, D. 1997. "(Not) Hanging on the Telephone: Payments Systems in the New Sweatshops," Working Paper 891, Centre for Economic Performance, London School of Economics (http://cep.lse.ac.uk/pubs/download/dp0390.pdf).

Frenkel, S. J., Tam, M., Korczynski, M., and Shire, K. 1998. "Beyond Bureaucracy? Work Organization in Call Centers," *International Journal of Human Resource Management* (9:6), pp. 957-979.

Gasson, S. 2006. "A Genealogical Study of Boundary-Spanning IS Design," *European Journal of Information Systems* (15:1), pp. 26-41.

Karsten, H., Lyytinen, K., Hurskainen, M., and Koskelainen, T. 2001. "Crossing Boundaries and Conscripting Participation: Representing and Integrating Knowledge in a Paper Machinery Project," *European Journal of Information Systems* (10:2), pp. 89-98.

Krishna, S., Sahay, S., and Walsham, G. 2004. "Managing Cross-Cultural Issues in Global Software Outsourcing," *Communications of the ACM* (47:4), pp. 62-66.

Richardson, H. J., and Howcroft, D. 2006. "The Contradictions of CRM—A Critical Lens on Call Centres," *Information and Organization* (16:1), pp. 56-81.

Star, S. L., and Griesemer, J. R. 1989. "Institutional Ecology 'Translations' and Boundary Objects: Amateurs and Professionals in Berkeley's Museum of Vertebrate Zoology, 1907-39," *Social Studies of Science* (19), pp. 387-420.

Taylor, P., and Bain, P. 1999. "'An Assembly Line in the Head': Work and Employee Relations in the Call Cente," *Industrial Relations Journal* (30:2), pp. 101-117.
Walsham, G. 2006. "Doing Interpretive Research," *European Journal of Information Systems* (15:3), pp. 320-330.
Yin R K. 2003. *Case Study Research: Design and Methods* (3rd ed.), Beverly Hills, CA: Sage Publications.

About the Authors

Paul Devadoss is a lecturer at the Department of Management Science, Lancaster University Management School, UK. He completed his Ph.D. in Information Systems at the School of Computing in the National University of Singapore. His research interests include enterprise systems and e-governments. In particular, he is interested in the social impacts of IT use in organizational settings and the managerial implications of technology use. He has previously published in journals such as *Decision Support Systems, MIS Quarterly Executive, Communications of the AIS, Information and Management*, and *IEEE Transactions on IT in Biomedicine*. Paul can be reached by e-mail at paul@davadoss.org.

Mike Chiasson is currently an AIM (Advanced Institute of Management) Innovation Fellow and a Senior Lecturer at Lancaster University''s Management School, in the Department of Management Science. Before joining Lancaster University, he was an associate professor in the Haskayne School of Business, University of Calgary, and a postdoctoral fellow at the Institute for Health Promotion Research at the University of British Columbia. His research examines how social context affects IS development and implementation, using a range of social theories (actor network theory, structuration theory, critical social theory, ethnomethodology, communicative action, power knowledge, deconstruction, and institutional theory). In studying these questions, he has examined various development and implementation issues (privacy, user involvement, diffusion, outsourcing, cyber-crime, and system development conflict) within medical, legal, engineering, entrepreneurial, and governmental settings. Most of his work has been qualitative in nature, with a strong emphasis on participant observation. Mike can be reached at m.chiasson@lancaster.ac.uk.

Part 5:

Panels

25 TURNING PRODUCTS INTO SERVICES AND SERVICES INTO PRODUCTS: Contradictory Implications of Information Technology in the Service Economy

Neil C. Ramiller
Portland State University
Portland, OR U.S.A.

Elizabeth Davidson
University of Hawaii at Manoa
Honolulu, HI U.S.A.

Erica L. Wagner
Cornell University
Ithaca, NY U.S.A.

Steve Sawyer
The Pennsylvania State University
University Park, PA U.S.A.

1 INTRODUCTION

Service industry sectors of modern economies are growing rapidly, in absolute size and in comparison to the manufacturing, agriculture, and other economic sectors. Implicit in the dramatic proclamations that have accompanied this worldwide shift to the service economy is a subtext about the displacement and subordination of products. The notion of displacement speaks, in a straightforward way, to the diminishing relative importance that material products play in production and consumption and, hence, in

Please use the following format when citing this chapter:

Ramiller, N. C., Davidson, E., Wagner, E. L., and Sawyer, S., 2008, in IFIP International Federation for Information Processing, Volume 267, Information Technology in the Service Economy: Challenges and Possibilities for the 21st Century, eds. Barrett, M., Davidson, E., Middleton, C., and DeGross, J. (Boston: Springer), pp. 343-348.

providing opportunities for employment. That information and communication technology (ICT) enabled innovation has played a crucial role in the emerging dominance of services is well known, for example, in all aspects of financial service sectors. Subordination is more interesting because it has to do with the changing and unsettled *relationship* between products and services and, indeed, ambiguity in the very definition of "service" itself. This subordination has taken a variety of forms:

➔ Producers and purveyors of products add information-intensive service dimensions in order to enhance relationships with customers and, in some cases, to generate network effects among the customers themselves (e.g., Amazon.com's online customer reviews of books and data-based book recommendations).

➔ Products themselves have become servicitized in the sense that a product comes with a service component that is crucial in fully constituting the product as a meaningful "solution" for the purchaser (Vandermerwe and Rada 1988). The delivery of these components is often ICT-enabled (e.g., call centers in support of consumer electronic products).

➔ We witness the servicitization of products in an alternative sense, where the purchaser never takes possession of the material product but rather subscribes to the service that the product provides, with the purveyor maintaining ownership of the physical asset (e.g., ASP models of computer or software use and outsourcing of certain other kinds).

This third service phenomenon is a central element in IBM's and others' emerging interest in "services science" and the proposed academic discipline of "service science, management, and engineering" (SSME) (Chesbrough and Spohrer 2006; Rust and Miu 2006; Sheehan 2006)[1]. As that particular discourse tries to craft accounts of the conversion of products into services, how the ownership of assets is situated appears to be an important issue. However, when we set ownership aside, we are left with the tantalizing implication that the true, extractable value of a product is only ever realized in the services that the product provides, even where the beneficiary holds title to the product. Accordingly, some researchers have gone on to argue that *all* economies are service economies, and that the division between services and products is not meaningful. Instead, it is the creation of value-added services that is the basis for exchange (Vargo and Lush 2004, 2008).

Formulations that portray products as amplified by, completed by, transformed into, or reconceptualized as services reflect important economic changes. However, in this panel we further argue that *services* are now sometimes taking on aspects of products. Accordingly, even as we witness the servicitization of products, we can also observe the *productization of services*. By this we refer especially to the introductions of technical artifacts into traditional service exchanges, and the implications that these new intermediaries have for the nature of service production. That is, if we consider the traditional

---

[1]For more about IBM's efforts to define, guide and advocate for SSME, see http://www.research.ibm.com/ssme/ for definitions, discussions, and links.

conceptualization of service as a change in the service consumer's state resulting from activities of the service provider that are jointly consummated, then productization might be evident in a variety of ways, for example,

- the inclusion of tangible and intangible goods, and reliance on ICT, as part of the service;
- the substitution of artifacts or artificial agents (with service knowledge standardized and encoded) for service provider or service consumer;
- removal of the temporal connection between production and consumption of the service, and hence the interaction of provider and consumer.

A simple example is a meal at McDonalds: the service effort—a meal—is distilled, measured, and packaged into numbered value-meals and sizes. Self-service fast food is productized and successful. A more relevant example is the bundling of iTunes with the iPod. The significance of the service is realized through its productization. More broadly, the ICT consulting activity: legions of capable young programmers and smooth business analysts swarming through your organization to create the information system of your dreams (or budget) reflects service via *service productization.*

Simply, ICT play a—if not the—crucial role in this reciprocal transformation. The productization of services is observed in the way that various kinds of artifacts based on ICT (much of this involving the Internet) have come increasingly to occupy and even to dominate the interaction space in which customers are called upon to cocreate service exchanges. These artifacts, acting as delegates for the organizations that offer the services, have in some cases replaced traditional human service providers; in other cases, they mediate a greater proportion of the interaction between humans; and in still other cases, they make service exchanges possible that had not previously been available or, perhaps, even envisioned.

This productization (which we might also call *artifactualization*) of services holds forth the promise of standardization, scalability, cost reduction, and geographic reach—outcomes of obvious appeal to managers in the provider firms. The results for service recipients may be more mixed, with convenience, speed, and reliability often being positive outcomes, but inflexibility and a lack of responsiveness also being possible. In particular, the novel, complex, recondite, obdurate, or exceptional service episode can place the customer at odds with an artifact-delegate's limited response set (and the provider's inscribed interests). In these situations, service recipients may have the means, with persistence and ingenuity, to transcend design assumptions and thus to create and experience a more robust and satisfying service encounter. However, we suggest that their interactions might be more akin to the manipulation of a product and exploration of its features, than to the negotiated exchange we more typically associated with service encounters. In this sense, what we as researchers and managers have learned about standardization, scalability, and cost reduction in the design, production, and distribution of products might apply to many ICT-enabled services.

Panelists will discuss the implications of artifacts-in-services in a set of industries/application domains, using a range of theoretical perspectives as noted in the next section. Building from these, the intent of the panel is to explore productization as the flipside of servicitization. And, in doing this, to examine the accompanying ambiguity in the semantics of *service*" (Bar 2001; Bresnehan and Greenstein 2001). Such an examination will surface interesting social, technological, and economic issues, confront both expectations and beliefs that the producers and consumers of services might encounter in developing these transformations, and challenge those who pursue research on services to conceptualize more coherently both servicitization and productization (e.g., Metka et al. 2006; Wolff 2002).

2 THE PANEL

Neil Ramiller will explore the mediating role of computer-based artifacts in distance education, in a context where maintaining a high level of instructor–student interaction is an expectation (in contrast, for example, to conventional computer-based training). His experience in teaching a distance-learning MBA core course in information technology will provide material for discussion. Actor–network theory will be used as the interpretive frame.

Elizabeth Davidson will examine how information technologies are being used in the delivery of healthcare services, not only to collect and distribute medical and patient information during service delivery, but as mediators in the interactions among medical professionals and with their patients, and even as substitutes for provider or patient (or both) in some interactions. She will consider how advancing technologies may challenge the institutionalized identities and legitimized roles of providers and patients, and the implications these changes may have for field-level change in healthcare.

Erica Wagner will focus on the hospitality sector, one of the largest services industries in the world, as the context for understanding the mutual shaping involved in the productization of service experiences. Using a social shaping lens, she will consider how the production of artifacts has been necessary to move toward the "self service" trend that is dominating airlines, lodging and restaurant segments. In turn, consumer responses to the productization of their service experience will be considered. Service recipients themselves are using web 2.0 technologies such as consumer-generated content sites like TripAdvisor to create delegates that have begun shaping the way hotels perform their services.

Steve Sawyer will draw from his ongoing work on the computerization of real estate to highlight two points of relevance to the artifactualization of services. Building on concepts of computerization he will use principles of economic sociology to make two points. First, he will highlight how different mobilizing rhetorics regarding the role of a service, and who participates, leads to conflicting views on who is providing which services to buyers and sellers. Second, in doing this he will contrast the social activities of real estate agent's work with the value-adding offerings of several online purveyors of real estate data.

References

Bar, F. 2001. "The Construction of Marketplace Architecture," in *Tracking a Transformation: E-Commerce and the Terms of Competition in Industries*, The BRIE-IGCC Economy Project Task Force on the Internet, Washington, DC: Brookings Institution Press, pp. 27-49.

Bresnehan, T., and Greenstein, S. 2001. "The Economic Contribution of Information Technology: Towards Comparative a User Studies," *Journal of Evolutionary Economics* (11), pp. 95-118.
Chesbrough, H., and Spohrer, J. 2006. "A Research Manifesto for Services Science," *Communications of the ACM* (49:7), pp. 35-49.
Metka, S., Jaklic, A., and Kotnik, P. 2006. "Exploiting ICT Potential in Service Firms in Transition Economies," *The Service Industries Journal* (26:3), pp. 287-299.
Rust, R., and Miu, C. 2006. "What Academic Research Tells Us About Service," *Communications of the ACM* (49:7), pp. 49-54.
Sheehan, J. 2006. "Understanding Service Sector Innovation," *Communications of the ACM* (49:7), pp. 42-47.
Vandermerwe, S., and Rada, J. 1988. "Servicitization of Business: Adding Value by Adding Services," *European Management Journal* (6:4), pp. 314-324.
Vargo, S., and Lusch, R. 2004. "Evolving to a New Dominant Logic for Marketing," *Journal of Marketing* (68), pp. 1-17.
Vargo, S., and Lusch, R. 2008. "Service-Dominant Logic: Continuing the Evolution," *Journal of the Academy of Marketing Science* (36:1), pp. 1-10.
Wolff, E. 2002. "Productivity, Computerization and Skill Change," *Federal Reserve Bank of Atlanta Economic Review*, pp. 63-78.

About the Panelists

Neil Ramiller is the Ahlbrandt Professor in the Management of Innovation and Technology at Portland State University's School of Business Administration. His primary research activities address the management of information technology innovations, with a particular focus on the role that rhetoric, narrative, and discourse play in shaping innovation processes and negotiating multi-party interests within organizations and across interorganizational fields. He also conducts work on the social construction of information technology scholarship, and the implementation of the "linguistic turn" in information technology studies. Neil has presented his work at a variety of national and international conferences, and published articles in a number of journals, including *Journal of the Association for Information Systems*, *MIS Quarterly*, *Information and Organization*, *Information Technology & People*, *Organization Science*, *Journal of Management Information Systems*, and *Information Systems Research*. Neil can be reached at neilr@sba.pdx.edu.

Elizabeth Davidson is the W. Ruel Johnson Distinguished Professor of Information Technology Management and Department Chair at the Shidler College of Business, University of Hawaii at Manoa. In her research, she has examined social structure change associated with computerized physician order entry in hospitals and barriers to adoption of electronic medical records by small physician practices. Elizabeth currently serves as research director for an action research program attempting to build a community of practice around implementation of health information technology and its use to improve patient outcomes among small practices in Hawaii. She also serves as an associate editor for *European Journal of Information Systems* and *MIS Quarterly* and on the editorial boards of *Information and Organizations* and *Information Technology & People*. Elizabeth can be reached at edavidso@hawaii.edu.

Erica Wagner is an assistant professor of Information Systems at Cornell University's School of Hotel Administration. She earned her Ph.D. from the London School of Economics and has an undergraduate degree in accounting. Her research interests focus on the ways software is "made to work" within different organizational contexts. The implication of this focus is theoretical development related to how technology is accepted within organizations, even when it is initially seen as problematic. Her research has been published in a variety of outlets including *Information and Organization*, *Journal of the Association for Information Systems*, *Communi-*

cations of the ACM, and the *Journal of Strategic Information Systems*. Erica can be reached at elw32@cornell.edu.

Steve Sawyer is a founding member and an associate professor at the Pennsylvania State University's College of Information Sciences and Technology. Steve holds affiliate appointments in the Department of Management and Organization; the Department of Labor Studies and Employer Relations; and the program in Science, Technology and Society. Steve does social and organizational informatics research with a particular focus on people working together using information and communication technologies. Steve can be reached at sawyer@ist.psu.edu.

26 INFORMATION SYSTEMS AND THE SERVICE ECONOMY: A Multidimensional Perspective

Steven Alter
University of San Francisco, U.S.A.

Uri Gal
University of Aarhus, Denmark

David Lipien
IBM Corporation

Kalle Lyytinen
Case Western Reserve University, U.S.A.

Nancy Russo
Northern Illinois University, U.S.A.

1 INTRODUCTION

This panel will examine the impact of the growth of the service economy on organizations and information systems from four perspectives: (1) internal changes in organizations, both service providers and service clients, in terms of their structures, processes, and competencies; (2) redefinition of interorganizational relationships and redrawing of organizational boundaries and identities; (3) the role of IS in enabling these new collaborative relationships; and (4) the possibility of designing better applications to enhance organizations' capacity to engage in service exchanges.

Please use the following format when citing this chapter:

Alter, S., Gal, U., Lipien, D., Lyytinen, K., and Russo, N. L., 2008, in IFIP International Federation for Information Processing, Volume 267, Information Technology in the Service Economy: Challenges and Possibilities for the 21st Century, eds. Barrett, M., Davidson, E., Middleton, C., and DeGross, J. (Boston: Springer), pp. 349-352.

2 PANEL DESCRIPTION

We live in a post-manufacturing world. Throughout most of the industrial era, the output of manufacturing has been product. The key to success was the ability of an organization to make more products cheaper than most other producers. In recent decades, many organizations are facing increased competition from producers that can make products as well as or better than they can and for lower prices. To stay competitive, many organizations have to redefine their business to emphasize the services they offer rather than the product they manufacture. This implies using the products they make for their customers as a platform to integrate a comprehensive set of services and processes to meet their clients' needs. These services represent a relationship between the provider and the client, comprised not just of technology but of people as well (Maglio et al. 2006). Today, nearly 80 percent of our economic activity consists of service jobs that comprise the *service economy* (Chesbrough and Spohrer 2006).

To become service oriented and increase service productivity, many organizations have reorganized their basic structure, created new organizational roles, and retrained their employees (or hired new ones). Additionally, they have reengineered their business processes, redefined their core competencies, and reconfigured their supply chains and relationships with key customers and suppliers. Indeed, changes to the provider–customer relationship are central to the emergence of the service economy. When products were the main focus of the exchange, they constituted a concrete mechanism, or boundary object, that facilitated interorganizational communication and reduced uncertainties. As products, their embedded technical standards, and their functionalities became well understood, suppliers did not need to understand the customers' business to become an exchange partner. Similarly, customers did not have to understand their providers' previous experiences and expertise as these were reflected in tangible products (Chesbrough and Spohrer 2006). The delivery of services is qualitatively different in that it lacks a concrete artifact that mediates the relationship. It involves, instead, an ongoing exchange of intangible assets between a provider and an adopter in which both parties play an active part. In this exchange, each party needs the other's knowledge in negotiating its involvement and role.

> The provider lacks the contextual knowledge of the customer's business, and how the customer is going to leverage the offering to compete more effectively....The customer lacks the knowledge of the full capabilities of the provider's technologies, and the experience of the provider from other transactions in assessing what will work best (Chesbrough and Spohrer 2006, p. 37).

Service exchanges, therefore, require a closer and more comprehensive familiarity between the involved parties and are often built around long-term relationships that last over the life time of an enterprise.

Successful organizations are able to integrate interorganizational and intra-organizational processes and configure them according to changing client expectations (Zhao et al. 2007). However, creating the information systems to support these processes is no simple matter. Traditional design and development tools were not constructed to model such complex systems, nor do we have standard theory or methods available for formal representation or evaluation of these systems (Maglio et al. 2006).

This raises a number of important issues that we wish to address in this panel.

Nancy Russo will open the panel with examples of new types of collaborative relationships between service providers and customers. She will highlight some of the main challenges organizations have faced when adjusting to these relationships and discuss how organizations have structured themselves internally to respond to these new conditions.

Kalle Lyytinen will address the following question that arises from the previous discussion: Do these collaborative relationships imply recharting organizational boundaries? If so, are there implications for the way organizations represent and define themselves, their goals, and their mission? Are there implications for organizational identities?

David Lipien will discuss the role of information systems in enabling these new collaborative relationships and examine their capacity to support broader and more comprehensive interorganizational interactions.

Steven Alter will inquire whether we can design better applications to enhance an organization's capacity to engage in service exchanges. Steve is the developer of the work system method, which has been proposed as a solution to the problem of how to model service systems.

Uri Gal will serve as the panel moderator.

References

Chesbrough, H., and Spohrer, J. 2006. "A Research Manifesto for Services Science," *Communications of the ACM* (49:7), pp. 35-40.

Maglio. P. P., Srinivasan, S., Kreulen, J. T., and Spohrer, J. 2006. "Service Systems, Service Scientists, SSME, and Innovation," *Communications of the ACM* (49:7), pp. 81-85.

Zhao, J. L. Tanniru, M. , and Zhang, L-J. 2007. "Services Computing as the Foundation of Enterprise Agility: Overview of Recent Advances and Introduction to the Special Issue," *Information Systems Frontiers* (9:1), pp. 1-8.

About the Panelists

Steven Alter is a professor of Information Systems at the University of San Francisco. He received a Ph.D. from MIT, taught at the University of Southern California, and was vice president of the manufacturing software firm Consilium before joining USF. His research for over a decade has concerned developing systems analysis concepts and methods that can be used by typical business professionals and can support communication with IT professionals. His latest book, *The Work System Method: Connecting People, Processes, and IT for Business Results*, is a distillation and significant extension of ideas in four editions (1992, 1996, 1999, 2002) of his information system textbook. His articles have been published in *Harvard Business Review, Sloan Management Review, MIS Quarterly, IBM Systems Journal, Communications of the Association for Information Systems*, and other journals and conference proceedings. Steve can be reached at stevenalter@comcast.net.

Uri Gal is an assistant professor of Information Systems at the Aarhus School of Business at the University of Aarhus. He holds a Ph.D. in Information Systems from Case Western Reserve University and an M.Sc. degree in Organizational Psychology from the London School of Economics and Political Science. His research takes a social view of organizational processes in the context of the implementation and use of information systems. He is particularly interested in the relationships between people and technology in organizations, and the changes in the nature of work practices, organizational identities, and interactions associated with the introduction of new information technologies. Uri can be reached by e-mail at urig@asb.dk.

David Lipien is a Senior Managing Consultant with IBM's Global Business Services Financial Services Insurance practice. He has over 10 years of IT systems development and leadership experience. He has expertise in all phases of the project life cycle and experience in numerous roles, including delivery executive, project manager, senior business analyst, and programmer. His specialties include Internet-based technologies, wireless, and object-based project methodologies. David is a PMI Certified Project Management Professional and holds a Bachelor of Science in Operations Management and Information Systems from Northern Illinois University and a Master of Business Administration from the University of Notre Dame. He has also completed an Executive Management Program at the Harvard Business School, and has published articles on the topic of enterprise software release management in IBM's international technical resource website, developerworks.com. David can be reached at lipien@us.ibm.com.

Kalle Lyytinen is the Iris S. Wolstein Professor at Case Western Reserve University, and adjunct professor at University of Jyväskylä, Finland. He serves currently on the editorial boards of several leading IS journals including *Journal of AIS* (currently as editor-in-chief*), Journal of Strategic Information Systems, Information & Organization, Requirements Engineering Journal, Information Systems Journal, Scandinavian Journal of Information Systems,* and *Information Technology and People,* among others. He is an AIS fellow (2004), the former chairperson of IFIP 8.2, and a founding member of SIGSAND. He has published over 150 scientific articles and conference papers and edited or written 10 books on topics related to nature of the IS discipline, system design, method engineering, organizational implementation, risk assessment, computer supported cooperative work, standardization, and ubiquitous computing. He is currently involved in research projects that looks at the IT induced radical innovation in software development, IT innovations in architecture, engineering, and the construction industry, design and use of ubiquitous applications, and the adoption of broadband wireless services in the U.K., South Korea, and the United States. Kalle can be reached at kalle@po.cwru.edu.

Nancy L. Russo is the Pavlović Professor of Information Systems at the Northern Illinois University's College of Business, and Partnership Director at Slobomir P University in the Republic of Srpska, Bosnia. She received her Ph.D. in Management Information Systems from Georgia State University. In addition to studies of the use and customization of system development methods, her research has addressed IT innovation, research methods, and IS education issues. Her work has appeared in *Information Systems Journal, European Journal of Information Systems, Information Technology & People*, and *Communications of the ACM.* Nancy serves as vice chair of the IFIP WG 8.2 (Information Systems and Organizations). She can be reached at nrusso@niu.edu.

27 THE SERVICITIZATION OF PEER PRODUCTION: Reflections on the Open Source Software Experience

Joseph Feller
University College Cork, Ireland

Patrick Finnegan
University College Cork, Ireland

Björn Lundell
University of Skövde, Sweden

Olof Nilsson
Mid Sweden University, Sweden

The concept of what Yochai Benkler called "peer production" as an alternative mechanism to traditional hierarchies and markets has captured the imagination of numerous communities in contexts ranging from t-shirt design to software to gold mining. While some question the suitability and potential longevity of this mode of production, others are focused on determining ways in which peer-produced products and services can be suitably packaged to meet the requirements of consumers. In particular, the mature peer production phenomenon known as open source software has emerged as a credible alternative to its proprietary counterpart and presents a compelling challenge to both industry and academia as we seek to understand how firms and other organizations can build sustainable business models leveraging the public commons of open source products and the collaborative engine that created them.

This panel will debate how intra- and interorganizational service offerings can be used to create and capture value from peer-produced intellectual property such as open source software. Drawing on their research in the secondary software, software services,

Please use the following format when citing this chapter:

Feller, J., Finnegan, P., Lundell, B., and Nilsson, O., 2008, in IFIP International Federation for Information Processing, Volume 267, Information Technology in the Service Economy: Challenges and Possibilities for the 21st Century, eds. Barrett, M., Davidson, E., Middleton, C., and DeGross, J. (Boston: Springer), pp. 353-355.

electronic business, and public administration sectors, the panellists will discuss how open source has been used to transform the value offering to consumers and citizens as well as the consequential business model changes for these organizations. In particular, the panelists will discuss the manner in which organizations

- utilize open source frameworks, tools and methods to build systems in an agile, rapid, and economically efficient manner
- build competitive strategies around delivering cocreated services that exploit open source
- adopt open source licensing structures to leverage the power of community-based development, promote particular platforms and standards, grow market- and mind-share, and steward public funds responsibly

In extrapolating from these experiences, the panellists will consider the need for private and public organizations to adopt sophisticated business strategies and models to

- offer services that exploit the various forms of peer-produced intellectual property
- manage the reliance on interorganizational/networked business dynamics
- maintain delicate relationships with peer production communities

The panel will include short position statements by the panellists and a dialogue between them on these positions, followed by a discussion with the audience. Significant time will be allocated to an open discussion with the audience, and those attending are invited to raise their own concerns/opinions in relation to developing service-offerings that exploit peer production.

About the Panelists

Joseph Feller (JFeller@afis.ucc.ie) is a senior lecturer in Business Information Systems, University College Cork, Ireland. He has written or edited three books on the topic of open source software and has published his research on the topic in a variety of international conferences and journals as well as practitioner publications. His paper (with Brian Fitzgerald), "A Framework Analysis of the Open Source Software Development Paradigm," was awarded Best Paper on Conference Theme at the 21st International Conference on Information Systems. He was program chair for the IEE/ACM workshop series on Open Source Software Engineering (2001-2005), program cochair of the Third International Conference of Open Source Systems, and has been a speaker/panelist on the topic of open source at academic conferences, industry workshops, and European Commission briefings and roundtables. He led, with Finnegan and Lundell, a work package in the EU FP6 CALIBRE project (2004-2006) and is currently a principle investigator in the Open Code, Content and Commerce (03C) Business Models Research Project, funded by the Irish Research Council for the Humanities and Social Sciences.

Patrick Finnegan (P.Finnegan@ucc.ie) received his Ph.D. from the University of Warwick, England, and is currently a senior lecturer in Management Information Systems at University College Cork, Ireland. His research on business models and interorganizational systems has been published in journals including the *Information Systems Journal*, *Information, Technology & People*, *Journal of Electronic Commerce*, and *Electronic Markets*. His interest in open source comes from two perspectives: (1) understanding the business models that are required/facilitated by open source, and (2) understanding the effective management of the inter-organisational cooperation needed to develop and exploit open source. He led, with Feller and Lundell, a work package in the EU FP6 CALIBRE project (2004-2006) and is currently a principle investigator in the Open Code, Content and Commerce (03C) Business Models Research Project, funded by the Irish Research Council for the Humanities and Social Sciences.

Björn Lundell (bjorn.lundell@his.se) has been a staff member at the University of Skövde, Sweden, since 1984, and has been researching the open source phenomenon for a number of years. He led, with Feller and Finnegan, a work package in the EU FP6 CALIBRE project (2004-2006) and is currently the technical manager in the industrial research project COSI (2005-2008), involving analysis of the adoption of open source practices within companies. His research is reported in a variety of international journals and conferences. He is a founding member of the IFIP Working Group 2.13 on Open Source Software, and the founding chair of Open Source Sweden, an industry association established by Swedish open source companies. He is the organizer of the Fifth International Conference of Open Source Systems (OSS 2009), which is to be held in Skövde, Sweden. In addition, his research has also included fundamental research on evaluation, and associated method support.

Olof Nilsson (O.Nilsson@ucc.ie) is a senior lecturer in Social Informatics at the Mid Sweden University and a Research Fellow at University College Cork, Ireland. His research focuses on access to public information systems, and he has taken an active part in two "triple helix" projects, developing open source applications for public authorities. His research on access has been published in *International Journal of Public Information Systems* and in a forthcoming edition of *International Journal for Humanistic and Social Computing*, and also in a number of international conference proceedings. He is currently researching how open source facilitates changes to the traditional approach to public service and government in Sweden.

28 eHEALTH: Redefining Health Care in the Light of Technology

Mike W. Chiasson
Lancaster University, Lancaster, U.K.

Donal Flynn
University of Manchester, Manchester, U.K.

Bonnie Kaplan
Yale University, New Haven, CT U.S.A.

Pascale Lehoux
Université de Montréal, Montreal, Canada

Cynthia LeRouge
St. Louis University, St. Louis, MO U.S.A.

1 INTRODUCTION

eHealth is the use of emerging Information and Communication Technology, especially the Internet, to improve or enable health and healthcare. (Eng 2004)

Information and communication technology is now the major enabler for healthcare organizations on many levels– national, regional and local–hoping to achieve structural and cultural change in healthcare provision; for example, the UK's NPfIT (National Program for IT) and the National Health Information Initiative in the United States. Final NPfIT costs are variously estimated from £12billion to £31 billion. Major initiatives are also underway in other developed as well as developing economies to address healthcare issues with eHealth technologies.

Please use the following format when citing this chapter:

Chiasson, M., Flynn, D., Kaplan, B., Lehoux, P., and LeRouge, C., 2008, in IFIP International Federation for Information Processing, Volume 267, Information Technology in the Service Economy: Challenges and Possibilities for the 21st Century, eds. Barrett, M., Davidson, E., Middleton, C., and DeGross, J. (Boston: Springer), pp. 357-362.

There are four main drivers of change in healthcare today:

- the new consumer (e.g., demanding information, control, choice, and service)
- the new science (of evidence-based medicine)
- the new technology (e.g., the Internet, monitoring devices and sensors, smart phone applications)
- the new focus on quality (e.g., the reduction of medical mistakes, comparative performance measurement, and focus on outcomes)

The drivers are combining to place new demands on healthcare organizations and providers who are faced with developing more efficient and effective healthcare delivery systems in the face of challenges such as ageing populations, who are expecting an extended life expectancy to be of high quality, and worldwide increase in chronic disease such as diabetes. Such systems are increasingly based on advances in ICT, changes in social demographics resulting in wider access to various technologies, and more informed patient demand. These changes are both contributing to and resulting in greater emphasis on community healthcare, patient self-management, and private healthcare provision. eHealth has thus both a technological as well as a social dimension, involving the use of ICT for health and a patient-centered approach to health.

While considerable advantages are expected for eHealth initiatives, these trends present both practice and research challenges. Considerable expertise is needed to design, develop, and evaluate ICT-based eHealth systems to suit a wide range of stakeholder perspectives in a variety of social contexts and changing patterns of health care delivery. Few health administrators and even fewer healthcare providers have the required mix of technological and social expertise. Medical informatics, management, and information systems researchers, too, face challenges. eHealth systems pose an interesting challenge to an organization-centric view of interacting social and technological systems. The different values and perceptions that individuals bring to bear on the patient–provider relationship, and the identification of social and cultural influences on patients' health beliefs and behaviors, including the use of ICT for eHealth purposes, pose policy, design, and deployment challenges. Another challenge is that, although qualitative methods are increasingly accepted, in many circles methodological standards in health care still regard the RCT (randomized control trial) as the gold standard and it is difficult to demonstrate that a particular element of an eHealth intervention, typically a complex, multi-element techno-social system, is a significant shaping influence on the chosen outcomes.

Thus, difficult and challenging issues arise as ICT is increasingly used in the expectation of coproducing health care and health care delivery. Examining what is occurring in healthcare points to emerging issues relevant to information systems researchers in other domains as well. This panel will address these issues from a variety of disciplinary and national perspectives by drawing implications for IS research from studies involving several different eHealth applications. A common theme is the nature and extent of change, individual and institutional, required for successful eHealth implementation.

Among the topics panelists will explore are:

- An overview of eHealth research areas, covering a range of stakeholder perspectives, with a focus on patients and healthcare providers. Advantages will be summarized, followed by a discussion of the evidence for improvements, leading to a summary of problems encountered in eHealth implementations.

- The redefinition of healthcare and its implications for patients (e.g., empowerment, culture, access) and the patient–provider relationship (e.g., nature of the relationship, provider education).
- The conflict between a range of institutional normative pressures and the values underlying eHealth initiatives, and the impact on evolving system designs.
- The extent and nature of changes required to traditional patient and provider roles and workflow and the different perspectives concerning a successful eHealth "encounter."
- From a technological innovation perspective, designers embed their assumptions about chronic illness into eHealth systems, and a narrow, efficiency-based managerial and clinical perspective can create as much as it can solve the problem of chronic illness.

2 PANELISTS AND TOPICS

Mike W. Chiasson is currently an Advanced Institute for Management Research (AIM) fellow and a senior lecturer in the Department of Management Science at Lancaster University's Management School. His research examines how social context affects IS development and implementation, using a range of social theories (actor network theory, structuration theory, critical social theory, ethnomethodology, communicative action, power-knowledge, deconstruction, and institutional theory). In studying these questions, he has examined various development and implementation issues (privacy, user involvement, diffusion, outsourcing, cyber-crime, and system development conflict) within medical, legal, engineering, entrepreneurial, and governmental settings. Mike can be reached at m.chiasson@lancaster.ac.uk.

Mike's discussion will focus on the various types of institutional pressures that affect development and use risks during eHealth initiatives aimed at reconfiguring patient–provider interaction (coproduction). His example will include the implementation of web-based systems to support communication between diabetes patients and their providers (nurses and physicians). While the institutional influences in health care are considered to some extent elsewhere, the timing and appearance of these influences and their management during eHealth projects, raise some potentially new insights. Four institutional pressures — hygienic, beneficial, effort-based, and equitable—will be discussed. He will also illustrate how the various eHealth initiatives were designed, shaped, and pitched in an attempt to address these institutional pressures.

Donal Flynn is a senior lecturer at the Manchester Business School and has research interests in the social and psychological processes that underlie the interaction between information systems and their human and organizational contexts. In 2005, he was a visiting research scholar in the Computer Information Systems Department, Robinson School of Business, Georgia State University in Atlanta. Currently, his focus is on eHealth, investigating interactive eHealth systems such as patient decision aids and web-based co-management systems for long-term conditions. Donal can be reached at donal.flynn@manchester.ac.uk.

Donal will present an overview of the different eHealth research areas, outlining their relationship to the different levels of stakeholders such as patient, healthcare provider, primary, secondary, and tertiary care institutions, healthcare planning and commissioning authority, and national epidemiological research institute. An information therapy viewpoint will be taken for the patient perspective while the provider perspective will outline clinical decision support, electronic health record, and CPOE, professional education and infrastructure systems. Having set the scene, the advantages (both predicted and realized) of eHealth will be summarized, followed by a discussion of the existing evidence for improvements resulting from eHealth. This will lead to a summary of the problems encountered in eHealth implementation and some important issues raised. Finally, some interesting points relevant to the 8.2 community will be noted.

Bonnie Kaplan is a lecturer in Medical Informatics at the Yale School of Medicine and Yale College, and is an adjunct clinical professor of Biomedical and Health Information Sciences at the University of Illinois at Chicago. A Fellow of the American College of Medical Informatics and recipient of the American Medical Informatics Association President's Award, she is chair-elect of both the American Medical Informatics Association Working Groups on People and Organizational Issues and on Ethical, Legal, and Social Issues. Bonnie co-chaired the IFIP 8.2 conference resulting in the book *Information Systems Research: Relevant Theory and Informed Practice* (Kluwer, 2004). Her research interests include evaluation, organizational, social, and ethical issues concerning new technologies, especially in health care. Bonnie can be reached at bonnie.kaplan@yale.edu.

The coproduction of service in health care represents a change to increased patient-centric care and collaboration between providers and patients. It is also seen as a way of redefining health care through a global network of information and services that increases patient access, autonomy, and empowerment.

Bonnie's presentation draws on years of research and experience in medical informatics and information systems to reflect on the research and ethical issues arising from the redefinition of the patient/health care consumer role in light of these emerging systems and changing processes in health care. In particular, the following will be explored: (1) consumer issues, such as what constitutes consent, empowerment, and autonomy—who is being empowered to do what; (2) what affects access and use; (3) organizational issues of reengineering healthcare globally, such as structural changes and role changes in health care delivery; and (4) work-life issues, such as those involved in changes in the patient–provider relationship, and in health care workers' roles and skill levels. All of these have implications for information systems research in domains other than health care. They raise concerns about design, such as how to take account of differences among individuals, populations, national structures, and stakeholder viewpoints, as well as the values embedded in technology and design approaches; empowering consumers and workers; and research ethics.

Pascale Lehoux obtained her Ph.D. in Public Health from the University of Montreal (Quebec, Canada) in 1996. An associate professor in the Department of Health Administration and a researcher with the Groupe de Recherche Interdisciplinaire en Santé (GRIS) at the University of Montreal, Pascale currently holds a Canada Research Chair on Innovation in Health (2005-2010). Her CRC program examines upstream factors that have an impact on the ultimate use and dissemination of health technologies (e.g., the design process itself, including needs analyses, design strategies, market constraints and opportunities, and group perceptions and practices guiding the innovation processes). She has published more than 50 papers examining the use of computerized medical records, telemedicine, scientific knowledge, home care equipment, and mobile and satellite dialysis units. Routledge published her book, *The Problem of Health Technology*, in 2006. Pascale can be reached at pascale.lehoux@umontreal.ca.

Pascale's presentation critically explores the role of technological innovation in the constitution of chronic states and illness. Drawing on the co-construction of technology and society perspective, the presentation focuses specifically on the way innovation designers envisage the enhancement of the chronically ill and build certain kinds of sociotechnical configuration to deal with chronic illness.

Using the case of "intelligent distance patient monitoring" as an illustration, Pascale argues that technology creates as much as it solves the problem of chronic illness. Technology is recursively embedded in chronic illness and it generates dual effects: it constrains and sustains users' daily practices. In a context where lack of financial and human resources dominates, managerial and clinical aims are steering the design process in a way that may first and foremost seek efficiency and provide a narrow response to chronic patients' concerns and preferences. Only by recognizing technology's duality, and eventually transcending it, will research and policy initiatives be able to deal creatively and responsibly with the design of our future health experiences.

Cynthia LeRouge is an associate professor in the Decision Sciences and Information Technology Management Department at St. Louis University. Cynthia has held various management roles in practice including roles in the software and healthcare industries prior to joining academe. Her current research interests relate to health care information systems, and in particular telemedicine and consumer informatics. She has over 60 publications including academic journal articles, edited chapters in research-based books, and peer-reviewed conference proceedings. Cynthia has co-chaired health care mini-tracks for various information systems conferences. She has also served as a special guest editor for the *European Journal of Information Systems*' special issue on Health Information Systems Research, Revelations and Visions. For the past year, she has actively worked as an executive board member of the Association for Information Systems special interest group for Healthcare Research and currently serves as the chair. Cynthia can be reached at lerougec@slu.edu.

There is a rising call in management literatures to revisit classic work design and process theories, given modern work technologies, practice, and roles in health care and other industries. Cynthia's presentation will cover the topic of coproducing new forms of technology-mediated service, specifically focusing on telemedicine (in the form of medical exams using high-end, real-time video conferencing and peripheral devices). These encounters may span the spectrum of contact from one-time exams with a specialist to enhanced encounters, where a patient and caregiver may meet via video conferencing on a periodic basis with multiple providers not colocated for the management of chronic diseases, such as AIDS. This phenomenon represents real-time coproduction of health service. Introducing technology as a mediator to the health delivery process initiates elements of change to the traditional roles of patients and providers, the medical exam workflow process, and the patient–physician communication process; ironically, it can also exacerbate some issues found in traditional service delivery. Although there is considerable overlap in what patients and providers define as key mechanisms that drive a successful medical video conferencing encounter process, there are differences that underscore the need for exploring multiple perspectives of health technologies in research and practice. The discussion will be based on a U.S. field study of medical video conferencing (telemedicine) and highlight the process, people, and mechanisms involved in coproducing the telemedicine encounter.

Part 6:

Workshop Paper Contributions

29 THE INFORMATION SERVICES VIEW

Matt Germonprez
Information Systems, College of Business
University of Wisconsin – Eau Claire
Eau Claire, WI U.S.A.

Dirk Hovorka
Information Systems, Leeds School of Business
University of Colorado – Boulder
Boulder, CO U.S.A.

ABSTRACT

The *information services view* engenders a conceptual shift from the provision of defined and predetermined services to an environment that enables users to select and integrate information services in the ongoing creation and recreation of unique information systems. The information services view (ISV) conceptualizes technology as an ensemble of facilities that perform an action or function on the users' behalf. The vision of ISV is the realization of user-enabled, real-time production of *ad hoc* information systems.

The ISV specifies that users of services are intelligent actors who are able to compute seamlessly across contexts and recognizes that services developers may not know how their services are going to be used but instead develop a reflective environment where users' thinking and redesign is supported. The ISV represents a dramatic shift in design from provision of a fixed, externally controlled service set to design of "a space of potential for human concern and action" (Winograd and Flores 1986, p. 37).

The ISV is a supplementary view of technology (Orlikowski and Iacono 2001) which focuses on the realization of flexible service development that engages users as secondary developers for which the technology bar was previously unacceptably high. Although new configurations for information systems can result from versioning or specific design changes, the evolutionary trajectory and evolving nature of systems is commonly the result of user-initiated mutability (Gregor and Jones 2007) or tailoring (Germonprez et al. 2007). The ISV embraces processes by which developers/providers expose information and allow user-initiated selection and configuration of services that fit "the idea of the arising of something from out of itself, or emergent properties, and behavior" (Gregor and Jones 2007, p. 326).

Please use the following format when citing this chapter:

Germonprez, M., and Hovorka, D., 2008, in IFIP International Federation for Information Processing, Volume 267, Information Technology in the Service Economy: Challenges and Possibilities for the 21st Century, eds. Barrett, M., Davidson, E., Middleton, C., and DeGross, J. (Boston: Springer), pp. 365-366.

The ISV distinguishes between the initial *design* and the *ways of doing design* and requires that attention be paid to the different experiences, perceptions, intentions, and goals that the user will use to recreate the design of the information system. It also aims to create a phenomenological *potential for action* in which the user tailors the information system and develops uses in new contexts or for new tasks (Germonprez et al. 2007). The ISV moves away from a dominant approach in systems design to over-engineer the information technology artifact through a restricted set of data structures, interfaces, and reporting systems, so that a limited range of work practices are allowed. By standardizing information gathering and presentation, many approaches produce and reproduce error by restricting the ability of users to reflexively and skillfully adjust their practices and computing systems to support changing goals, use patterns, and tasks. The ISV requires the support of classes of tasks, use patterns, recognizable conventions and components, and metaphors that the end user reflects on and engages during use. The ISV suggests that it is incumbent on designers to build a flexible, holistic picture of what services are and how they can create novel recombinant information systems.

References

Germonprez, M., Hovorka, D., and Collopy, F. 2007. "A Theory of Tailorable Technology Design," *Journal of the Association for Information Systems* (8:6), pp. 315-367.

Gregor, S., and Jones, D. 2007. "The Anatomy of a Design Theory," *Journal of the Association for Information Systems* (8:5). pp. 312-335.

Orlikowski, W., and Iacono, C. 2001. "Desperately Seeking the 'IT' in IT Research –A Call to Theorizing the IT Artifact," *Information Systems Research* (12:2), pp. 121-134.

Winograd, T., and Flores, F. 1986. *Understanding Computers and Cognition: A New Foundation for Design*, Norwood, NJ: Ablex Publishing Corporation.

About the Authors

Matt Germonprez is an assistant professor of Information Systems at the University of Wisconsin – Eau Claire. He received his Ph.D. at the University of Colorado in Boulder in 2002. His research interests are in the domain of human–computer interaction with a secondary interest in Information Systems theory. He is a member of IFIP 8.2, ACM, and AIS, and currently serves on the executive council for the AIS Special Interest Group for Human–Computer Interaction. He has published work in *Journal of the AIS, Communications of the AIS, Organization Studies, International Journal of IT Standards and Standardization Research, Designing Ubiquitous Information Environments: Socio-Technical Issues and Challenges* (IFIP 8.2), and *Information Systems Research: Relevant Theory and Informed Practice* (IFIP 8.2). Matt can be reached at germonr@ uwec.edu.

Dirk S. Hovorka is currently a Scholar in Residence at the Leeds School of Business, University of Colorado at Boulder. He attended Williams College in Massachusetts for his B.A., holds an M.S. in Geology and an M.S. in Interdisciplinary Telecommunications, and received his Ph.D. in Information Systems from the University of Colorado. His research includes the philosophical foundations of IS research, the development of design theory, the evolving role of information systems in science, and the influences of social networks on knowledge management. He has published research in the *Journal of the AIS, European Journal of Information Systems, Communication of the AIS, Business Agility and Information Technology Diffusion* (IFIP 8.6), and *Perspectives on Information Management: Setting the Scene*. Dirk can be reached at dirk.hovorka@ colorado.edu.

30 AN EPISTEMOLOGY OF ORGANIZATIONAL EMERGENCE: The Tripartite Domains of Organizational Discourse and the Servitization of IBM

Michelle Carter
Department of Management
Clemson University
Clemson, SC U.S.A.

Hirotoshi Takeda
Computer Information Systems
Georgia State University
Atlanta, GA U.S.A.
Centre de Recherche en Management et Organisation
University of Paris Dauphine
Paris, France

Duane Truex
Computer Information Systems
Georgia State University
Atlanta, GA U.S.A.

ABSTRACT

This paper draws from 21 years of discourse to examine a narrative about IBM's transition to a service-oriented company. Covering three leadership eras during a period of sweeping change for IBM and the information technology industry, this discourse, found in the IBM Corporation's annual reports, in illustrates the emergence of policy, technology, and business models in one of the largest and most influential IT companies in the world. Our purpose in drawing from these texts is twofold: (1) to provide a more

Please use the following format when citing this chapter:

Carter, M., Takeda, H., and Truex, D., 2008, in IFIP International Federation for Information Processing, Volume 267, Information Technology in the Service Economy: Challenges and Possibilities for the 21st Century, eds. Barrett, M., Davidson, E., Middleton, C., and DeGross, J. (Boston: Springer), pp. 367-370.

thorough discussion of the notion of "emergence" in IT organizational settings, and (2) to introduce a fuller process model of how emergence is manifest in organizational discourse than is currently present.

In much of the information systems literature, the term *emergence* has been informally used in describing organizational contexts and the process of IS development (Markus and Robey 1988; Orlikowski 1996; Pfeffer and Leblebici, 1977). In three papers, Truex and his colleagues formally describe and situate a theory of emergence in the discourse on ISD methods (Truex and Baskerville 1998, Truex, Baskerville, and Klein 1999; Truex, Baskerville, and Travis 2000). They liken ISD to "emergent grammars" in a linguistic system. However, they stop short of developing a full epistemology of the notion and provide little more than analogical and descriptive examples grounded in linguist Paul Hopper's (1987, 1988) emergent grammar hypothesis. The incomplete development of the epistemology and an ontology of the emergence construct has proven problematic for scholars attempting to apply emergence theory in practice (Bello et al. 2002). While researchers or practitioners might find the idea of emergent organizations inviting, without descriptive and explanatory models, the concept is difficult to use in the practice or study of information systems. Accordingly, this paper seeks to contribute to the development of a theory of emergence.

We draw from the organizational communication and organizational discourse literature. In a subset of this community, scholars have advanced theories on the nature of organization as a discursive construction. For them, discourse is the very foundation on which "organization" is built (Fairhurst and Putnam 2004; Heracleous 2006; Heracleous and Barrett 2001; Taylor and Robichaud 2004; Taylor and Van Every 2000). Using this meta-theoretical framework, we explore how emergence arises through an examination of IBM's annual reports and industry-level discourses, which were, in turn, influenced in part by the IBM declarations and subsequent behavioral changes.

We introduce a new process model of organizational emergence by extending and addressing shortcomings in a set of current perspectives in the literature. The tripartite domain model identifies three domains—context, task, and negotiation-at-hand—as integral components of any concrete occurrence of discourse. To test its efficacy, we apply the tripartite domain model *post hoc* to a longitudinal set of IBM Corporation data. The tripartite domain model provides a lens to examine the servitization of IBM and, in the process, illustrates the emergent discourse on the notion of "service" and on the evolution of the meaning of "customer" in the IBM dataset.

References

Bello, M., Sorrentino, M., and Virili, F. 2002. "Web Services and Emergent Organizations: Opportunities and Challenges for IS Development," in *Information Systems and the Future of the Digital Economy: 10th European Conference on Information Systems*, S. Wrycza (ed.), Gdansk, Poland, June 6-8.

Fairhurst, G. T., and Putnam, L. 2004. "Organizations as Discursive Constructions," *Communication Theory* (14), pp. 5-26.

Heracleous, L. 2006. *Discourse, Interpretation, Organization,* Cambridge, UK: Cambridge University Press.

Heracleous, L., and Barrett, M. 2001. "Organizational Change as Discourse: Communicative Actions and Deep Structures in the Context of Information Technology Implementation," *Academy of Management Journal* (44), pp. 755-778.

Hopper, P. 1987. "Emergent Grammar," *Berkeley Linguistics Society* (13), pp. 139-157.

Hopper, P. 1988. "Emergent Grammar and the a priori Grammar Postulate," in *Linguistics in Context: Connecting Observation and Understanding*, Vol. XXIX, D. Tannen (ed.), Norwood, NJ: Ablex Publishing, pp. 117-134.

Markus, M. L., and Robey, D. 1988. "Information Technology and Organizational Change: Causal Structure in Theory and Research," *Management Science* (34), pp. 583-598.

Orlikowski, W. J. 1996. "Improvising Organizational Transformation Over Time: A Situated Change Perspective," *Information Systems Research* (7), pp. 63-92.

Pfeffer, J., and Leblebici, H. 1977. "Information Technology and Organizational Structure," *The Pacific Sociological Review* (20), pp. 241-261.

Taylor, J. R., and Robichaud, D. 2004. "Finding the Organization in the Communication: Discourse as Action and Sensemaking," *Organization*, (11), pp. 395-413.

Taylor, J. R., and Van Every, E. J. 2000. *The Emergent Organization: Communication as Its Site and Surface*, Mahwah, NJ: Lawrence Erlbaum Associates.

Truex, D. P., and Baskerville, R. 1998. "Deep Structure or Emergence Theory: Contrasting Theoretical Foundations for Information Systems Development," *Information Systems Journal* (8), pp. 99-118.

Truex, D. P., Baskerville, R., and Klein, H. 1999. "Growing Systems in Emergent Organizations," *Communications of the ACM* (42), pp. 117-123.

Truex, D. P., Baskerville, R., and Travis, J. 2000. "Amethodical Systems Development: The Deferred Meaning of Systems Development Methods," *Accounting Management and Information Technologies* (10), pp. 53-79.

About the Authors

Michelle Carter is pursuing her Ph.D at Clemson University. She has Master' degrees in Computer Science (Anglia Polytechnic University, UK) and Management Information Systems (Georgia State University). Her research interests, informed by a decade in IS development, include managing project risk through requirements validation, how organizations emerge in IS-related communicative actions, and technology adoption in developing economies. Her research has appeared in the proceedings of the SAIS, AMCIS, IFIP OASIS, and IFIP WG.8.2. Michelle can be reached at mscarte@clemson.edu.

Hirotoshi Takeda is a Graduate Advancement Program Fellow and Ph.D. student in Computer Information Systems at Georgia State University and the University of Paris Dauphine. He has degrees in electrical engineering and computer science from the University of California, Irvine, a Master's of Electrical Engineering from Georgia Institute of Technology, and an MBA from Southern Methodist University. His research interests include discourse analysis, mobile computing, bibliometrics and knowledge management. His research has appeared in the proceedings of the SAIS, UKAIS, ISECON, IFIP WG 8.2, ICIS SIG-ED, and SIG MIS. Hirotoshi can be reached at htakeda@cis.gsu.edu.

Duane Truex, an associate professor in the Department of Computer Information Systems and the Institute for International Business at Georgia State University, is interested in the social impacts of IS on social organizations and how emergent properties of organizations may be reflected in emergent ISD. Duane is active in the IFIP WG 8.2 and 8.6 communities, is an associate editor of *Information Systems Journal*, has edited for *Database for Advances in Information Systems*, and serves on several other journal editorial boards. His work has appeared in the *Communications of the ACM, Accounting Management and Information Technologies, Communications of the AIS, DataBase, European Journal of Information Systems (EJIS), Information Systems Journal, Journal of the AIS, Journal of Arts Management and Law, IEEE Transactions on Engineering Management, Scandinavian Journal of Information Systems, Le Journal Systè mes d'Information et Management,* and more than 60 IFIP transactions and edited books and conference proceedings. Duane can be reached at dtruex@gsu.edu.

31 ORGANIZATIONAL LEARNING IN HEALTH CARE: Situating Free and Open Source Software

Gianluca Miscione
Margunn Aanestad
Department of Informatics
University of Oslo
Oslo, Norway

ABSTRACT

Free and open source software (FOSS) has been attracting the interest of organizations involved in the development and implementation of information and communication technologies (ICTs) in developing countries for years. ICTs for development initiatives often have public sector orientations, as governments' ICT policies are expected to shape and support socio-economical development. *The usual mismatch between formal bureaucracies' functioning, the usual top-down software development schemes, and the actual trajectories of development initiatives* (mostly run by international agencies) *provides a promising empirical field.* This paper intends to discuss the *connection between FOSS and organizational learning* in contexts where the usual assumptions about them cannot be taken for granted. It is argued that the relevance of open technologies as public goods is in allowing organizational learning in public administration. Such a focus on the organizational aspects would complement existing studies on the economical relevance of FOSS.

The argument is built by addressing FOSS-related emphatic expectations for emancipation in the "knowledge society" on one side (Government of Kerala 2002), and implementation and use, on the other. Then, a meso-level between global trends and local specificities is identified as crucial in situating FOSS for development potentialities. Empirically, this level is between the two usual poles in information systems studies: decision makers (public administrators and software developers, both oriented by a top-

Please use the following format when citing this chapter:

Miscione, G., and Aanestad, M., 2008, in IFIP International Federation for Information Processing, Volume 267, Information Technology in the Service Economy: Challenges and Possibilities for the 21st Century, eds. Barrett, M., Davidson, E., Middleton, C., and DeGross, J. (Boston: Springer), pp. 371-373.

down approach to systems design) and the ground of implementation (usually sensitive to a variety of contexts). As it is unusual in developing contexts to have spontaneous voluntary participation, the software development process needs to be designed and carried out in a way that allows local organizations to "indigenize" FOSS.[1] Its fluidity allows inscribing a variety of context-bound socio-technical arrangements (De Laet and Mol 2000), and also can cause avoidance of path-dependencies and vendor lock-ins (Weerawarana and Weeratunge 2004).

The case of a health information system being implemented in Kerala, as part of an international initiative, is presented. We describe, on one side, the principles and views supporting the network and local politics and, on the other side, aspects of the implementation dynamic, which is underestimated in the common approach to ICT for development (Avgerou 2007).

The project presented in the paper has significant links both at the global level (participating in a broad and heterogeneous network of trend-setting organizations like universities and research centers, international donors, ministries of different countries) and local levels (where systems are piloted and implemented, capacity building is carried out, requirements for further developments are collected). Empirical exploration showed that the (formal and informal) institutional constraints, which FOSS implies and relies on, are fragmented or absent, whereas others can be relevant. Nevertheless, *FOSS narrative proves to be present and effective both in negotiations between stakeholders, and in facilitating local participation to information system development.*[2]

The meso-level position of this initiative shows the distance between the two ends, descriptively. Prescriptively, it suggests how possible bridges can allow interorganizational relations across a variety of actors rarely involved in the same FOSS initiative. The interactions around local technical skills improvement and *the increased ability for organizations to formulate, express, negotiate, and inscribe their needs in technology is proposed as a chance for organizational learning.* In contexts of multiple accountabilities (Suchman 2002), we claim that *the relevance of FOSS emerges from negotiating alliances,* and does not inhere in FOSS itself. FOSS facilitates learning as far as its openness is allowed by software development processes, and enacted by brokering activities to relate dispersed practices (Gherardi and Nicolini 2002).

References

Avgerou C. 2007. "Information Systems in Developing Countries: A Critical Research Review," Working Paper Series, Innovation Group, London School of Economics and Political Sciences, October.

Camara G., and Fonseca F. 1997. "Information Policies and Open Source Software in Developing Countries," *Journal of the American Society for Information Science and Technology* (58:1), pp. 121-132.

---

[1]Camara and Fonseca (2007) relate modalities of participation to code writing and software modularity.

[2]Myths and narratives are discussed by Czarniawska (1997) in neoinstitutional terms. The legitimizing role of myth is clearly presented by Noir and Walsham (2007), also through a case from Kerala.

Czarniawska B. 1997. *Narrating the Organization: Dramas of Institutional Identity*, Chicago: University of Chicago Press, Chicago.

De Laet, M., and Mol, A. 2000. "The Zimbabwe Bush Pump: Mechanics of a Fluid Technology," *Social Studies of Science* (30), pp. 225-263.

Gherardi, S., and Nicolini, D. 2002. "Learning in a Constellation of Interconnected Practices: Canon or Dissonance," *Journal of Management Studies* (39:4), pp. 419-436.

Government of Kerala. 2007. "Information Technology Policy: Towards an Inclusive Knowledge Society" (http://www.keralaitmission.org/web/main/ITPolicy-2007.pdf)

Noir, C., and Walsham G. 2007. "The Great Legitimizer: ICT as Myth and Ceremony in the Indian Healthcare Sector," *Information, Technology & People* (20:4), pp. 313-333.

Suchman, L. 2002. "Located Accountabilities in Technology Production," *Scandinavian Journal of Information Systems* (14:2), pp. 91-105.

Weerawarana, S., and Weeratunge, J. 2004. *Open Source in Developing Countries*, Swedish International Development Cooperation Agency, January (http://www.eldis.org/fulltext/opensource.pdf).

About the Authors

Gianluca Miscione received his Ph.D. in Information Systems and Organization from the Sociology Department of the Universityof Trento (Italy) with a dissertation focused on the interplay between information and communication technologies and health care change in "developing" contexts. At the University of Oslo, his research focuses on information infrastructures and organizational studies. Gianluca can be reached at gianluca.miscione@gmail.com.

Margunn Aanestad is a researcher at the Department of Informatics, University of Oslo. She worked in the health care and telecommunications industries before her doctoral study of surgical medicine. Margunn's research interests are broadly related to large-scale information infrastructures, specifically in health care. She can be reached at margunn@ifi.uio.no.

32 UNDERSTANDING THE EXCHANGE INTENTION OF AN INDIVIDUAL BLOGGER

Wee-Kek Tan
Department of Information Systems
National University of Singapore
Singapore

Chuan-Hoo Tan
Department of Information Systems
City University of Hong Kong
Hong Kong

Hock-Hai Teo
Department of Information Systems
National University of Singapore
Singapore

ABSTRACT

This research explores the issue of how a blogring, a circle/community of blogs with a common theme (Xanga 2008; Chua and Xu 2007), could be utilized to form a loosely distributed exchange of products for monetary and nonmonetary returns. Specifically, we present a research model that identifies factors influencing a blogger's intention to participate in a *commercial exchange*, a commercial activity involving a blogger and a reader within a blogring.

A blogring is conceptualized as a *natural segmentation of user-alike* whereby members (i.e., bloggers and readers) converge toward common interests reflected in the blog contents (Kumar et al. 2004; Gumbrecht 2004). The formation of this social network provides bloggers with the social capital to engage in collaboration and cooperation with other members (Putnam 2000; Peece 2002). Economizing on this social capital, bloggers could engage in product, service, and/or money exchanges. We term this process the

Please use the following format when citing this chapter:

Tan, W-K., Tan, C-H., and Teo, H-H., 2008, in IFIP International Federation for Information Processing, Volume 267, Information Technology in the Service Economy: Challenges and Possibilities for the 21st Century, eds. Barrett, M., Davidson, E., Middleton, C., and DeGross, J. (Boston: Springer), pp. 375-378.

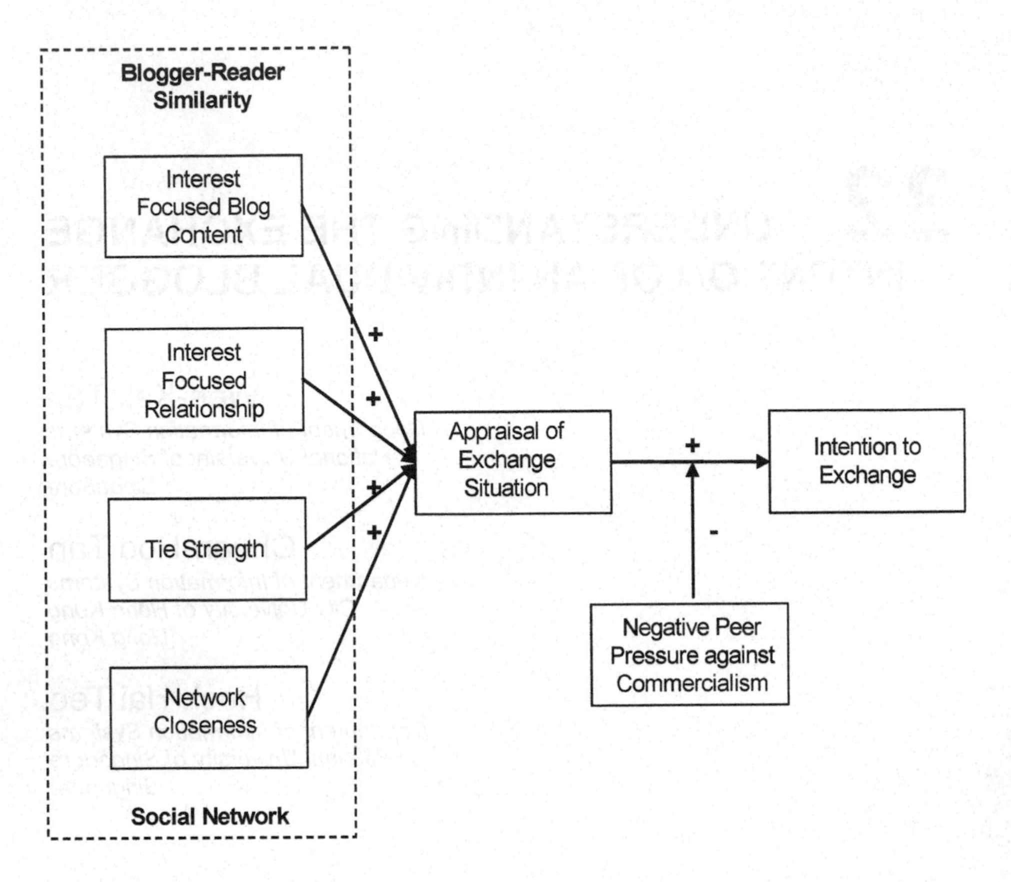

Figure 1. Research Model

economic leverage of personal blog (Balasubramanian and Mahajan 2001). Such an economic leverage includes selling, barter trading, and outright giving away items that can be identified or associated with the shared interests of the blogring. In other words, any exchange between bloggers and readers need not necessarily be for monetary returns but could also involve an exchange of knowledge or expertise in return for approval-related utility (Balasubramanian and Mahajan 2001). With this understanding, we next present our research model.

The research model (Figure 1) draws from three streams of literature. First, we reference the similarity attraction theory (Byrne 1971), which posits that people prefer to interact with like-minded others. For instance, salespeople generally prefer to sell to prospects sharing characteristics (e.g., gender and age) with themselves (Dwyer et al. 1998). Hence, we conjecture that bloggers could be more inclined to commercial exchange with readers who have similar interests. We term this *blogger-reader similarity*. Second, the general premise of the social network dictates that members could place higher priority on relationships and ties with other members within a blogring compared with individuals' personality attributes. Hence, we could posit that the stronger the tie (Frenzen and Davis 1990; Marsden and Campbell 1984) between a buyer and a seller, the higher the likelihood for purchase (Frenzen and Davis). Additionally, the closer a reader is to the blogger, the greater the ease of engaging in an exchange due to greater influence exerted by the blogger over the reader (Burkhardt 1994).

Third, building on the decision-making framework proposed by Luce, Bettman, and Payne (2001), we hypothesize that the decision of whether to engage in an exchange depends on the appraisal of the exchange situation, which in turn is influenced by blogger-reader similarity and social network. A positive appraisal should lead to a higher exchange intention from the blogger. However, any negative peer pressure against commercialism exerted by other blogring members could negatively moderate the appraisal (Luce et al. 2001), thus leading to lower exchange intention.

References

Balasubramanian, S., and Mahajan, V. 2001. "The Economic Leverage of the Virtual Community," *International Journal of Electronic Commerce* (5:3), Spring, pp. 103-138.

Burkhardt, M. E. 1994. "Social Interaction Effects Following a Technological Change: A Longitudinal Investigation," *Academy of Management Journal* (37:4), August, pp. 869-898.

Byrne, D. 1971. *The Attraction Paradigm*, New York: Academic Press.

Chau, M., and Xu, J. 2007. "Mining Communities and Their Relationships in Blogs: A Study of Online Hate Groups," *International Journal of Human-Computer Studies* (65:1), pp. 57-70.

Dwyer, S., Richard, O., and Shepherd, C. D. 1998. "An Exploratory Study of Gender and Age Matching in the Salesperson–Prospective Customer Dyad: Testing Similarity-Performance Predictions," *Journal of Personal Selling & Sales Management* (18:4), Fall, pp. 55-69.

Frenzen, J. K., and Davis, H. L. 1990. "Purchasing Behavior in Embedded Markets," *Journal of Consumer Research* (17), June, pp. 1-12.

Gumbrecht, M. 2004. "Blogs as 'Protected Space,'" presentation at the WWW 2004 Workshop on the Weblogging Ecosystem: Aggregation, Analysis and Dynamics, New York May 18.

Kumar, R., Novak, J., Raghavan, P., and Tomkins, A. 2004. "Structure and Evolution of Blogspace," *Communications of the ACM* (47:12), December, pp. 35-39.

Luce, M. F., Bettman, J. R., and Payne, J. W. *Emotional Decisions: Tradeoff Difficulty and Coping in Consumer Choice*, Chicago: The University of Chicago Press.

Marsden, P. V., and Campbell, K. E. 1984. "Measuring Tie Strength," *Social Forces* (63:2), December, pp. 482-501.

Preece, J. 2002. "Supporting Community and Building Social Capital," *Communications of the ACM* (45:4), April, pp. 37-39.

Putnam, R. D. 2000. *Bowling Alone: The Collapse and Revival of American Community*, New York: Simon & Schuster.

Xanga. 2008. "Xanga Help Blogrings FAQ," (http://www.xanga.com; accessed March 25, 2008).

About the Authors

Wee-Kek Tan is a graduate student at the National University of Singapore. His research interests include social computing, online decision aid design, and information systems development and education. He has published in conferences, such as ACM SIGMIS CPR. He can be reached by e-mail at tanwk@comp.nus.edu.sg.

Chuan-Hoo Tan is an assistant professor of Information Systems at the City University of Hong Kong. His research interests include agent design, online market institutions, and IT innovation adoption. He has published in reputable journals such as *IEEE Transactions of Engineering*

Management and **Communications of the ACM**, and conferences, such as the International Conference on Information Systems. He can be reached by e-mail at chuantan@cityu.edu.hk.

Hock-Hai Teo is an associate professor of Information Systems at the National University of Singapore. His research interests include IT innovation adoption, assimilation and impacts, information privacy, and electronic market institutions. Dr. Teo has published in many journals including *MIS Quarterly, Journal of MIS,* and *IEEE Transactions on Engineering Management*. He can be reached by e-mail at teohh@comp.nus.edu.sg.

33 TOWARD UNDERSTANDING THE CAPABILITY CYCLE OF SOFTWARE PROCESS IMPROVEMENT: A Case Study of a Software Service Company

Yu Tong
Lingling Xu
Shanling Pan
Department of Information Systems
National University of Singapore
Singapore

ABSTRACT

The emergence of a service economy facilitates proliferation of software service companies (SSCs), small firms offering various software services such as software maintenance, testing, and customization. Given the unique characteristics of SSCs (i.e., small-scale projects, limited resources, strong customer dependency, etc.), software process improvement (SPI) in SSCs faces additional challenges. Based on Helfat and Peteraf's framework of a dynamic capability lifecycle, this study aims to understand how dynamic capabilities are developed to facilitate SPI implementation, and how these capabilities can be maintained and transformed in a changing context. An in-depth interpretive case study was conducted in a SSC, namely SGSC, which is the an offshore software service center of a leading U.S. document management company, Xerox Corporation. Drawing insights from the framework, this study demonstrates the evolution of key resources and dynamic capabilities as well as their impacts on the success of the SPI project.

Responding to the ownership transfer from Xerox to Fuji Xerox group in 2005, the center decided to implement new SPI process. As shown in Figure 1, three capabilities

Please use the following format when citing this chapter:

Tong, Y., Xu, L., and Pan, S., 2008, in IFIP International Federation for Information Processing, Volume 267, Information Technology in the Service Economy: Challenges and Possibilities for the 21st Century, eds. Barrett, M., Davidson, E., Middleton, C., and DeGross, J. (Boston: Springer), pp. 379-381.

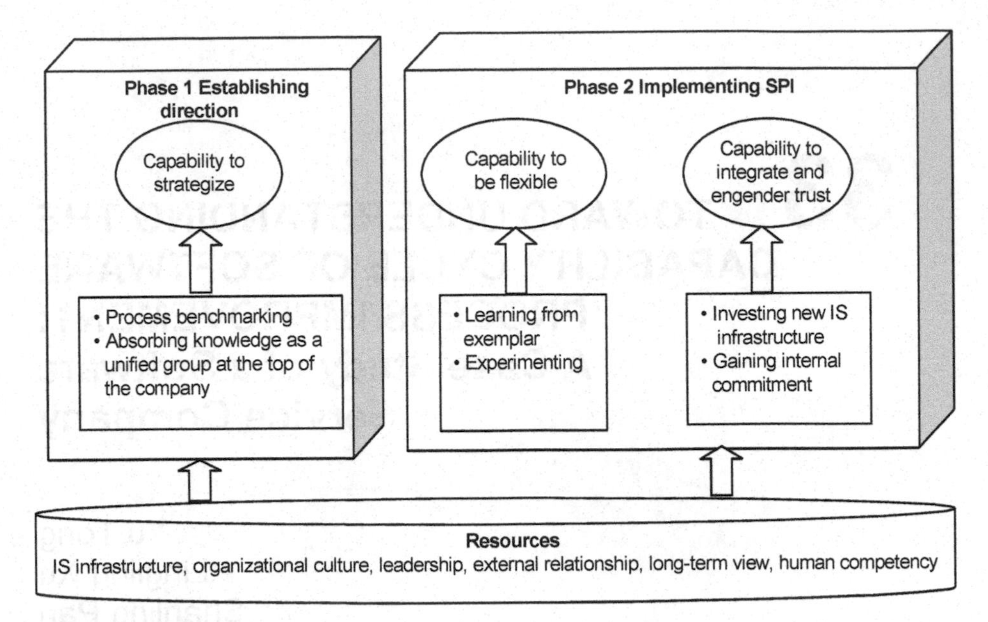

Figure 1. Capability Development

(i.e., *capability to strategize*, *capability to be flexible*, and *capability to integrate and engender trust*) were built up during the first two phases (i.e., establishing direction and implementing SPI). Development of each capability was accomplished through some actions with the support of various resources.

After development, capabilities can evolve over time (Zollo and Winter 2002). Two capabilities (*capability to strategize* and *capability of integrate and engender trust*) were maintained through regular exercise, which helps to refresh the company's memory and facilitate the creation of embedded knowledge. Besides being maintained, capabilities may also branch into different forms when external factors have a strong impact to alter the current development trajectory (Helfat and Peteraf 2003). At SGSC, ownership transfer resulted in a dramatic change in terms of external relationships and internal structure. With changes in the supporting resources, capabilities built up in previous phases were further transformed into different branches. *Capability to be flexible* was renewed because of better communication and sharing resources with external partners. *Capability to strategize* was also renewed after obtaining a long-term view from the new parent company. Figure 2 summarizes the life cycle of three capabilities in this study.

By integrating theoretical perspective with empirical evidence, this paper contributes to both researchers and practitioners. For researchers, it constitutes one of the first empirical studies to extend Helfat and Peteraf's general framework of the capability life cycle by demonstrating how capabilities can be transformed through altered resources over time. Moreover, this study advances the SPI literature by suggesting a conceptual framework for SPI implementation in a small-scale, service-oriented company. Practically, this study provides invaluable suggestions to managers on how firms can successfully implement SPI strategy when facing dramatic changes in the external environment or internal structure.

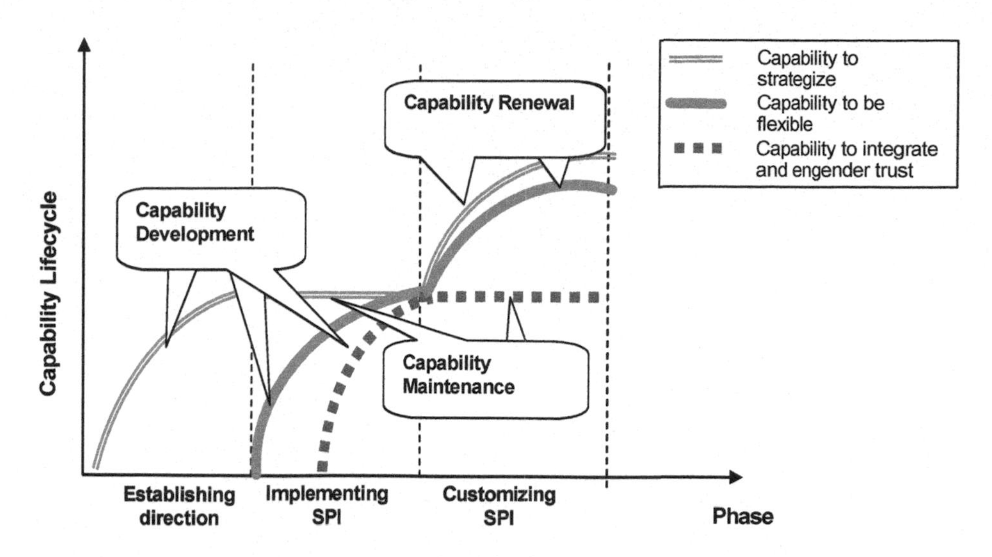

Figure 2. Capability Life cycle

References

Helfat, C. E., and Peteraf, M. A. 2004. "The Dynamic Resource-Based View: Capability Life-cycles," *Strategic Management Journal* (24:10), October, pp. 997-1010

Zollo, M., and Winter, S. G. 2002. "Deliberate Learning and the Evolution of Dynamic Capabilities," *Organization Science* (13:3), pp. 339-351.

About the Authors

Yu Tong is a Ph.D. candidate in the Department of Information Systems at the National University of Singapore. Her research interests include IT adoption and assimilation in organizations, virtual team management, and IT-mediated marketing communications. She can be reached by e-mail at tongyu@comp.nus.edu.sg.

Lingling Xu is a Ph.D. candidate in the Department of Information Systems at the National University of Singapore. Her research interests include human–computer interaction, social commerce, and IT adoption and implementation. She can be reached by e-mail at xulingling@comp.nus.edu.sg.

Shanling Pan is the coordinator of the Knowledge Management Laboratory in the Department of Information Systems, School of Computing, National University of Singapore. Dr. Pan's primary research focuses on the recursive interaction of organizations and information technology (enterprise systems), with particular emphasis on issues related to work practices, culture, and structures from a knowledge perspective.

34 A CASE STUDY APPROACH TO EXAMINING SERVICE INFORMATION REQUIREMENTS

Rachel Cuthbert
Paris Pennesi
Duncan McFarlane
Distributed Information and Automation Laboratory
Department of Engineering
University of Cambridge
Cambridge, UK

ABSTRACT

In this paper, we propose a case study approach to examine and assess the information required to underpin services for particular industrial service offerings. The focus of this paper is on the means by which service information requirements may be extracted and understood, as opposed to on how service information requirements are subsequently used.

The term *service information requirements* refers to a set of information needed to support the delivery of a service to a customer. The area of service information requirements is a new, and a less researched, area compared with product information requirements. Service is important for both product and service delivering organizations as companies move toward the provision of integrated solutions. Within manufacturing, one of the most significant trends is toward servitization, in particular for high-value, complex goods, where the focus of the product and service providers is on the associated service delivered.

Information is important in service as a means of enhancing decisions. The information has no direct value, but the impact of improved information quality can reduce costs or enhance service decisions. In the context of product servicing, the information can provide details about the condition and usage of the product. In a service delivery context, information provides the specification of the customer to enable service delivery decisions to be made.

Please use the following format when citing this chapter:

Cuthbert, R., Pennesi, P., and McFarlane, D., 2008, in IFIP International Federation for Information Processing, Volume 267, Information Technology in the Service Economy: Challenges and Possibilities for the 21st Century, eds. Barrett, M., Davidson, E., Middleton, C., and DeGross, J. (Boston: Springer), pp. 383-385.

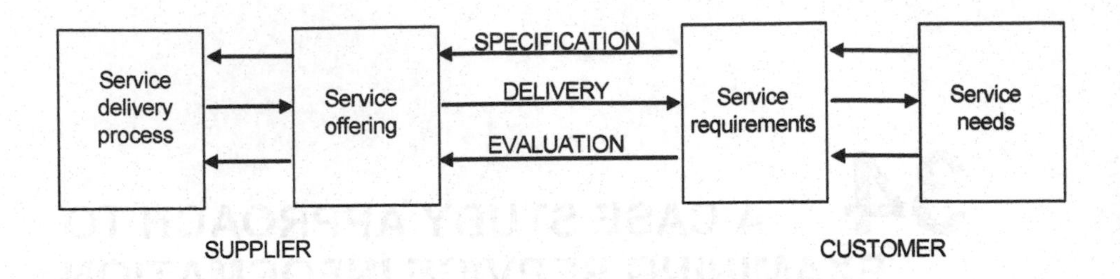

Figure 1. Service Information Model

This research proposes a process of obtaining service information requirements in order to determine the needs of the customer and to determine the information required for the supplier to deliver this service to the customer. To achieve this approach, we propose a service information model (see Figure 1).

The first stage of this approach is to identify the *specification process* or the information needed to translate the customer need into requirements in the form of a performance-based contract.

Having gained an understanding of the requirements of the service, the actual process for the *delivery* of the service is mapped to identify the information needed to deliver the service.

The *evaluation* stage determines the information needed to assess the service provided against that specified, and the means by which the measurements of service delivery are achieved.

Analysis of the approach should assess the role of information and its impact on the performance of the service delivery process as well as an understanding of the consequences of a variation in information quality throughout the requirements, delivery, and evaluation processes. From this analysis, two questions need to be answered.

(1) What is the impact of the information on the service performance?
(2) What is the current quality of the information?

The proposed approach to service information requirements collection, detailed in the paper, will test the effectiveness in determining the information requirements for service provision. This will be achieved through a series of cross-sector case studies. A further extension to this could assess the industry status of information requirements.

About the Authors

Rachel Cuthbert graduated in 2000 with an M.A. and M.Eng. in Engineering from Cambridge University with a specialization in fluid mechanics and thermodynamics. Following this, she successfully completed the Advanced Course in Design, Manufacture and Management (ACDMM, now ISMM) in 2001 at the Institute for Manufacturing, Cambridge University. She has gained significant industrial experience from her work in the chemical process and inkjet industries from research anf development, manufacturing, and supply chain roles. Rachel is a

Chartered Engineer and a Member of the Institution of Mechanical Engineers (2004). She is a Research Associate in the Distributed Information and Automation Laboratory. Her current research focus is on service information requirements and service supply chains. Her e-mail address is rc443@eng.cam.ac.uk

Paris Pennesi obtained his Laurea in Electronic Engineering (2002) with honors from the Universita' degli Studi di Ancona, Italy, and his Ph.D. in Artificial Intelligent System (2006) from the Universita' Politecnica delle Marche. During the same period, he also created JEF, a consultancy company, and worked in several software and industrial projects with Italian SMEs. He held a visiting scholar appointment with the College of Engineering at Boston University, from September 2004 to December 2005. His current research interests lie in the fields of stochastic distributed systems and control, optimization and information architecture. The main application areas he is targeting include service supply chain, dynamic pricing, capacity management, inventory control, and multi-agent coordination. His e-mail address is pp300@eng.cam.ac.uk

Duncan McFarlane (University of Cambridge) has been Professor of Service and Support Engineering at the Cambridge University Engineering Department since 2006, and head of the Distributed Information and Automation Laboratory (DIAL) within the Institute for Manufacturing since 1995. He is currently the research lead on the BAE Systems Service and Support Engineering Programme at Cambridge, and founder of the Cambridge Service Systems Forum. He is also Research Director of the Aero ID Programme, involving 20 companies from the aerospace sector examining the role of ID systems within their operations. Until March 2007, he was Director of the Cambridge Auto ID Lab. Between 2000 and 2003, he was the European Research Director of the Auto-ID Center, a program that has driven the industrial adoption of RFID on a global scale. Since 2001, he has been a co-investigator in the EPSRC funded Innovative Manufacturing Research Centre based in the Institute for Manufacturing, currently investigating the role of information in complex asset tracking. He has been a principal investigator on a significant number of UK, EU, and industrially based research programs and now leads a research center of more than 20 researchers and doctoral students. His specific research interests can be summarized as valuing industrial information, information quality assessment, ID and sensor integration methods, distributed information architectures, lifecycle information management, distributed automation, and decision making and reconfigurability. Duncan has developed a research profile which straddles both technical and operations management issues and works closely with researchers in policy, strategy and operations areas. His e-mail address is dcm@eng.cam.ac.uk.

Index of Contributors